UP FROM THE MUDSILLS OF HELL

UP FROM THE MUDSILLS OF HELL

THE FARMERS' ALLIANCE, POPULISM, AND PROGRESSIVE AGRICULTURE IN TENNESSEE, 1870–1915

CONNIE L. LESTER

THE UNIVERSITY OF GEORGIA PRESS ATHENS AND LONDON

Acknowledgments for previously published material appear on page xi.

Athens, Georgia 30602

Designed by Mindy Basinger Hill and Erin Kirk New
Set in 10.5/13.8 Adobe Caslon by Wordstop Technologies (P) Ltd.
Printed and bound by Maple-Vail
The paper in this book meets the guidelines for
permanence and durability of the Committee on
Production Guidelines for Book Longevity of the
Council on Library Resources.

Printed in the United States of America
10 09 08 07 06 C 5 4 3 2 1

Library of Congress Cataloging-in-Publication Data
Lester, Connie L., 1946–
Up from the mudsills of hell : the Farmers' Alliance,
populism, and progressive agriculture in Tennessee,
1870–1915 / Connie L. Lester.
p. cm.
Includes bibliographical references and index.
ISBN-13: 978-0-8203-2762-4 (hardcover : alk. paper)
ISBN-10: 0-8203-2762-X (hardcover : alk. paper)
1. Agriculture—Economic aspects—Tennessee. 2. Agriculture,
Cooperative—Tennessee. 3. Populism—Tennessee. 4. National Farmers'
Alliance and Industrial Union. I. Title.
HD1333.U62 T465 2006
338.1'09768'09034—dc22 2005022127
British Library Cataloging-in-Publication Data available

FOR ROBERT AND SARAH THE JOY OF MY LIFE

CONTENTS

ILLUSTRATIONS

Tables

Maps

ACKNOWLEDGMENTS

Scholarship is at once an individual pursuit and a collaborative effort. Long hours in the archives and in front of a computer screen are supported by family, friends, colleagues, and professionals who listen, question, and encourage until the project is completed. This book began as a dissertation, and I am especially grateful to my major professor and mentor, James C. Cobb, for his wisdom, his enthusiasm for the profession, his guidance on this project, and his continued friendship. I also thank the rest of my committee, Paul H. Bergeron, W. Bruce Wheeler, Susan D. Becker, and Anne Mayhew, for their capable instruction and helpful advice.

I owe a debt of gratitude to the librarians and archivists at the Tennessee State Library and Archives in Nashville; the Special Collections at the Hoskins Library, University of Tennessee; the McClung Collection, Lawson-McGee Library, Knoxville; Mississippi Valley Collection at the University of Memphis Library; Mitchell Memorial Library at Mississippi State University; and the Southern Historical Collection, Chapel Hill. The historical societies of Obion, Gibson, and Marshall counties gave me access to their archives and provided much-needed assistance in tracing the development of the Agricultural Wheel and the Farmers' Alliance. Among the individuals who shared their knowledge and their resources were R. C. Forrester, Betty B. Wood, John Brannon, Fred Culp, and Thomas Pierce.

I would also like to thank the benefactors of the Bernadotte Schmitt Scholarship and the Lee Verstandig Scholarship for providing financial assistance during the dissertation phase of this work. I am also grateful to the Departments of History at Mississippi State University and the University of Central Florida for their assistance as this work evolved into a book.

Sections of this book were presented in papers at the Southern Historical Association Annual Conferences (1995, 1998) and at a meeting of the Rural Women's Studies Association (2000). The helpful suggestions made by panel commentators and the questions raised by the session audiences improved my analysis of insurgent actions. Some of the work on rural women was previously published as "'Let Us Be Up and Doing': Women in the Tennessee Movements for Agrarian Reform, 1870–1892," in the *Tennessee Historical Quarterly* and in

Tennessee History: The Land, the People, and the Culture (see bibliography). The Tennessee Historical Society graciously gave permission to use that work in this book.

The University of Georgia Press demonstrated enthusiasm for the project and provided technical and editorial guidance. Fred Arthur Bailey and an anonymous reviewer provided critical and insightful comments. Thank you for your careful consideration of my work; any errors or interpretive flaws are mine alone. A special thanks goes to Andrew Berzanskis, who answered my e-mails promptly, patiently fielded the most arcane questions, and lifted my spirits when they sagged.

Students and colleagues at Middle Tennessee State University and Mississippi State University pushed me to rethink and rewrite, and thereby to improve this book. Special thanks are due Kathy Barbier, Brendan Buttimer, Mark Cheathem, Rebecca Conard, Derek Cronin, Lorenzo Crowell, Richard Damms, Wanda Dodson, Stanly Godbold, Matthew Hale, Janice Leone, Charles Lowery, John Marszalek, Peter Messer, Rebecca Montgomery, Sara Morris, Susan Myers-Shirk, Elizabeth Nybakken, William Parrish, Jason Phillips, Stephanie Rolph, Mallory Scholl, Jamie Sewall, Fred Smith, Thad Smith, Amy Staples, and Martha Swain.

Family and friends appeared at special moments to relieve the pressure of work and study, offer advice, debate esoteric points of Populist ideology, and laugh at it all. I will remain ever grateful to Margaret Binnicker; Jenny Brooks; Brian Crowson; Jayne DeFiore; Tim and Pat Ezzell; Mary Hoffschwelle; Anne-Leslie Owens; Toni McDaniel; Jim, Jane, Jimmy, and John Phillips; Dan Pierce; Edie Quin; Lisa and Scot Sheffield; and Carroll Van West for their support. With loving gratitude, I thank my mother, Margie Lester, for her patience and confidence. The rest of my family, DeeGee, Rick, Jenny, Molly, and Joe Lester, Steve Pilgreen, and Ted, Esther, and Dennis Gowder, never faltered in their belief that dissertations and manuscripts do come to an end. Most important, I owe thanks to my children, Robert Gowder and Sarah Gowder Pilgreen, who gave perspective to everything. You have always been the joy and hope of my life, and it is to you that I dedicate this book.

Last, I would like to honor the memory of several friends and family members who saw the beginning of this journey but not its end: my father George Cuthbert Lester, Dennis Floyd and Lillie Beulah Whatley Gowder, Susan Lester McVey, W. B. Lester, Florine Baker Lester, Imogene Lester Hinson, Theo Lester Adcox, Lettie Sue Lamb Gowder, Richard W. Hepler, and Shannon Mallard. In countless ways, you encouraged me to pursue this course, and I hope you would have been pleased with the results.

UP FROM THE MUDSILLS OF HELL

INTRODUCTION

During the bitter gubernatorial campaign of 1892, newspaperman and Democratic stalwart Edward W. Carmack held a rally in Obion County, the heartland of the Agricultural Wheel and the Farmers' Alliance in Tennessee. In his harangue against the reelection of Alliance governor John P. Buchanan, Carmack referred to the Populists who supported the incumbent: "If you dig down under the mudsill of Hell," he intoned, "you would find Populists down there!"[1]

Carmack's impassioned denunciation of Populists demonstrated how seriously Democrats took the threat posed by Populists, as former Democratic political adversaries put aside their divisions and committed themselves to party unity and victory against the agrarian insurgents. Throughout the 1880s, while the Bourbon and New South wings of the Democratic Party continued politics as usual, the state's farmers had organized to address the problems of credit, transportation, and marketing that stymied agrarian efforts to shape commercial agriculture to fit the hopes and expectations of the yeoman class. In rallies across the state, rural leaders preached cooperativism in grassroots organizations that offered the greatest benefits to propertied, small-scale, commercial-minded farmers. The Agricultural Wheel and then the Farmers' Alliance initially resisted pressures to create an agrarian political party and convinced a significant number of rural voters to set aside partisanship in favor of candidates who promised to represent agriculture's interests. Largely ignored or dismissed by Democratic and Republican leaders, farmers added to their legislative strength with each biennial election, winning the gubernatorial office and a significant minority of the General Assembly seats in the 1890 election.

Acting on what one newspaper editor called the "Tennessee Plan," the governor and the farmer-legislators proceeded cautiously with the expectation of retaining the executive office for a second term and increasing the Alliance numbers in the 1893 general assembly, thereby positioning farmers to name Tennessee's next U.S. senator. The Democratic Party, which harbored the majority of the insurgents, reacted to that possibility with alarm. The Bourbons, who held both Senate seats, worried that tentative agrarian efforts to bridge both racial and lingering sectional divisions threatened white supremacy. New South proponents feared their industrial goals would lose support if the farmers gained access to national office. Indeed, former governors, legislators, congressmen, and both U.S. senators depicted the intraparty struggle as class warfare and mounted a campaign to purge the agrarians from the Democratic Party, thus ensuring the political demise of the Alliance.

Farmers might have agreed with Carmack's claim that they occupied a place in hell, but they would not have agreed with him on how they got there. As agricultural producers farming some of the richest land in the South, they nevertheless believed they faced bankruptcy and ruin. Proud inheritors of a national mythology that enshrined the yeoman farmer as the backbone of the republic, they recognized that their votes were counted out, and their demands for reform fell on deaf ears. Drawn into a cash nexus dominated by the power of capital, they feared becoming the handmaidens of industry. As the national economy industrialized, Tennessee's farmers struggled between the traditional view of agriculture as a way of life, with its attendant localism and communalism, and the modern view of agriculture as a livelihood with its increasing pressures to adopt business practices, incorporate scientific methods, and replace exchange networks with capital and technology.

While avoiding Carmack's inflammatory language, historians of Gilded Age agrarianism likewise relegated the Populists to some nether world as they paradoxically concluded that the Tennessee farmers offered little serious challenge to the political status quo. Most scholars agreed that Tennessee agrarianism was relatively inconsequential, a "reluctant rebellion" that failed to secure Wheel, Alliance, and Populist demands. Although briefly arousing the fears of entrenched Democratic elites, the agrarians proved to be "conservative" in their demands and actions, offering few impediments to railroad and industrial development.[2]

Scholars assessing Tennessee's agrarianism viewed the Populists as backward-looking and out of touch. Impelled by their fears, these small-town merchants and farmers were seen to be engaged in demagoguery in a futile effort to regain their lost status through political office. An interpretive school more sympathetic to agrarian movements depicted farmers as rational but beleaguered men who ran for political office in order to humanize industrial capitalism. In this second view, the failure of Populism was the failure of democracy. Both analyses viewed election to political office as the critical measure of organizational success and seemed to imply that control of the legislative process represented the only avenue whereby farmers might achieve their goals and influence economic and social change.[3]

A closer look at the Tennessee experience suggests a new thesis with broad implications for reevaluating agrarianism regionally and nationally. Tennessee agrarianism provides a microcosm for understanding the complexity and evolutionary nature of rural cooperative organizations as they transitioned from localism to political insurgency to business-minded progressive farmers. In the course of those multiple transitions, farmers developed and refined the mechanisms for achieving their goal of securing a place for family agriculture in the emerging industrial economy. At times, control over votes proved necessary, but local community action and bureaucratic organization were also paths to success. At every stage of development, the agrarian movements threatened the existing hierarchies of power as they challenged the class structure with their demands for social and economic justice.

Tennessee's agrarian movement grew out of a history of class antagonism, and farmers joined the Grange, the Agricultural Wheel, the Farmers' Alliance, and the Farmers' Union during a protracted period of political upheaval and economic crisis. Class differences predating the Civil War widened as middling-level farmers became increasingly fearful that the economic schemes of New South proponents ignored the problems encountered by small producers and focused instead on industrial development. Similarly threatened by the economic control exerted by the Bourbon Cotton Kings who dominated the cotton warehousing and agricultural furnishing businesses, farmers constructed purchasing cooperatives as an alternative to the crop lien system of agricultural credit and formed marketing cooperatives in the expectation of obtaining higher prices for their commodities. Organizing through existing networks of friendship and kinship, agrarian leaders tapped into an extant rural religious critique of social hierarchies and wealth. Perhaps most disturbing to southern elites, farmers forged alliances with the more politicized African American cultivators and opened their ranks to women. Only a threat on so many levels could unite the Bourbon and New South Democrats in destroying the political insurgency of the agrarians. Although defeated politically, agrarianism survived to fight another day, most effectively as farmers entered the new agricultural bureaucracies and shaped the agricultural institutions and the farm legislation of the early twentieth century.

Understanding the multiple transitions requires an approach that incorporates social, economic, political, and environmental history. Analysis of the agrarian movement must begin with the farmers and the conditions of Tennessee agriculture in the late nineteenth century. Though they generally spoke collectively and through organizational resolutions or the reform press, farmers articulated a rational and increasingly sophisticated response to the economic, political, and social displacement produced by industrial capitalism. To fully understand the agrarian uprising's movement from the barnyards to the statehouse and the agricultural bureaucracies, we should examine not just the perceptions of their opponents, but the words of the farmers themselves.

In speeches, campaign pamphlets, letters, poems, and newspaper articles, rural men and women engaged in a public dialogue that grappled with the contradictory perceptions of farming as a livelihood and farming as a way of life. The shift from the relative safety of a traditional emphasis on home production to commercial agriculture simultaneously opened country life to the social and economic benefits of industrialization but put farm families at economic risk as farmers struggled with fluctuating prices, costly technology, market standardization, and the increasing need for capital to make improvements in methods of production. Farm families clung to traditional exchange networks, subsistence production, and intense localism as bulwarks against the vagaries of modernity. As the Gilded Age transitioned into the Progressive Era, the tension between the opposing positions intensified with enormous social consequences for rural life.

The agrarian movement did not begin with the Farmers' Alliance in 1886 and end with the defeat of William Jennings Bryan in 1896: The issues that dominated rural insurgency began with the reconstruction of the farms and rural institutions following the Civil War and continued through the early years of Woodrow Wilson's administration. The creation of state and federal agencies and farm programs in the Progressive Era provided the mechanisms for institutionalizing the rural voice and addressing the social, economic, and political demands farmers had made for forty-five years: rural uplift, repeal of the national banking system, access to farm credit, stabilization of farm prices, regulation of modern industrial and consumer corporations, revision of the tax system, and popular election of U.S. senators.

Although cooperativism remained central to agrarian organization, the mechanisms for implementing it evolved over time. Farmers first organized their cooperative society around friendship and kinship networks that emphasized personal bonds of self-help and rural community. As the limitations of localism in the battle against highly capitalized commercial and industrial adversaries became apparent, agrarians looked to state and federal governments as the arbiters between capital and labor. Farmers moved into politics only reluctantly in order to protect their claims of agriculture as a livelihood *and* as a way of life.

Historians suggest by their silence that the collapse of the Farmers' Alliance and the political defeat of the Populist Party sent insurgents home where they abandoned their demands and retreated into a self-destructive, one-crop economy. The evidence reveals a more complex history of rural transformation and intransigence and extends evolutionary cooperativism into the Progressive Era.

Having weathered the worst economic depression (1893–1897) in American history, many farmers retreated to a position that eschewed risk as they continued to produce first for home consumption. Marketing the surplus to pay taxes and buy the few goods considered essential, these farmers generally managed to hold onto their land, but with considerable long-term adverse consequences. Unable to isolate themselves from market changes, they remained vulnerable to the power of commodity buyers, warehousemen, and transportation consolidations. Unwilling to finance local improvements in road construction, schools, and public health, they fell further behind as the new century advanced. When pressed, these farmers protected their perceived interests in violent confrontations with those who threatened their traditional communities.

A second, smaller group, which included many former Alliancemen and Populists, developed in more progressive directions. Allying themselves with reinvigorated rural organizations and the state and federal institutions and bureaucracies that developed during the Progressive Era, these farmers adapted cooperativism to modern business practices and abandoned the mass political culture of the Gilded Age. The 1891 separation of the Tennessee Bureau of Agriculture, Immigration, and Mines into two agencies, one for agriculture and the other representing industrial and mining interests, provided the mechanism for the emergence of a state agency devoted to farming and rural life. The new department became the voice of progressive agriculture linking farm organizations with the research and educational programs of the University of Tennessee and the U.S. Department of Agriculture.

Progressive farmers pressed their demands by conforming to the bureaucratic and professional order of the day. They joined the Farmers' Union or organized as horticulturalists, livestock and dairy farmers, as well as cotton and tobacco producers. They attended Farmers' Institutes, enrolled in short courses at the University of Tennessee, organized marketing cooperatives, subscribed to farm journals, campaigned for good roads, supported consolidation of schools, promoted agricultural education, and testified before congressional committees investigating rural life and farm legislation.

Joining progressive leaders in county seats, farmers campaigned for prohibition, woman suffrage, and the country life movement. By its own admission, the Tennessee Bureau of Agriculture was riddled with Farmers' Union men, whose commitment to modern business practices and improved agricultural technology did not subsume its organized purpose: "To establish justice, to secure equality, and to apply the Golden Rule."[4]

The first three chapters of this book analyze Tennessee agriculture and the mechanisms for uniting farmers with disparate histories to challenge the economic and political powers. Chapter 1 surveys Tennessee farms in the Gilded Age. As Tennessee farmers surveyed their barnyards, they saw themselves in various related contexts: as dwellers in a landscape that defined their opportunities and shaped their history, as producers who carefully weighed the risks and benefits as agriculture inexorably moved toward commercial farming, and as voters who sought to protect their interests and advance their position through favorable government action. These farmers assessed their positions individually, but they also took their cues from the world around them.

Farm size, mortgage indebtedness, and tenancy were individual and communal measures of stability and potential that rural families understood. Decisions about what to plant, the availability of labor for planting and harvesting, and access to capital influenced the productive capacities of individuals and regions. Farmers' private decisions in turn influenced regional development. Fearing debt more than they desired advancement, some retreated to an isolationist, near-subsistence agriculture. Others envisioned more commercial farming if the obstacles to their advancement could be removed or mitigated. A third, politically and economically powerful group believed they benefited from the status quo and so fought against measures that potentially altered the established hierarchies of power.

Chapter 2 traces the organization of farm societies from the Grange through the Wheel and Alliance. Tennessee farmers joined national organizations that promised to promote their agricultural interests through cooperativism. While upholding the demands forged in regional and national conventions, the Tennessee experience also reflected the unique social and political characteristics of the state as agrarian leaders struggled to unite the Wheel and Alliance, establish an official newspaper, fund the work of reform, smooth the differences between former Confederate and Union soldiers, and incorporate women and African Americans into the larger movement.

Chapter 3 explores the communal and cooperative ties that bound the members of the Agricultural Wheel and Farmers' Alliance together. Friendship and kinship ties that included church memberships and social celebrations provided a foundation for organization. Likewise, Wheel and Alliance cooperativism built upon the exchange networks that sustained farm families through shared labor and communal assistance in hard times. However, the cooperative commonwealth envisioned by agrarian organizers transcended local communities and demanded a commitment to the greater good that proved difficult to sustain. Building a state agency, organizing the jute boycott, and establishing a tobacco warehouse tested the limits of Tennesseans' commitment to cooperativism.

Chapters 4 and 5 analyze the transition from local cooperativism to Populism as Tennessee farmers organized for political action, first as nonpartisan, antiparty critics of the status quo, then as insurgent captors of the Democratic Party, and finally as organizers of the People's Party. Chapter 4 situates the Wheel and Alliance within the context of the most vigorous two-party system in the South. Agrarian reformers adopted nonpartisanship out of a commitment to the principles of their organizations and as an effective mechanism for overcoming the historic political divisions that separated the state's rural electorate. Anti-partyism proved the insurgents' most powerful tool for mobilizing farmers to effect political change. Marshaling grassroots anger against "courthouse rings" and "political bosses" tapped into longstanding local fears and subsumed other potentially divisive issues.

With the merger of the Wheel and Alliance in 1888–1889, the enlarged agrarian movement became increasingly involved in political campaigns and tested the perseverance or stickability of farmers as they battled for elective office at the state and national levels. Chapter 5 explores the agrarian political insurgency from the successful execution of the "Tennessee Plan" to the collapse of Populism in 1896. Alliancemen plotted their political advance on two fronts: control over state government and election of congressional candidates who measured up to the Alliance "yardstick." Farmers interpreted their 1890 political victory as a harbinger of things to come, but once in office they found governing more difficult than they had imagined. The perseverance they demonstrated in the canvass evaporated as their divisions became apparent under the pressures of legislative work.

In subsequent elections, Bourbons and New South men forced farmers to choose between their politics and their economic interests. Pushed into a weak People's Party, the farmers made pragmatic alliances with the Republican Party in order to maintain their ideological integrity and gain offices. The inability to effect outright fusion hinged as much on GOP factionalism as Populist inexperience. Paradoxically, the strongest examples of fusion occurred in congressional races in Middle and West Tennessee, where a weaker Republican presence made collaborative effort more viable. The relentless agrarian pressure in the congressional races produced no electoral triumphs but forced both major parties to acknowledge rural demands and reshape their party platforms.

Chapter 6 traces the fortunes of agricultural reform throughout the remainder of the nineteenth century and the first decade of the twentieth century. The political defeat of Populism did not provide a solution to the agrarian crisis, and farmers did not abandon their efforts solve the problems. Faced with renewed consolidation by industrial processors of agricultural products and increased pressure to engage in commercial agriculture, farmers responded in both progressive and traditional ways. With the political avenue blocked, many returned to localism and communal cooperativism to defend their interests. More locally focused without the need to bridge the divisions so evident in all aspects of Tennessee agriculture, these associations of farmers grappled with issues of race and class and resorted to violence in order to maintain traditional order in the communities of Black Patch and Reelfoot Lake.

A second group that included a number of well-known former Wheelers, Alliancemen, and Populists pursued a more business-oriented course. They organized in the Farmers' Union and the Farmers' Institutes, where their view of rural cooperativism combined the modern tenets of professionalism and efficiency with their longstanding demand for social justice. Uniting with county seat progressives, these farmers pushed for better roads, consolidated schools, and rural uplift. Abandoning the mass political culture of the Gilded Age, they exerted considerable influence over the agricultural legislation of the twentieth century as they inserted themselves in state and federal bureaucracies. Largely unnoticed in agrarian scholarship, these farmers soon faded from the scene as they were replaced by professional agriculturalists educated in the land-grant colleges. Nevertheless, in the small window of opportunity afforded men with practical experience, they shaped agricultural legislation for the first half of the twentieth century.

Active capital to provide suitable labor and tools, and to enable them to hold their crops for the best prices, is probably one of the greatest needs of the farmers of Tennessee.

—JOSEPH B. KILLEBREW, *Introduction to the Resources of Tennessee*

CHAPTER ONE SURVEYING THE BARNYARDS TENNESSEE FARMERS IN THE AGE OF AGRARIAN REFORM

Unlike most farmers of his day, Archelaus M. Hughes kept a diary. The father of young children, he had little help in the fields and generally limited his diary entries to days when bad weather drove him inside. He tended to the morose, with rain seeming to add to his melancholy but bad weather did not fully account for the gloom of his agricultural forecasts and his own prospects for retaining his hold on the land and providing for his family.[1]

Like most farmers of his time, Hughes believed society had an agricultural foundation. Farmers produced the food and fiber that sustained the nation; rural family life anchored the social structure in a patriarchal hierarchy that guaranteed civic order; and Thomas Jefferson himself had identified the yeoman farmer as the bedrock of the republic, the incorruptible pillar of democracy. An honest man, Hughes sought neither power nor wealth as he fulfilled his duties as husband, father, farmer, and citizen.

In several ways Hughes stood apart from his neighbors, a difference of which he was fully aware. He always voted the Republican ticket and displayed a lively interest in state and national politics. Although money was scarce in the Hughes household, he subscribed to several newspapers and committed his opinions to letters for publication in the Nashville and Cincinnati papers. During Reconstruction he taught school and harbored the view that black and white children should be educated alike. He confided to his diary that he was drawn to Universalism in his religious views. If his beliefs marked him as more liberal than his neighbors, he was nonetheless accepted as a citizen of the community and served on juries, performed his share of roadwork, and was elected master of his local Grange.

Although Hughes' opinions and beliefs set him apart from other southern cultivators, he shared with them an increasingly perilous existence as a farmer. It was an economic dilemma born of the national transition into industrial modernity, and neither hard work nor good weather alleviated its symptoms. Farmers, like other Americans of the day, optimistically believed that the technological and industrial advances of the Gilded Age would ease the burdens of life and solve the problems that beset humanity generation after generation. But industrial capitalism gave with one hand and took with the other, and producers like Hughes were soon struggling in a cash economy where the exchange networks that had traditionally sustained the rural community no longer sufficed. Hughes planted carefully, worked hard, scrambled to supplement his meager crop income, exchanged goods and services with other farmers, and scrimped or did without the manufactured goods that lined merchants' shelves. And in the end, Hughes lost his farm.

With the return of peace in 1865, the region's leading men had predicted a rosier future, promising stability and prosperity in which the farmer would play a starring role. "There is no State in the Union which possesses greater natural advantage and which opens a theater of greater enterprise and wealth than the State of Tennessee," J. D. B. DeBow confidently wrote in 1866. An unfailing promoter of the South, DeBow's enthusiasm for Tennessee's future in postwar America seemed neither misplaced nor a product of mere boosterism at the time. Although congressional estimates placed the state's Civil War losses in slave and nonslave property at a staggering $185 million, expectations of Tennessee's recuperative power rested on a solid history of productive capacity.[2]

DeBow's enthusiasm was matched by that of Tennessee's foremost agriculturalist, Joseph B. Killebrew, a successful tobacco planter and the first commissioner of the Volunteer State's agricultural bureau. As the publisher of the *Rural Sun* and the compiler of the massive *Introduction to the Resources of Tennessee* (1874), Killebrew promoted a program of scientific agriculture and rural boosterism promising a happy marriage between idyllic country life and modern industrial development. He envisioned a countryside dotted with small, independent farms that would feed the miners and supply the raw materials for the mills of a new South. The immigration of hardy European peasants into a revitalized South would offset the out-migration of shiftless blacks and landless whites. The frugal practices of the new residents would transform marginal land into productive use. Immigrants and native Anglo-Saxon yeomen would together create the ideal blend of agriculture and industry.[3]

DeBow's and Killebrew's expectations for prosperity resonated with farmers even if they recognized the exaggeration. In part, the promises of prosperity propelled weary men and women forward in the work of agricultural reconstruction. Likewise, the dream of well-tended farms revitalized the centrality of the yeoman farmer in the life of the nation: men who owned and worked their own land had a stake in government and the good order of society. Even men who had embraced the Confederacy could resume their prior roles in American life as economic contributors and patriots. Farmers looked to their barnyards as a means of confirming both their place in history and their role in the future.

If DeBow's and Killebrew's New South represented one possibility for Tennessee's farmers, an equally powerful elite provided an alternative. The Bourbons upheld the mythology of the Old South to advocate a rejuvenation of the cotton and tobacco economy and the use of black labor in conditions little removed from slavery. As in antebellum times, wealthy planters expected to dominate the social and political hierarchies, with lesser local offices going to men of the middling yeoman class. Ironically, Bourbon agriculturalists envisioned a role for yeomen farmers that was surprisingly similar to that proposed by New South advocates like Killebrew. In the minds of the elites (both New South and Bourbon), middling-level farmers were best suited to home production, marketing their surplus meat, milk, eggs, and vegetables to local buyers. While New South agriculturalists advocated crop diversification as a way to end the region's cotton and tobacco monoculture, Bourbons championed King Cotton but sought to limit access to the "monarch's" court to the largest planters.

Both views promised a hierarchical social order, an economy that offered modest benefits for white farmers, and a government that would uphold such a world. Farmers were soon demanding a reevaluation as the limitations of both the Old South and the New became increasingly apparent. Adherence to the traditions of the Old South demanded a provincialism that commercial-minded farmers found increasingly untenable. Access to markets, capital, and technology required a southern reentry into the national economy and a setting aside of sectional prejudices. The success of the New South scheme depended on rapid industrial and urban development: farmers were expected to supply foodstuffs to industrial cities, but such cities were slow in coming. The alternative, of course, was to supply distant northern markets, a more viable option as transportation and technology expanded into the South, but one that produced a different set of problems, including enhanced demands for capital, reliance on monopolistic rail service, and compliance with increasingly standardized markets.

Caught between the economic promises and the social and political limitations of both the Old and New South, Tennessee farmers assessed the boundaries and the potential of their farms within a complex calculus of geography, history, and hard-nosed reality. Climate, soil fertility, farm size, availability of labor, access to capital, and dependable transportation figured into their barnyard calculations. They balanced the practical side of the agrarian equation by less quantifiable factors of cultural expectations and historical experience. To understand Tennessee farmers, the choices they made, and the demands they promoted, it is necessary to know what they grew, where they planted it, and how they interacted with their neighbors in shaping a political and social community in which to enjoy the fruits of their labors.

A Landscape Divided by Geography and History

Grasping Tennessee's complex physical, social, and political divisions is mandatory to understanding the agrarian movements that rocked the state in the last third of the nineteenth century. When Killebrew amassed his data on Tennessee's resources, he utilized paradigms that characterized the state from its inception and remain the standard for modern analyses. He assessed the state's barnyards according to three organizational yardsticks: history, geography, and civil polity. Collectively and individually, farmers also evaluated their situations within the confines of these divisions.

Historically, Tennessee was, and is, a divided state, and no analysis can disregard the complexity of those divisions. The most obvious division and the one most frequently used in analysis is a geographic one, the three grand divisions of East, Middle, and West Tennessee. These geographic divisions shaped the barnyards, defined the culture, and determined the politics.

Geographers recognized nine physiographic regions that further divided the state and shaped economic development. Soil type, topography, and access to transportation provided the advantages some farmers needed to succeed in the emerging agricultural markets. Absence of those advantages fostered a more limited, risk-averse approach to farming.

Finally, as farmers recognized the need to protect their economic interests, they looked to the state and federal governments to regulate banking and currency, ensure fair transportation rates, and break up monopolies and trusts. As voters, they were divided into ninety-six counties and ten congressional districts that reinforced some geographic and historical divisions and overlapped other. All three divisions—historic, geographic, and political—affected the barnyards of individual farmers and shaped the potential of agrarian movements.

Tennessee's long history of enthusiastic expectations for agricultural profits began in the late eighteenth century when settlers from the Virginia and Carolina backcountry crossed into the rich valleys of upper East Tennessee to establish the first permanent white settlements, before moving into the Cumberland region. Quickly laying claim to fertile lands still in the possession of the seaboard states or local Native American tribes, Tennessee's founding families gained control over what would become, in 1796, the sixteenth state in the new Union. When the Chickasaw (or Jackson) Purchase completed the state's territorial claims in 1819, Tennessee measured 432 miles at its longest east–west axis and 112 miles at its greatest north–south distance. The trapezoidal state borders eight other states of the upper and lower South (see map 1).[4]

From its earliest history to the present, Tennesseans self-consciously identified themselves according to the three distinct regions. East Tennessee, which stretches from the Unaka Mountains bordering North Carolina and across the Great Valley, also incorporates the eastern slopes of the Cumberland Mountains. Middle Tennessee, which spans the region from the Cumberland Plateau to the northward flowing arm of the Tennessee River, includes the eastern and western Highland Rim surrounding the rich farmland of the Central (or Nashville) Basin. West Tennessee extends from the western arm of the Tennessee River to the Mississippi River at the state's westernmost border.

The altitude drops from 6,643 feet at Clingman's Dome in the east to less than 200 feet above sea level on the Mississippi Flood Plain of West Tennessee. Rich in water resources, Tennessee encompasses several river systems, including the Tennessee, Cumberland, and Mississippi rivers. These waterways benefited the transportation needs of early farmers in Middle and West Tennessee, but a series of navigational problems on the Upper Tennessee River limited economic potential for East Tennessee farmers, as both rivers and mountains initially isolated them from regional and national markets.

Changes in soils and climate that accompanied altitude differences also shaped commercial production and slaveholding patterns. A shorter growing season, rockier soil on the hillsides surrounding the Great Valley, and limited access to transportation prompted East Tennessee farmers to practice more self-sufficiency or safety-first agriculture, which included corn and free-ranging hogs. Slaveholding in East Tennessee was the lowest in the state, calculated at 15 percent in 1860. In the rolling hills of Middle Tennessee, the production of corn, tobacco, and cotton vied with the raising of livestock to create a unique region that supported both slavery (25 percent in 1860) and a more diversified agriculture. Parts of West Tennessee, especially the counties in the southwestern corner and along the Mississippi River, closely resembled the Deep South in

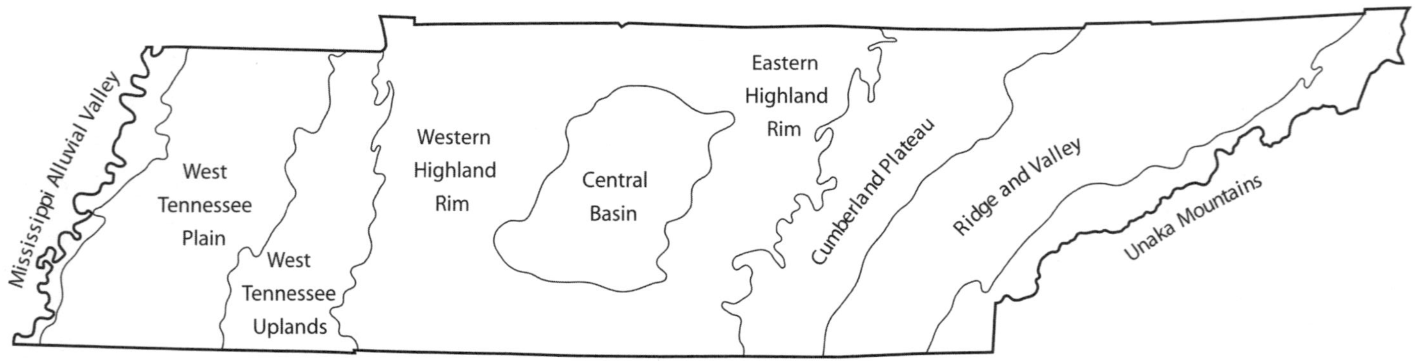

Map 1. Major Physiographic Regions of Tennessee. University of Tennessee Cartographic Services. Redrawn from: Carroll Van West, editor, *Tennessee Encyclopedia of History and Culture.*

climate, soil fertility, slaveholding patterns, and the production of cotton. As it did everywhere in the antebellum South, West Tennessee's cotton culture supported slavery, estimated at 40 percent in 1860. On the eve of the Civil War, Tennessee's barnyards encompassed agricultural production typical of a number of southern states stretching from the self-sufficiency of Appalachia to the tobacco and mixed agriculture of the Upper South and the cotton fields of the black belt and Delta regions.[5]

Overall, antebellum Tennessee boasted a solid economic history and seemed well on the way to separation from regional dependence on staple crop monoculture as it moved toward mixed agriculture and industry. Planned or completed railroads promised to link Tennessee farms and manufacturing to New Orleans, Mobile, Atlanta, Charleston, Chicago, and New York, providing ready access to regional, national, and international markets. With abundant resources in iron and coal, antebellum investors anticipated a dominant role in southern industrialization. By the 1840s the state's reputation for diversified crop and livestock production prompted the editor of *Niles Weekly Register* to proclaim, "Of all the states in the union[,] that which probably has, and admits of in the future, the most various productions . . . is Tennessee."[6]

Planters in the Deep South regarded the state as a regional breadbasket and a reliable source for the purchase of livestock, corn, mules, and swine to supplement their own production. One historian of Mississippi agriculture noted that, even as planters in that state expanded the use of horses and mules on their plantations, they remained reluctant "to purchase the stallions and jacks needed for home breeding purposes," so they "continued the practice of buying their mules and horses from Tennessee and Kentucky." Likewise, South Carolina plantations relied on the annual hog drives organized by East Tennessee farmers to supplement their supply of pork. Tennessee ranked sixth in the nation in corn production, tenth in wheat, third in tobacco, and eighth among the fourteen cotton-producing states. According to agricultural historian Donald L. Winters, "no other state ranked as high in so many different agricultural goods."[7]

Yet, despite the generally optimistic picture of economic growth and market strength, recent research suggests that pockets of decline in the late antebellum period pointed to future problems. In particular, slaveholders and commercial-minded yeomen farmers expressed their concern that a rising number of poor whites portended trouble ahead. In East Tennessee, an expanding population exerted increasing pressure on the thin, hillside soils of farms outside the Great Valley. Elsewhere the presence of landless and tenant-farming whites threatened racial solidarity as impoverished men and women disregarded social norms and established an illegal economy with slaves and free blacks to exchange goods and services. Some historians have even suggested that economically marginal white farmers became enmeshed in a credit arrangement similar to the postbellum crop lien system.[8]

Agricultural Production, 1870–1900

Whatever Tennessee's economic potential in 1860, the state lay in ruins by 1865. Secession and Civil War exacerbated any antebellum problems experienced by farmers and spawned new ones. Divided in its support for southern secession, Tennessee became the bloodiest battleground of the western theater, fighting more major battles than any state except Virginia. Parts of West Tennessee and upper Middle Tennessee quickly fell under Union control and remained in federal hands throughout the conflict. The rest of the state was not so fortunate. East Tennessee and southern Middle Tennessee suffered extensive losses as the result of the warfare and occupation that wracked both regions. Parts of Middle Tennessee repeatedly experienced the ruinous effects of the movements of large armies as both sides fought for control of the major transportation arteries into the Deep South. War left no part of the state unaffected, and in its aftermath, the geographic grand divisions harbored populations of men with bitter memories of the historic conflict.[9]

The protracted postwar recovery rested on factors ranging from the destruction of land, fences, buildings, tools, and work animals to the loss of capital and labor, and the lack of physical strength for sustaining the reconstruction and modernization of farms. Contemporary observers, hoping to entice immigrant settlers to the state, exaggerated the speed of recovery and suggested that most of the problems associated with the Civil War fell on the former slaveholding planters, leaving smaller farmers in a more advantageous position. Killebrew assured prospective investors that the small farmers of Middle and East Tennessee were "both prosperous and happy." He asserted that owning farms of 100 to 200 acres allowed these farmers to "luxuriate in an abundance of all the necessaries and many of the luxuries of life." The dry statistical evidence presented another story.[10]

By 1890, twenty-five years after the war, the total valuation of farms remained more than $27 million below the 1860 assessment even though Tennessee farmers had put an additional 2.6 million acres under cultivation. Census reporters valued land, fences, and buildings at $242.7 million, a figure $28.7 million less than 1860 figures. The value of implements and machinery rose almost $1.5 million over the earlier valuation, but livestock assessments remained virtually unchanged. Although Tennessee farmers now led the southern states in numbers of mules per farm and rated third in horses per farm and sixth in cattle per farm in the regional enumerations of 1889, the number of swine recorded in the state continued to lag behind 1860 figures, and sheep counts fell to nearly their prewar level. A similar story emerged for crops. Cotton and corn production finally reached or surpassed 1860 levels in the 1880 census. Tobacco did not reach the 1860 crop of 43 million pounds for the remainder of the nineteenth century. Such mixed results illuminated the difficulty of recovery.[11]

Even without the disruption of Civil War and Reconstruction, agriculture faced a chaotic and uncertain future. The transformation of the United States from an agrarian, rural economy to an urban, industrial society affected the agricultural sector by drawing increasing numbers of farmers into a cash-based market economy that depended less on local exchanges of goods and services and tied the value of farm products to distant markets. Agricultural modernization demanded more than the sale of surplus production. Lack of capital to finance production changes and improvements in technology stymied efforts to incorporate "scientific" methods to enhance agricultural output. Despite popular claims of general economic growth, low farm prices, shrinking farm size, and a general perception of rural stagnation or decline pervaded the agricultural sector in the Gilded Age. Although risk-averse farmers championed home production with the marketing of surplus, most rural families found such agrarian isolationism untenable in postwar America. Reluctantly or enthusiastically, Tennessee farmers surveyed their barnyards anew and readjusted their sights to accommodate the transformation into modern farming and marketing.

Farming has always been risky, perverse in its rewards and penalties. Bountiful harvests produce low prices, and meager crops, rescued from drought, flood, or pests, reward the lucky or the persistent with higher prices. By the last decades of the nineteenth century, the familiar complaints against weather and prices took on a new intensity as farmers across the South struggled against what they perceived as man-made challenges: national and international markets controlled by futures trading and industrial trusts, differential transportation costs, the agrarian dependence on store credit to survive rather than access to capital to institute farm improvements, and rising tax burdens to fund the progressive changes associated with an industrializing New South.

Farmers complained that hard work no longer produced its just rewards, only indebtedness and failure. Believing that the farmer of the first half of the century benefited from the largely agrarian economy, they argued that debt, taxes, and transportation costs combined with downward spiraling prices to produce a situation in which "the financial demands made today are not so

Table 1. Estimated Value of Farm Products in Tennessee Counties, 1889

Group 1 *>$1,000,000*	*Group 2* *$500,000–$999,999*	*Group 3* *$250,000–$499,999*	*Group 4* *<$250,000*
Bedford	Blount	Anderson	Bledsoe
Davidson	Carroll	Benton	Clay
Fayette	Crockett	Bradley	Cumberland
Gibson	Dyer	Campbell	Fentress
Giles	Franklin	Cannon	Grundy
Lincoln	Greene	Carter	Hancock
Madison	Hardeman	Cheatham	Houston
Maury	Hardin	Chester	James
Montgomery	Hawkins	Claiborne	Johnson
Obion	Haywood	Cocke	Lewis
Robertson	Henderson	Coffee	Morgan
Rutherford	Henry	Decatur	Pickett
Shelby	Humphreys	DeKalb	Polk
Tipton	Jefferson	Dickson	Scott
Weakley	Knox	Grainger	Sequatchie
Williamson	Lauderdale	Hamblen	Trousdale
Wilson	McMinn	Hamilton	Unicoi
	McNairy	Hickman	Van Buren
	Marshall	Jackson	
	Monroe	Lake	
	Sevier	Lawrence	
	Smith	Loudon	
	Sullivan	Macon	
	Sumner	Marion	
	Warren	Meigs	
	Washington	Moore	
		Overton	
		Perry	
		Putnam	
		Rhea	
		Roane	
		Stewart	
		Union	
		Wayne	
		White	

Source: U.S. Bureau of the Census, 1890 (11th) census, 228–229.

easily met as in 1850, or thirty years ago." Debts paid with ten bales of cotton in 1870 required almost twenty by 1890. The benefits of the new industrial society remained out of reach for rural families, prompting farm women like Melia Wilburn to ask, "Why is it, that in the great middle class to which we belong as compared with twenty years ago[,] we find ourselves so much lower down in the scale of real possessions?"[12]

Wilburn's observation was mirrored in the speeches of state and national political figures and in the sermons of country ministers as they contemplated the transformation of farm life. The most obvious change came with the advancement of rail lines into previously isolated counties. Access to regional and national markets encouraged farmers to abandon traditional marketing of surplus production and shift to commercial farming. Modern urban and industrial markets imposed demands for standardization in agricultural production that required investment in land, tools, fertilizer, and shipping to insure delivery of a uniform and reliable product. Initially more flexible, the demands gained in rigor and cost over time as railroads and processing mills enforced uniformity, and state and federal government agencies acted to protect consumers with health and sanitation regulations.

Once committed to commercial agriculture, farmers struggled to maintain their economic independence as farm prices fell over the course of the 1880s and 1890s. In order to meet mounting financial obligations, many increased the acreage devoted to commercial crops, simultaneously reducing their home production and accelerating the downward spiral of prices by adding to the already glutted markets.

Variations in soil and climate meant that Tennessee's farmers produced for several different markets, with their individual demands and opportunities. Aggregate estimated crop valuations compiled by the enumerators of the eleventh census provided a rough picture of agricultural production. The figures show that Tennessee's production divided into four distinct groups, ranging from the most commercially oriented farming to the most risk averse. Table 1 and maps 2 and 3 show the 1889 distribution of estimated values of farm products for the ninety-six counties.

Farmers living in the counties in group 1 (above $1 million) were the most commercially oriented in the state. Located in Middle and West Tennessee, these commercial counties formed a broad north–south band that spanned four distinct soil regions. Middle Tennessee claimed ten of the most productive counties: eight in the fertile Central Basin and two on the northern extension of the western Highland Rim. Nashville (Davidson County) and Clarksville (Montgomery County) provided important market centers for the farmers in these areas. The remaining seven top producing counties were located in West Tennessee, three in the alluvial plain that bordered the Mississippi River and four in the brown loam tablelands that encompassed the middle counties of West Tennessee. Memphis (Shelby County) was the most important market center in the region, with Jackson (Madison County) serving as a secondary market. Not surprisingly, farmers in group 1 also reported the highest investments in farm tools, machinery, and livestock. By the input and return records of the day, these farmers were the most market oriented in the state, with all the benefits and risks that position entailed.

Group 2 included a broader span of counties that stretched down the Great Valley of East Tennessee, across the rolling hills of Middle Tennessee, and included most of the remaining counties in West Tennessee. In the east, Knoxville (Knox County) anchored the surrounding farm communities and provided access to markets as well as consumer goods. Subsequent enumerations in 1900 and 1910 showed considerable upward movement among the five Middle Tennessee counties and eleven West Tennessee counties in this group. Improvements in farm prices and agricultural methods occasionally permitted these sixteen counties to move into group 1.

The counties in group 3 hovered on the geographic periphery of the market-oriented counties in groups 1 and 2. Located in three concentrated areas, these counties occupied the hillier regions surrounding the Great Valley of East Tennessee, the eastern and western Highland Rim of Middle Tennessee, and the western banks of the Tennessee River in West Tennessee. These farmers could not afford the risks associated with commercial agriculture. They pursued the more prudent course of providing for home needs and marketing their surplus.

The poorest farmers lived in the counties comprising group 4. Isolated on the Upper Cumberland Plateau and in the mountains of East Tennessee, they contributed little to regional markets. Without rail transportation and dependent upon unreliable water routes, farmers in these counties pursued the most self-sufficient course.

In 1890 corn was Tennessee's principal crop, both in value and in number of acres planted. However, most corn was consumed by humans or fed to livestock on site, and therefore figured into the farm income only indirectly. Tennesseans also planted more than 9 percent of their improved acreage in wheat and 6 percent in hay. Orchards, fruits, and vegetables previously grown for local consumption increased in economic value when refrigerated box cars opened distant industrial cities as markets.[13]

Unlike commercial farmers in the Deep South, Tennesseans were not wedded to cotton production: only 8 percent of the state's improved acreage was

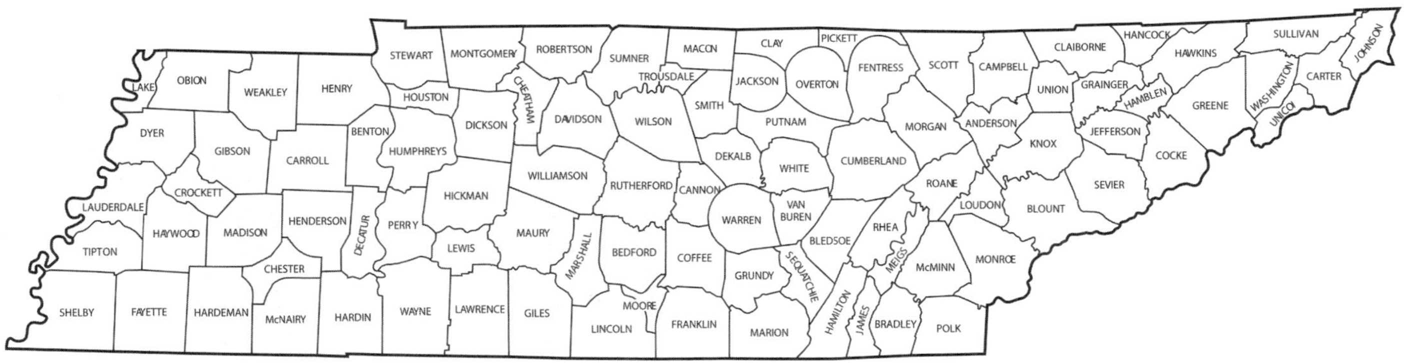

Map 2. The Counties of Tennessee.

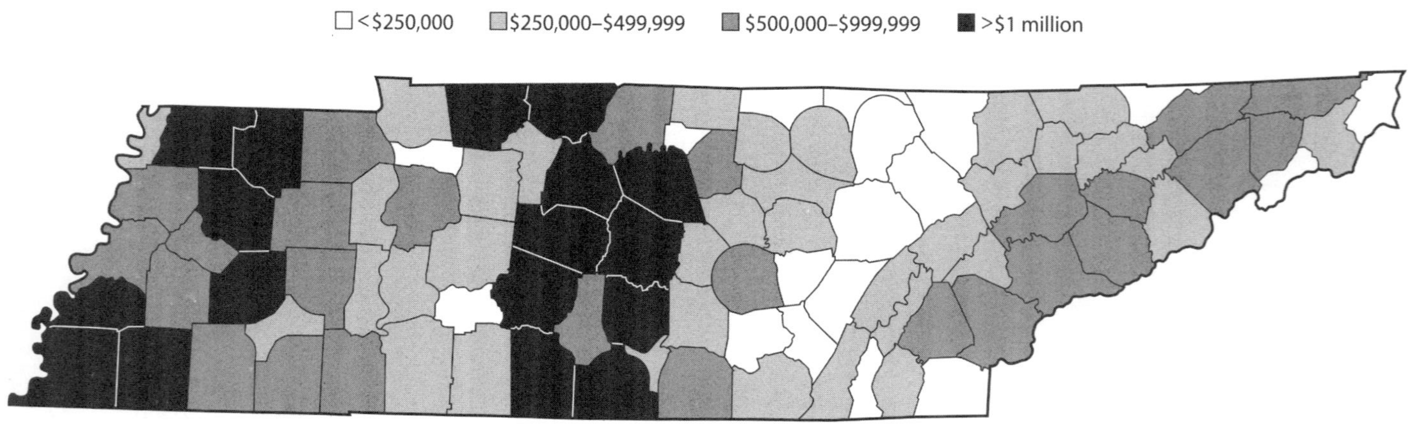

Map 3. Distribution of Value of Farm Products, 1889.

planted in cotton in 1889. Twenty-eight counties produced no cotton, and five-sixths of the crop was grown in the area between the western arm of the Tennessee River and the Mississippi River. Six counties contributed nearly 50 percent of the total cotton acreage and produced almost 56 percent of the crop. Three counties in West Tennessee—Tipton, Fayette, and Shelby—planted more than 50,000 acres each in cotton in 1890. In those counties, agricultural census takers noted, "there is greater density of the productive area than is found in any county in Texas, Arkansas, Louisiana or North Carolina, and one that is exceeded only in a few counties in Georgia, Alabama, Mississippi, and South Carolina." Another eight counties reported 25,000 to 50,000 acres in cotton, while ten counties planted 10,000 to 25,000 acres, and twelve counties had 1,000 to 10,000 acres in cotton. Due to poor weather conditions, the 1889 crop fell below average at 190,579 bales (averaging 0.26 bales/acre). By comparison, the 1879 crop produced 330,621 bales (averaging 0.46 bales/acre).[14]

Tennessee farmers generally held that the break-even point for those who could produce a half bale of cotton per acre was 8¢ per pound. At prices below this level, the production of the fiber was a losing proposition. Cotton prices began falling in the early 1870s, moving from a high of 20.5¢ per pound in 1870 to 10.66¢ by 1880. Prices averaged 8.87¢ per pound in the 1880s and stood at 8.59¢ in 1889. Barely above the break-even point as the new decade began, prices plummeted to a low of 4.59¢ per pound in 1893. Only one year in that decade, 1899, would find cotton selling above the break-even point. Although agricultural experts repeatedly advised destitute farmers to reduce cotton acreage and plant alternative crops, many sought salvation in devoting additional land to the production of the fiber. As John E. Gwin of McNairy County explained, "we are growing poorer every year, but have to raise it [cotton] as a money crop to pay taxes and on debts."[15]

Like cotton, tobacco production centered in a few counties. Tennessee farmers raised three varieties of tobacco—dark-fired, White Burley, and Yellow—with the heavy, dark-fired tobacco dominating the market. Dark-fired tobacco was grown in the Middle Tennessee counties of the Clarksville District and a secondary region that encompassed the northern tier of counties in West Tennessee. This so-called Black Patch extended far into Kentucky and accounted for a significant portion of the region's economy. Warehouses in Springfield and Clarksville, Tennessee, and Fulton, Kentucky, housed the Tennessee crop and attracted buyers who bid on the leaf offered by individual farmers in competitive sales. As historian Christopher Waldrep noted, tobacco farmers interpreted the prices offered for their crops as a reflection of their skills as growers; price, therefore, connoted status within the agricultural community.[16]

Farmers produced the other two varieties of tobacco on a smaller scale. In East Tennessee, farmers planted Yellow tobacco, which was almost exclusively used by domestic manufacturers. Yellow tobacco had been transplanted from North Carolina, and Tennessee's tobacco expert, Joseph Killebrew, believed it could be successfully grown in a number of areas where poor soils unsuited the land for other commercial crops. By the mid-1890s, Greeneville had emerged as a marketing center for the eastern producers. The third tobacco, White Burley was grown on the Upper Cumberland Plateau. By the early 1890s, Trousdale County had become the most important producer of Burley. Plateau cultivators shipped their leaf to Louisville, Kentucky.[17]

Successive commissioners of agriculture touted tobacco as the best cash crop for small planters even as they recognized its labor-intensive production requirements. A single man could cultivate four acres of what amounted to a thirteen-month crop. A year's work began with the preparation and planting of tobacco beds to start the young plants before transplanting them to the field in late spring. Tobacco required repeated hoeings, then removal of horn worms, and topping before harvest began in middle August. Curing of the tobacco occurred in specially constructed tobacco barns, where the leaf was transformed through the use of curing fires fueled by smoldering hickory logs. Once cured, the tobacco was sorted and prepared for market.[18]

Killebrew estimated that tobacco returned approximately $60 per acre, providing the average farmer with $240 for his year's work. For most farmers, tobacco was an "extra crop" planted in addition to other activities that might include livestock raising or food crops. Killebrew acknowledged that the labor-intensive nature of tobacco often meant farms took on a neglected look since the usual time for clearing fence corners was the busiest period for tobacco farmers. He also addressed the negative public perception of tobacco, but dismissed critics as "supersensitive moralists" and countered with an argument that proclaimed tobacco a "moral agent." In a defense that rings strangely in modern ears, he claimed that "tobacco-using people are rarely opium eaters or opium smokers, nor do they use hasheesh or cocaine." He further argued that tobacco "represses the desire for narcotics that destroy the mind, soul, and body."[19]

In addition to cotton and tobacco, Tennessee farmers also produced peanuts for the commercial market. Peanuts were grown in most counties, but as a market crop, red and white Virginia varieties were cultivated in five counties of the western Highland Rim and five counties of the western Tennessee River Valley. Three counties—Benton, Humphreys, and Perry—marketed more than three-fourths of the 747,668 bushels produced in 1889. Peanut farmers vigorously supported agrarian organization and worked to create a regional peanut alliance.[20]

In portions of Middle and West Tennessee, access to rail transportation made strawberries an important crop that added income and, when combined with tomatoes, offered an alternative to staple crop farming. J. W. Rosamon introduced strawberry farming to West Tennessee in 1867. The area around Humboldt and Gadsden devoted thousands of acres to the crop and shipped more than 115,000 crates in 1892. Strawberries averaged 50 crates per acre and generated cash income for day laborers, who received 2¢ per quart (generally around $2 a day) for their efforts. Tennessee strawberries traveled in refrigerated box cars to cities as far north as Chicago, Milwaukee, and St. Paul. In anticipation of the 1892 Columbian Exposition in Chicago, West Tennessee farmers doubled their acreage. An African American farmer living near Fruitland in Gibson County developed a highly praised early variety strawberry. Named Columbian, his berry won the premium at the 1892 meeting of the West Tennessee Horticulture Society, where R. H. C. Mitchell, president of the society, described it as "the best early variety for the market."[21]

Strawberries were marketed in summer as fresh fruits and made into preserves for later sales. Nannie Williams found that spring was a busy time, when in addition to other chores, she "made berry boxes every possible moment." Her usual litany of work included a new complaint, "Monday—yesterday—and to day and the rest of this week and as long as the berries last we will be busy busy." But the rewards proved worth the effort to this cash-strapped family, and Williams later proclaimed, "I am only too glad to have the opportunity to make something. Today the proceeds have reached $82.00." More money was yet to come because Williams also busied herself "making preserves to sell," an endeavor that she proudly noted "made nearly a hundred dollars."[22]

Poultry production provided another avenue for diversification. Long the domain of farm women, and a form of production many farmers had previously regarded as "beneath the dignity of a grown man," it gained in appeal as cotton prices continued to slide. The relatively low cost of entry coupled with the growth of cities and the availability of rail transportation encouraged farm households to undertake poultry and egg production more seriously and systematically. Women's butter, egg, and poultry operations were a source of income or barter for household needs, but, as interest in hens and eggs spread during hard times, it attracted the attention of "prominent poultry men," who initiated investments in improvements that eventually removed production from the domestic sphere into the realm of big business.[23]

Professional agriculturalists in the agricultural bureau, farmers' institutes, and at the University of Tennessee promoted dairy farming and sheep raising as alternatives to row crops. Although dairy and sheep advocates claimed low start-up costs, farmers generally shied away from what they perceived as a costly and speculative venture when compared to the certainty of cotton and tobacco. To gain maximum benefit from either dairy herds or sheep required investment in blooded stock—a significant barrier to cash-strapped farmers. The poor state of veterinary medicine and the lack of protection against roving packs of dogs added to the risk of maintaining herds and flocks, dissuading most farmers from that investment until the early twentieth century.[24]

Tennessee's greater degree of diversification may have cushioned the blows of the prolonged agricultural depression, but many farmers believed themselves at a disadvantage in virtually every market. They looked enviously at regions whose production dwarfed that of the Volunteer State and worried that Tennessee's small market share placed the state's farmers at the mercy of monopolies. An important antebellum wheat-producing region in the Tennessee Valley disappeared as harvesting and milling technology shifted the bulk of U.S. wheat production to the Plains states. In 1890 the Meigs County Farmers' Alliance petitioned Governor Robert L. Taylor to "prohibit the importation of dressed beef," complaining that "the cattle growers of our state cannot compete with the cattle kings of the West." Madison County farmer Robert Cartmell worried about Tennessee's decreasing share of the cotton production. "Texas will soon make it all," he declared. "It [Texas] is an empire unto itself."[25]

As map 3 shows, the location of the most market-oriented counties in discrete clusters in Middle and West Tennessee appeared to offer a solid base for the organization of a strong agrarian insurgency. But in this case appearances were deceiving. The largest producers were divided between cotton and tobacco counties and between plantation-based and small-farm economies. Moreover, potential insurgents were physically separated from like-minded East Tennessee farmers by mountains and divided from their probable allies in the east by historic barriers that included the Civil War and political and cultural differences. Similar to producers regionally and nationally, Tennessee farmers found as many reasons to divide as they did to unite on behalf of their common interests.

Survey of Landholding Patterns

By the mid-1880s, farmers were anxious as commodity prices plummeted and angered as monopolies, trusts, and pools consolidated markets. As Tennesseans surveyed their barnyards, they worried about the security of individual farms—the ability to hold onto the land, make improvements, and pass the land to future generations. By the mid-1880s, many farm families no longer believed they would be able to pass their land to sons and grandsons.[26]

As farmers gravitated toward radical programs aimed at bringing markets and middlemen under control and increasing the availability of money, they looked to their state legislators and congressmen to support the agrarian agenda. Familiarity with the general farm problems in their own counties and congressional districts shaped individual views of their economic future. Tennessee's ninety-six counties were divided into ten congressional districts that remained stable over the course of the 1880s and 1890s (see chapter 4 for a discussion on reapportionment in this period). Map 4 provides a breakdown of the congressional districts.[27]

The scatter-shot placement of highly commercial agricultural counties meant that no one congressional district dominated the farm vote. District ten was home to the cotton counties with the strongest ties to antebellum plantation production. District six encompassed the heart of the dark-fired tobacco region. District nine, with its fertile soil and extensive rail networks, contained highly commercial farms of modest size growing cotton, tobacco, and fresh produce. Close behind these three districts in crop value were districts eight and five. Congressmen in every district heard mounting concerns about the security of farms and landholding as local farm organizations petitioned for investigations of tenancy and mortgage rates.

Farmers did not need to be told that agriculture was in trouble: they could see their future in the shrinking size of farms and the forced sales of land. Not everyone viewed the declining size of the state's farms with alarm. Indeed, postbellum promoters of scientific agriculture counseled intensive land use and routinely chastised farmers for retaining large tracts of unimproved acreage. Killebrew asserted that "the attempt to cultivate too large a breadth of land is probably as fatal to the prosperity of the farmer as the want of means." He assured farmers that "an acre well cultivated can be made to yield as much as three half cultivated." Rural newspaper editors frequently pointed to excessive land holding among farmers as an extravagance they could no longer afford.[28]

Despite their perception of extensive landholding, however, by 1880, if not earlier, "small farms had become the hallmark of Tennessee agriculture." With few exceptions, each federal census recorded the diminishing size of the region's farms, and Robert Tracy McKenzie's recent survey concluded that "One of the most conspicuous developments in southern agriculture . . . was the rapid disappearance of the plantation system of production . . . and a drastic, widespread decline in the scale of agricultural operations." In the old plantation district in southwestern Tennessee, median farm size dropped by two-thirds, while in East and Middle Tennessee, median farm size fell by nearly one-half during the

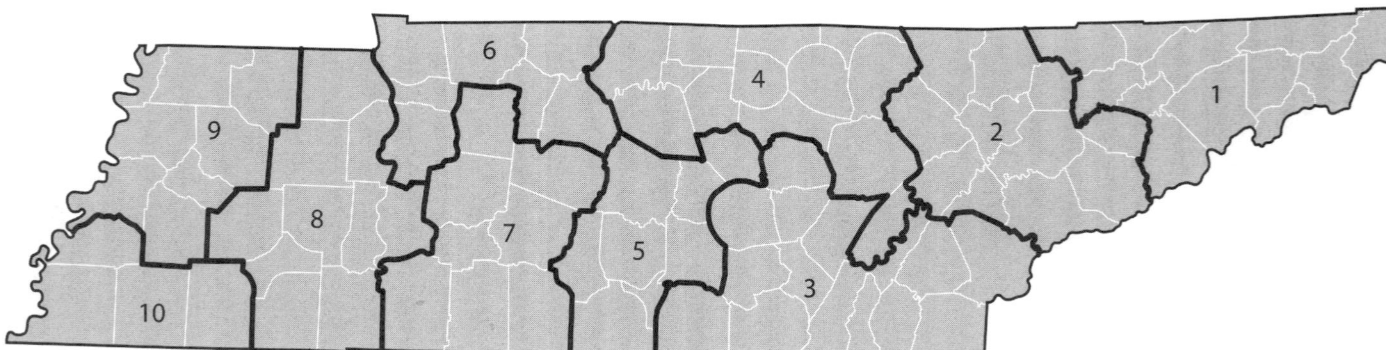

Map 4. Tennessee's Congressional Districts, 1890. Redrawn from Chas. A. Miller, *Official and Political Manual of the State of Tennessee* (1890).

same period. Both farm size and number of improved acres per farm generally declined across the state.[29]

With slight variations, the pattern of decline is evident, as the average farm size in Tennessee dropped from 124.8 acres in 1880 to 81.5 acres by 1910—a decline in farm size of almost 35 percent over thirty years. Although Tennessee's land losses were comparable to the decrease in farm size of other Upper South states, there were obvious pockets of concern, particularly in the twelve counties of the ninth and tenth congressional districts. In those counties, average farm size was 22–30 percent smaller than the state figures in the period 1880–1910. With its cotton plantation history and large African American population, the tenth district closely resembled the black belt region of the Deep South. In the ninth district, however, good soil, a yeoman farming history, and access to both northern and southern markets seemingly offered the best opportunity for small-farm commercial agriculture. But, as commercial-minded farmers in this district watched their opportunities fade, Tennessee's agrarian reform movement gained its first adherents and raised up its first and most enduring leaders.[30]

The importance of farm size in shaping perceptions and actions rested on Jeffersonian political agrarianism, social expectations in post-emancipation Tennessee, and market changes that rewarded specialization. In the minds of late-nineteenth-century farmers, land ownership, social order, citizenship, and economic security remained firmly bound together. Shrinking farm size gave physical dimensions to the frequent complaints that the views of farmers no longer carried weight in legislative halls. Landholding became an important measure of political status and confirmed white racial superiority over black tenant farmers and sharecroppers. In the views of many Tennesseans, the passage of land to sons and grandsons insured family continuity of place and guaranteed social order. Finally, by the last decades of the century, farm size exerted an important influence on decisions to purchase expensive farm technology and incorporate scientific agricultural practices to maximize profits in the market economy. Recent scholarly research suggests that declining farm size, even when accompanied by expanding ownership, represented a step backward. The long-term effects included overworked land, a reliance on low-wage farm labor, and regional poverty.[31]

Although farm size continued to shrink, the majority of Tennessee farmers owned the land they plowed and harvested. In the decade that spawned the agrarian movements, farm ownership expanded from 62.8 percent in 1880 to 69.2 percent in 1890, placing the state near the national land ownership figure of 71.6 percent.[32]

Historian James Tice Moore attributed the rising farm ownership to Tennessee's low (6 percent) interest rate, but an alternative explanation can be found in the population and mortgage figures. Between 1870 and 1890, the state's white farm population increased by 40 percent. By the end of that period 77.3 percent of the total land surface had been claimed for farms, making Tennessee the most "settled" of the former Confederate states. Population density rose from 30.1 per square mile in 1870 to 42.3 per square mile in 1890, with the ninth congressional district experiencing the largest nonurban growth. In the wake of this population boom, the number of owner-operated farms increased, but the percentage of mortgaged farms remained low despite the presence of a presumably favorable interest rate. Data on farm indebtedness collected for the 1890 census put the percentage of farms with outstanding mortgages at 3.2 percent, suggesting that a significant portion of the rise in ownership represented hereditary or familial transfers.[33]

Following the economic depression of the 1890s the consequences of small farm growth became apparent, and the pattern changed in an alarming way. Although the number of farms increased by 10.7 percent between 1890 and 1900, the percentage of farm owners fell to 59.4 percent. By 1910, despite a further increase in the number of farms by 8.6 percent, ownership stood at 58.9 percent. Moreover, the percentage of mortgaged farms rose from 3.2 percent to more than 17 percent by 1910, with mortgages heavily concentrated in the northwestern corner of the state and in Middle Tennessee. Clearly, farm ownership had not protected rural families against the vagaries of the new industrial economy.[34]

The steady decline in farm size and percentage of owner-operators mirrored a corresponding rise in tenancy. Statewide, the rate of tenancy rose from 34.5 percent in 1880 to 41 percent in 1910. Uneven distribution of African Americans across the state meant that tenant farmers in East Tennessee were predominantly white. Indeed, whites accounted for the majority of the tenants in every congressional district, with the exception of the tenth, which was made up of the four counties (Fayette, Hardeman, Shelby, and Tipton) that most closely mimicked the plantation South. In 1880 Tennessee posted tenancy rates comparable to those elsewhere in the South; by 1910 the state's tenancy rates were smaller than those reported regionally. Nevertheless, Tennessee farmers found little reason for celebration.[35]

In eight of the ten congressional districts, tenancy rates showed modest declines in the 1890 census, representing a shift of approximately four thousand farms away from tenancy. At least three possibilities could account for the apparent short-lived improvement in tenancy rates. The most optimistic explanation ties the reduction in tenancy to the growth of farm ownership, suggesting a modest advance up the agricultural ladder. If this explanation holds true, the decade of the 1890s proved harsh indeed as tenancy rates rebounded and surpassed previous levels by 1900.[36]

A second explanation emerges from scattered anecdotal evidence documenting a growing concern over the number of young men abandoning the farm to seek employment in towns and cities. In a letter to the *Weekly Toiler*, the state's agrarian newspaper, Paul F. Kefauver of Monroe County in East Tennessee complained that "agriculture has become a very uninviting occupation for the ambitious youth of the present day." Declaring themselves "sick and disgusted with farming," young college men vowed "not to have anything to do with agriculture." Kefauver claimed an exodus to Chattanooga and Knoxville so large that "the daily application of country boys for employment has become a positive nuisance." The concerns of Nannie Williams for her own son mirrored the observations of Kefauver. After a "particularly hard year" on their family farm, Williams reluctantly permitted her sixteen-year-old son to accept employment at the Aetna Furnace, where he was promised a monthly wage of $25. The need for young men to find employment reflected declining opportunities and the demands of farm families for cash. For many farm youths, the migration into the towns and cities became a permanent relocation as they discovered better opportunities for advancement than farming offered.[37]

Finally, many farmers and their families alternated periods of tenancy with land ownership or industrial labor in an effort to improve their economic position or provide their children with better education. An account of Dorie Woodruff Cope's life in the mountains of East Tennessee documents the transition from farm to industrial labor that characterized the Appalachian experiences of many families as they alternately farmed their own land, rented land from others, and accepted work in cotton mills and logging woods. Although life in the mills provided consumer goods that remained out of reach of most farm families, rural life held the greater attraction for Cope, as it did for many others.[38]

In the Upper Cumberland, where tenancy rates stood at more than 27 percent in 1890, Bob Gwaltney's career also exemplified the movement between farm and wage labor. In 1892, Gwaltney married a woman who had inherited a quarter interest in a Smith County farm near that of his parents. From 1897 to 1899, Gwaltney worked for the McCormick Harvester Company before returning to farming, where he remained until 1906. In that year, he and his wife moved to Nashville to provide better education for their daughters. After a year of odd jobs, the desire to return to farm life brought the family back to Smith County. In 1920, when he told his story to a local newspaper, Gwaltney attributed "any success" to hard work, a good partner, and "studying an economical way of doing things." He might also have added his willingness to abandon the farm as necessity dictated to his list of characteristics for "success."[39]

Like Gwaltney, J. H. Bittick of West Tennessee's Obion County temporarily abandoned his own farm to provide a better education for his children. Unlike Gwaltney, Bittick rented his land to tenants and became a tenant himself in order to acquire a farm close to Union City, where his children enjoyed advantages not found in the one-room country schools. The Copes, Gwaltneys, and Bitticks, like many Tennessee families, periodically relocated to maximize their opportunities.[40]

Roughly a quarter of all farmers in East Tennessee rented land or farmed on shares. In the middle portion of the state, the rate rose to one-third of all farmers as production shifted to cotton, tobacco, and peanuts. By 1900, West Tennessee, the most market-oriented section of the state, recorded tenancy rates of 50 percent or higher. Black tenancy raised no particular alarms: it had long served as a mechanism for producing staple crops and for exerting social control. It was the growth of white tenancy alongside that of black that caused concern. The growth in white tenancy appeared especially problematic in the ninth congressional district. Although tenancy declined modestly in the decade from 1880 to 1890, the ninth district contained 9,344 mostly white tenant farmers, the largest number in the state. This was particularly troubling when compared to the cotton-producing tenth district, which recorded 7,993 mostly black tenant farmers. More than forty years later, W. J. Cash understood the social implications of the rise in white tenancy and viewed the "mighty and always multiplying horde of the landless," whose livelihood now depended on contracts negotiated with landed neighbors as the end to the frontier individualism of past days.[41]

Recent scholarship supports this concern. Harold Woodman suggests that in the southern states, "an increasing *proportion* of both races worked as permanent tenants on plots that remained consistently small." McKenzie argued that in postwar Tennessee "landless farmers found it consistently more difficult to acquire farms and landowning farmers found their hold on the land more tenuous." Moreover, the fertile soils of the western counties attracted landless migrating whites, who swelled the tenancy figures in both the former plantation region and the northwestern counties.[42]

The distribution between sharecroppers and tenant farmers among whites demonstrates important regional differences. In East Tennessee, where the percentage of African Americans ranged from 7.2 percent in district one to 17.4 percent in district three (which included Chattanooga), most landless farmers cropped on shares. Moving westward across the state, the percentage of white sharecroppers declined as the percentage of African Americans in the total population increased. In the old plantation districts, landless white men farmed as cash tenants, and blacks sharecropped.

In West Tennessee, where tenancy rates were the highest, the movement away from owner-operated farms constituted an undeniable "revolution" by 1880. It was a revolution that raised new questions about the "relation of master and man, patron and client" and was rapidly "pouring over into the taboo confines of white men" by threatening white supremacy. White tenants generally rented land for cash in the western counties of the state, distinguishing themselves from black sharecroppers in economic as well as social status. Nevertheless, white planters associated tenancy with emancipation and the need for black labor in the days following the Civil War. Not surprisingly, some farm owners attributed the same negative work characteristics to white tenants that they ascribed to blacks and occasionally assumed similar paternalistic attitudes toward white tenant farmers during periods of acute economic distress.[43]

Robert Cartmell, a Madison county farmer, recorded a litany of labor woes among his hired hands and tenants. He especially resented the Saturday evenings laborers of both races spent in town, concluding: "There is not as much difference between white and black as is supposed. A good deal of the same *kind* of *Human Nature* about *both*." After a particularly trying episode with his "man" Reeves, Cartmell decided that "there is but little difference between the white laborer, such as we have, and the negro." Indeed, he preferred a black laborer, "when he is disposed to work at all."[44]

J. F. Frank, a merchant and planter, was seventy-three years old when he testified in Memphis before the U.S. Senate Agricultural Committee on the condition of cotton. In addition to farming, he acted as a cotton broker and furnishing merchant. He had acquired eight or ten farms in the Mississippi Valley when the owners proved incapable of paying their debts, but he denied that market forces influenced the loss. He argued instead that farmers' own mismanagement was the cause and asserted that "if a man does not make money at his business, it is his own fault." Frank operated his farms entirely with hired labor because he believed tenancy was a losing proposition, vowing, "if I can't have control of my labor, I quit." When the committee pressed Frank on the possibility of similarities between the economic problems of poor whites and blacks, he responded by labeling blacks "an ignorant people" who required constant supervision. By contrast, he concluded, "a white man ought to have more sense than a negro." Clearly the increasing numbers of whites who struggled with the consequences of tenancy experienced a loss of social status in the eyes of more successful farmers and planters as they were blamed for their own losses and equated with less highly regarded black labor. If farmers in East Tennessee still clung to tenancy as a temporary rung on the agricultural ladder, farmers in the western portion of the state operated under no such illusions. For them, tenancy increasingly represented a visible symbol of poverty and social decline.[45]

Occasionally, elites exhibited a sympathetic—if paternalistic—concern for tenant farmers. The disastrous cotton crop of 1889 followed a year of droughts, torrential rains, and invasions of worms and insects. Everyone felt the pinch of hard times, prompting the editor of the *Milan* (Gibson County) *Exchange* to remind readers that "a number of poor renters and croppers will likely suffer if their creditors press them." The newspaperman expressed the hope that "no man with a heart in him will take the last barrel of corn of a poor man." Although the price of cotton rose to a reported 11¢ per pound as a result of the shortfall, the "small crops naturally hurt the small farmer." While sympathetic to the plight of tenants and sharecroppers, the editor left the source of relief to the good will of the elites, thereby reinforcing the sense of hopelessness that marked landless men, who presumably lacked the means to better their situation.[46]

Farm Debts and Capital Needs

Increasingly, farmers blamed farm debt for agricultural decline and increasing tenancy. Potentially the most devastating form of debt was the mortgage. New South advocate Henry Grady characterized mortgages as "the bulldog of obligations—a very mud-turtle for holding on." Farmers understood Grady's analogy and penned poignant reminders about the tragedy of mortgage debt. In a poem titled "The Mortgage," Will Carlton depicted mortgage as a worse threat than any storm or pest.

> We worked through spring and winter, through summer and through fall;
> But the mortgage worked the hardest and the steadiest of all.
> It worked all night and Sunday; it worked each holiday
> It settled down among us and never went away.
> Whatever we kept from it seemed always a theft;
> It watched us every minute, and ruled us right and left.
> The rust and blight were with us sometimes, and sometimes not,
> The hard-browed, scowling mortgage was forever on the spot.
> The weevil and the cut-worm they went as well as came,
> The mortgage staid forever, eating hearty all the same.

I nailed up every window, stood guard at every door,
And happiness and sunshine made their home with us no more.
Till with failing crops and sickness we got stalled upon the grade,
And there came a dark day on us when the interest wasn't paid;
And there came a sharp foreclosure, and the farm was cheaply sold.
The children left and scattered when they hardly yet were borne;
My wife she pined and perished and I found myself alone.
What she died of was a "mystery," the doctors never knew;
But I knew she died of mortgage—just as well as I wanted to.
If to trace a hidden sorrow were within a doctor's art,
They'd a found a mortgage lying on that woman's broken heart.
Worm or beetle, drought or tempest, on a farmer's land may fall,
But for first class ruination, trust a mortgage gainst them all![47]

Some observers held out the prospect of a new start farther west for victims of foreclosure. Archelaus Hughes painted a grim picture of families sneaking off in the night to Kansas or Texas, leaving behind their unpaid debts. The poem "Mortgage on the Farm" told a similar story of years of struggle against debt that ended in foreclosure and the move west.

. . . But I built to the home an "addition,"
And made things all cosy and warm;
And I bought forty acres of "timber,"
And for it we "mortgaged" the farm.

Then hard times grew harder than ever,
Misfortune was followed always,
With failure on failure of harvest,
That no mortal foresight could stay,
The mortgage is closed and our homestead
Is gone for its half and is sold
No help, for it's law, so it's justice,
And avarice clusters his gold.

So out—heaven help us—we wander,
Our youth and our labors are lost,
Ah, little we dreamed as we signed it,
The anguish that mortgage would cost.
Out over the prairie tomorrow,
A white-covered wagon will roam,

And eyes that are misty and blinded,
Shall take a last look at the home.[48]

Many southerners believed the consequences of mortgage debt and foreclosure and the resulting slide into tenancy and sharecropping would reduce American agriculture to a plight comparable to that of impoverished Irish farmers. In a speech to farmers at Elberton, Georgia, Henry Grady argued that money paid in interest flowed out of the South to enrich foreign investors in the same way that Irish tenant farmers paid "every tithe of their earnings to deepen the purple luxury of London." Such concern was not limited to the leaders of the urban New South. John McDowell, the editor of the *Weekly Toiler*, the Tennessee Wheel and Alliance newspaper, warned that if farmers did not rise up in protection of their liberties they would succumb to "a condition of common servitude, similar to that which exists in Ireland."[49]

Mortgage debt raised such fears in Tennessee that farmers demanded statistical evidence of the extent of the problem. Farmers' organizations across the state passed resolutions calling upon their congressional representatives to enact legislation to include mortgage data in the 1890 federal census. Moore County farmers resolved to petition their congressman and senators to demand passage of legislation "to collect the data, in order to show in the next census what percentage of the people in this country occupy their own homes and farms." The Wilson County Alliance wanted to know "what proportion [of farmers] have their property free from debt" as well as the number of "farms and homes mortgaged." The pressure exerted by local and county alliances and wheels produced the desired results. Benjamin A. Enloe of the eighth congressional district and James D. Richardson of the fifth district introduced a bill into the House of Representatives to include mortgage debt questions in the census.[50]

When the results of the census became known, agricultural critics complained that the severity of the problem was not fully reflected in the data. The national Farmers' Alliance newspaper, the *National Economist*, labeled the study "the most disappointing document yet received" from the census bureau and calling it "a compilation of deductions and percentages more complicated and bewildering than anything yet produced by government experts or economic theorists." Like his farmer subscribers, the editor believed the census bureau had been reluctant to undertake the question, fearing it would support the agrarian view that "money and the non-producer were obtaining the lion's share, while the producer and tribute-bearer were forced to be content with the crumbs." As a result, the questions and the manipulation of statistics meant that "every form of mortgage indebtedness possible was eliminated [and] every point that would tend to lessen the amount was utilized." Of particular concern was the failure to record the amount of mortgage indebtedness incurred before 1880. This eliminated from consideration the "deluge of mortgages that followed the panic of 1873, many of which were still burdens to cash-poor farmers." Despite the perceived drawbacks to the study, the editor concluded that the report "furnishes abundant evidence as to the correctness of the position taken by the farmers and laborers."[51]

Without the 1890 manuscript census, it is difficult to assess the mortgage indebtedness reported by individual Tennessee farmers. Nevertheless, one important fact emerges from the published statistics: farmers in the ninth congressional district had more mortgages than farmers in any other part of the state. Over a ten-year period, they gave a total of 8,791 mortgages, small by Midwestern terms, but more than twice the Tennessee average. Thus, in every measure of economic viability—farm size, tenancy, and mortgage indebtedness—district nine appeared as the state's most significant problem. Its average farm size was the smallest in the state, while mortgage indebtedness and tenancy topped the other districts.[52]

The average indebtedness per mortgaged farm in Tennessee rose from $777 in 1880 to a high of $1,535 in 1887. It stood at $1,286 per farm in 1890. The average mortgage for the decade was $1,037. By comparison, the average mortgage for Alabama was $877; Arkansas, $583; Georgia, $552; Kentucky, $950; Mississippi, $759; and North Carolina, $493. Of the eleven former Confederate states, Tennessee's mortgage debt per acre was the highest at $6.64 per acre, though it was lower than the national average of $8.04 per acre. Men forty years of age or older were more likely to own farms (68 percent of the farms were owned by men in this age range), making them more likely to have a mortgage on their land. Fifty-nine percent of the mortgages in 1890 were made on land farmed by men age forty or older. They were seasoned farmers, who now faced uncertainty and the threat of foreclosure and ruin.[53]

Economic historians suggest that foreclosure remained unlikely even in the states with the highest mortgages. Nevertheless, the fear of foreclosure and bankruptcy permeated the countryside and dominated the concerns of men and women unfamiliar with the exigencies of the cash nexus. The circulation of mortgage notes in cash-poor rural communities meant that virtually every farmer was aware of the extent of indebtedness present throughout the open country. With mortgage indebtedness concentrated in the most commercial-minded counties, it only required a few foreclosures to raise the level of apprehension among all nearby farmers.[54]

Concerns about the extent of mortgage indebtedness merged with complaints about interest rates. At 6 percent, Tennessee boasted one of the lowest interest rates in the country, but farmers claimed they actually paid rates as high as 12 percent. Lenders skirted the law by adding the additional interest to the face value of the loan or by claiming the added interest as an initiation fee. West Tennessee cotton farmers, who struggled with the inequities of the crop lien system, joined some New South proponents in advocating a legal rise in the interest rate to 10 percent. They argued that the higher rate offered the advantage of honesty in the negotiation of loans and created an attractive climate for outside investment in the state—something the farmers felt would improve the availability of money and credit. Support for an increase in interest rates dried up rather quickly after farmers from other parts of the state launched a vigorous attack against the scheme. The whole episode points to the differing needs of farmers in the three geographic regions as well as the lack of competition among banking interests.[55]

In 1886, R. G. Dun and Company listed seventy banks of all types, scattered across Tennessee. National banks (of which Tennessee had thirty-three) were in the major trading centers, but they were prohibited by law from making mortgage loans. Most of the state and private banks were located in secondary markets or county seats. East Tennessee had eight national banks, eight state banks, and three private banks. The forty-one counties of Middle Tennessee were served by twenty-two national banks, three state banks, and eleven private banks. Banks in the midstate were concentrated in a narrow band of counties north and south of Nashville, as befitted the rising commercial and industrial influence of the state capital. West Tennessee, with its cotton market, supported the fewest number of banks. Three national banks, eight state banks, and seven private banks served the region.[56]

Banks were by no means the only or even the principal sources of capital and credit for farm families. Crossroads stores extended credit to farmers and tenants who produced cash crops, which were sold through the merchant. The farmer agreed to purchase all, or part, of his grocery needs from the store that provided credit. The crop was the collateral, and interest rates rose significantly above the legal banking level of 6 percent. The evidence for rates of interest charged by merchants is too sparse to make a definitive statement for Tennessee, but ledger book evidence from Georgia indicates that the credit price for corn averaged 53 percent above the wholesale price. Anecdotal evidence supports a similar claim for credit prices in Tennessee. J. H. White of Tipton County estimated the markup for credit "from twenty five to fifty and even to one hundred percent above the actual value." Rural reformers explained the necessity of establishing and maintaining a state cooperative agency for buying farm equipment, seeds, and large household needs such as sugar by holding out the expectation that buying in bulk "cuts off 25 to 50 per cent retail profits," making "a little money go a long ways."[57]

In addition to institutional loans and store credit, small loans could sometimes be arranged between friends or among family members. Archelaus Hughes's diary chronicles the IOUs exchanged and redeemed in place of scarce cash as the circulating medium for the town of Dresden, and indeed throughout Weakley County. Under such circumstances, trust—and therefore access to future loans—rested on a public reputation for economic stability and personal honesty. On more than one occasion, Hughes expressed astonishment when the insolvency of a prominent man of the community was revealed by his death and the paucity of the estate. His own overwhelming preoccupation with repaying his debts and his attention to maintaining the appearance of solvency testified to the personal and communal tensions created under simultaneous conditions of rising debt and declining agricultural prices. In the privacy of his diary, Hughes recorded his personal fear (and presumably the fears of other farmers) that attended the possibility of public exposure of insolvency and the humiliation associated with losing the land.

> In vain may you preach reformation. I must needs do as my neighbors do, or it will be whispered around "He is tight. He is broke. He can't pay his debts" and whenever such opinions as these are entertained in any neighborhood about any one of the neighbors that poor suspicioned neighbor begins to lose confidence in himself. His neighbors at first look coldly at him, then neglect and finally slight him. . . . Every man that he owes is ready to pounce upon him and seize and sell everything he has liable to seizure and sale.[58]

Nor was Hughes's situation unique; the "pursuing phantom" of debt haunted many agricultural households as farm prices fell. In the changing economy of the Gilded Age, farm families devoted more of their resources to market production, increasing their dependency on industrially produced goods and making them vulnerable to demands for payment by agents of a growing number of agriculturally related businesses, who were less receptive to the excuses for nonpayment that previously satisfied local merchants. Before the Civil War, all but the most commercial-minded farmers produced virtually everything for family consumption and marketed the surplus. Confederate veterans remembered that antebellum farm wives preserved food and spun the thread, wove the cloth, and sewed the garments the family wore.[59]

By the last decades of the century, farmers marketed a larger share of their production and scrambled to find new ways to earn cash to pay for goods they once produced themselves. Farm wives still made the family clothing, but now those garments were sewn from cloth bought at the crossroads store or from traveling peddlers rather than homewoven. Commercially oriented farmers demanded improved seeds, blooded stock, factory-made plows, and fertilizers, all of which had to be purchased with cash or bought on credit. In order to earn the necessary cash, farmers experimented with new crops, especially strawberries and tomatoes, and cut timber for use by the railroads. As Anne Mayhew noted in her analysis of the changing economic picture, "as agriculture was commercialized, the farmer was drawn into the rapidly developing commercial-industrial economy and was no longer exempt from the pressures and consequences of changes in that economy."[60]

Nannie Williams worried endlessly about the cost of maintaining her family of ten children and stepchildren. After years of living with her "white ghost" of concern, Williams described herself as "pinched and sour" and complained that she felt "entirely surrounded sometimes." Mrs. Williams herself had negotiated the $1,300 note—the largest portion of the family indebtedness—with her brother-in-law, and her husband Henry signed it. As the family fell "distressingly behind in [their] *accounts*," Williams fretted that "Henry leaves all these important moves to be made by me which causes me such frightful anxieties at times that I cannot be myself." Clearly, although Williams had no legal standing as a debtor, her familial ties proved critical to the economic stability of the farm.[61]

The burdens of debt eventually affected family life, prompting Williams to describe the aged and careworn disposition of Henry, sadly noting that "he is getting fretful and uncertain in his temper," having lost "the proper forbearance with young children." As agricultural prospects continued to decline, the older Williams children left home to find employment. When the girls found work in Nashville, Williams consoled herself with the observation that "this is an industrious age," when "no one is expected to sit and fold their hands." But the situation did not improve without an act of unexpected generosity. In October 1889, Williams's brother-in-law returned the note and seven years' interest as a gift to the family. "*Debt* has at last been vanquished," Williams cheered, and she optimistically calculated that "next year if we have a good crop of Berries We can pay every cent we owe & which does not exceed three hundred dollars and then oh then—Free—Free—Free." Few agricultural debtors could expect the generosity experienced by the Williams family, but many knew the mental anguish that accompanied overwhelming debts with their attendant family tensions and strictures of poverty.[62]

In a period of declining agricultural prices, as the prospects of debt repayment shrank and the likelihood of foreclosure and tenancy increased, contemporaries advised farmers to practice what historian Steven Hahn called "safety-first" agriculture—self-sufficiency in food and household needs, with market sales limited to surplus production. In a speech before the Giles County Farmers' Alliance, President J. F. Black reminded listeners that Alliancemen "believe[d] in the doctrine 'pay as you go,'" and he urged farmers to abandon the use of credit as soon as possible. His words echoed those of B. A. Tucker of Decatur County, who encouraged farmers to be "independent by raising our own food, and if need be, our clothing." Tucker voiced a common belief that farmers held the upper hand in the war against capital, insisting that "if the capitalist wish to corner on wheat," farmers could "raise our biscuit or eat corn cake." Likewise, referring to the sugar trust, he assured farmers if they "wish to corner sugar, let us eat molasses." Turning to history, Tucker compared the farmers' struggles against credit and debt to the Boston Tea Party, and like that event, he urged farmers to "heave it [credit] overboard and let them learn a lesson from the farmer." In his report to the Senate Agricultural Committee, Attwood Pierson described the condition in Lauderdale County as "one of Debt, poverty, and despair," which he attributed to farmers "who have nothing but the hopes of a crop of cotton and good prices to pay their debts." While he judged most of the county's farmers to be "hopelessly insolvent," Pierson reported that "the owner of land who has raised supplies of all kinds at home and sold to the unfortunate or indigent who did not is in much the best fix."[63]

Despite assurances that self-sufficiency held the key to prosperity, farmers slipped deeper into debt. Even frugal men found themselves in need of credit as prices spiraled downward year after year. Robert Cartmell, a Madison County farmer, summed up the year of 1890 with a crop report in which he noted that two years of short crops left many in the country hungry. Cartmell believed the problem was not a scarcity of wheat, oats, or corn, but "many did not *have* money to buy with & sometimes credit is not at hand." By the end of January 1891 he had sold the last of his cotton crop bringing his total income from cotton to $364.05. A year of hard work and careful living did not meet his expenses, and early in February, he recorded, "gave my note to R & Sons for my acct.—payable at one the Banks 1st Dec. next." As if to justify his need for credit, he asserted, "I have tried to avoid that [credit] for the last year or so, but it seems a hard matter to do it. Short crops & low prices does the work—and crops or no crops, expenses keep going on."[64]

Fears of mortgage debt, concerns over plummeting commodity prices, and the specter of a decline into tenancy merged with worries over transportation costs as farmers looked to state and federal government to arbitrate between industrialists and more traditional producers. The railroads presented a problem produced by the confluence of the desire of farmers and town dwellers for economic expansion and an emerging need for regulation to protect the public interests against what many perceived to be the growth of "evil monopolies." Men and women in small towns and rural counties experienced the rise of corporate America primarily through their dealings with the railroads. The often ruthless and impersonal nature of business in an era of industrial building and merger struck many Americans as a threat to the very fabric of the nation. Regulation of the railroads gained widespread popularity, and by 1880 fully half the states had established railroad commissions. At the same time, small businessmen, town boosters, and farmers recognized the economic potential that accompanied railroad growth, and they encouraged it in a variety of ways, including tax exemptions and public bond issues. The confluence of the problems and the potentials associated with railroads sometimes created the paradoxical situation in which the same men who demanded regulation cheered wildly when the railroad finally reached their neck of the woods.[65]

Farmers and Railroads

The railroad "bug" bit the Volunteer State early in the 1830s as schemers and visionaries foretold a future tied to that of the coastal cities of Charleston and Baltimore. Railroad building fell short of meeting the immediate expectations of the state's leaders, but by the beginning of the Civil War Tennessee had developed rail connections to the Atlantic coast, as well as to New Orleans and Mobile.[66]

The destruction caused by the war and the changing needs of post-war Tennessee produced a rush of railroad projects—many poorly conceived and undercapitalized. Too much railroad construction was undertaken with public money, which was lost when economic hardship and/or fiscal reality led to bankruptcy. As a result, much of the turmoil surrounding the state debt controversy of the 1870s centered on the problem of railroad bonds. The divisiveness generated by the state debt issue and the role played by the railroads in creating the debt left many voters disillusioned.[67]

In the last two decades of the nineteenth century, Tennessee railroads expanded substantially to more than 3,100 miles (map 5). At the same time the multitude of small locally owned lines merged into three national rail lines: the Southern Railway controlled by J. P. Morgan, the Louisville and Nashville Railroad (L&N), and the Illinois Central. Rail lines in all three systems lost their regional identification and operated under northern leadership. The most contentious transfer of control occurred in 1880, when Vernon K. Stevenson, founder and first president of the Nashville and Chattanooga Railroad, secretly negotiated its sale to the L&N, ending the existence of an important independent transportation link in Nashville.[68]

The railroads worked hard to promote a good public image. Many rail companies included an honored former Confederate general on the board of directors, and former governors also found a welcome home in the corporate structure. The railroads were especially solicitous of their most vocal critics, the farmers, who complained bitterly about rate differentials for local and long hauls. The Nashville, Chattanooga, and St. Louis Railroad (formerly the Nashville and Chattanooga) hired former State Agriculture Commissioner, Joseph Killebrew, as one of its agents and encouraged his promotion of Tennessee agriculture, mining, and immigration efforts. Agents of the railroad aided agricultural fairs, distributed seeds, and encouraged the development of agricultural societies in the expectation that improved and diversified farm production would certainly benefit transportation.[69]

When the Agricultural Wheel and the Farmers' Alliance organized in the 1880s, the railroads provided reduced rates to state and national meetings. In July 1888, when the State Wheel met at Dunbar's Cave in Montgomery County, the L&N offered transportation for Wheelers and their families at excursion rates. The railroad likely justified the offer as "good business," since the Wheel hoped to attract several thousand members to the combination picnic–rally–business meeting. Profit may not have been the only motive for the decision to provide transportation at a reduced rate for delegates to the national meeting of the Farmers' Alliance in Memphis in November 1892. There agrarian leaders debated the future of the Alliance and the newly established People's Party, and presumably plotted strategies for consolidating their recent electoral gains in order to promote regulation of monopolies. In addition to providing cheap transportation for agricultural leaders, the railroads bought weekly advertising space in the cash-strapped Wheel and Alliance newspaper, a potential conflict of interest that no one seemed to notice.[70]

There were several examples of the merging of railroad and agricultural interests. In 1889, the *Toiler* reported that both the L&N and the Nashville and Chattanooga Railroad had granted permission to Alliances in Giles "and one or two other counties" to construct warehouses "on the road's right of way, free of charge." Exalting in this "exhibition of good will," McDowell claimed that there should be "no antagonisms between us and any legitimate industry." The editor exhorted farmers to remain "firm in demanding our rights," but to "grant equal justice to all other honorable avocations," including railroads.[71]

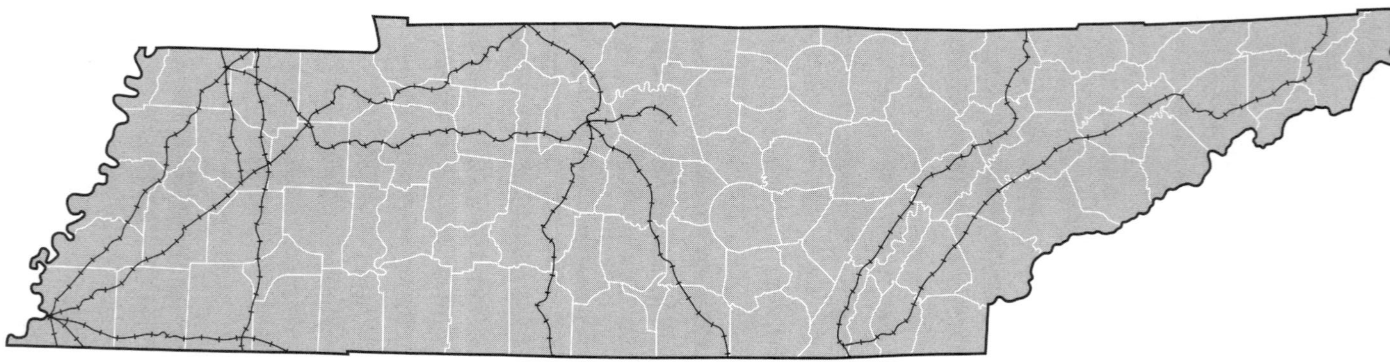

Map 5. Tennessee Railroad System, 1876. Redrawn from Joseph B. Killebrew, *Introduction to the Resources of Tennessee.*

A second example of close relations between railroads and farmers occurred in the town of Gates in Lauderdale County. The town was established in 1883 with the completion of the Memphis and Paducah Railroad. One historian of the area asserts that the town was named for Colonel Robert Gates, an assistant agricultural commissioner. As an agricultural leader, he "look[ed] after the railroad interests in the legislature, and in a general sense promot[ed] good will between the people and the railroads in Tennessee." Both as an advocate of farm institutes and as an agent of the Ohio Valley Railway and the Tennessee Midland Railroad, Gates preached diversification from cotton to truck and fruit farming. The completion of the railway and the establishment of the town of Gates fulfilled the expectations of many area farmers who accepted the "gospel" of diversification. Without the railroad, transportation over country roads made cotton the only realistic market crop because perishable fruits and vegetables required ready access to distant markets. With the coming of the railroad and the availability of refrigerated box cars, the area around Gates added strawberries and tomatoes to its production of cotton. Production increased yearly, and in 1917 the town reportedly shipped 107 railroad cars of strawberries.[72]

The experiences of individual farmers with the railroads rocked between enthusiasm for new spur lines and conflict over property rights. Working on railroad construction and harvesting timber to sell for railroad ties provided new economic opportunities for farmers along the tracks that extended into previously isolated counties. Railroads also enhanced communication between farmers historically separated by geographical obstructions. Although Wheel and Alliance organizers expressed their willingness to travel anywhere, realistically the areas with rail transportation received the most attention. Thus, the most inaccessible areas, including the Upper Cumberland Plateau, the Western Highland Rim, and the mountainous regions of East Tennessee, remained largely unorganized. Isolated from commercial production and bypassed by the railroads, farmers in these areas eschewed the agrarian movement.[73]

While they praised some aspects of railroad accessibility, farmers across the state expressed deep reservations about the unabated growth of railroad power. From the standpoint of small businessmen and farmers, competition provided the surest guarantee of reasonable transportation rates. As the number of independent railroads shrank, the pressure grew for state regulation. In 1881, J. M. Coulter of Gibson County, a future leader in the Wheel and Alliance, introduced a bill in the Tennessee House of Representatives to regulate railroad rates. This weak measure was easily defeated, but the issue was far from dead, and in the next legislative session both the house and the senate introduced regulatory bills.

Agitation for the passage of railroad regulation embraced the concerns of small producers who believed differential rates unfairly pitted "our poor land in direct competition with the alluvial prairies of the West, stripping us of the privileges of a local market, which has been developed and matured by the advantage of toil and proximity to market." The East Tennessee Farmers' Convention complained about the cost of shipping farm implements, which averaged 50 percent higher than rates through Knoxville to points farther south. Many voters found the practice of distributing free railroad passes to state legislators a despicable breach of ethics and justified the creation of a railroad commission on the grounds of ending this questionable arrangement. Railroads complained that they had become the victims of a "fierce attack by all classes of citizens," and pointed to the high costs of operation and the frequency with which railroads fell into the hands of receivers. They warned of the potential losses to their investors and begged for a more "thoughtful and considerate" law that would be "of mutual advantage to the railroads and the people." Despite the pressure brought to bear on the legislature by the railroad lobbyists, the measure to create a regulatory commission passed in 1883. Governor William B. Bate appointed John H. Savage, G. W. Gordon, and J. A. Turley to the three-man commission. The commission tackled its duties with a zeal that alarmed the railroads and, no doubt, facilitated the quick demise of the agency.[74]

Demands for rate schedules by the commissioners and subsequent rate regulation drove the railroads to seek relief from the federal courts. In February 1884 an injunction against the commission was issued by the U.S. District Court. The court found that the act was "a regulation of inter-state commerce, and to that extent an intrusion upon the exclusive legislative authority of Congress." Democrats promised reform of the measure, but Republicans agitated toward abolition of the agency.[75]

The Railroad Commission became the central issue in the 1884 election and doomed regulation with the election of three commissioners who promised to get rid of the agency. When the general assembly met in 1885, repeal of the Railroad Commission easily passed in both chambers despite an appeal by Governor Bate to reform rather than abolish the Commission. John H. McDowell, not yet editor of the *Toiler*, voted for the commission as a member of the house in 1883 and for the repeal as a member of the senate in 1885. He explained his actions by reaffirming his desire for a railroad commission, but, "considering the fact that commissioners have been elected who are opposed to the principle . . . I favor repeal of the present commission law, and there to let the matter rest until the next session of the General Assembly."[76]

The failure to achieve state regulation of the most visible corporate entity confirmed agrarian views that modern business practices of influence buying and political pressure rendered the states helpless to deal with the situation. The U.S. Supreme Court's decision in *Wabash v. Illinois* (1886) that denied states the right to regulate interstate rates further dampened enthusiasm for state relief. No doubt some put faith in the 1887 Interstate Commerce Act, though the *Knoxville Tribune* predicted their expectations were misplaced. "It is not possible," wrote the editor, "that the new Interstate Commerce law can correct all the abuses of railroad operation. There are hundreds of cases not touched by the law." Indeed, the editor expected the railroads to take advantage of the abuses not prohibited, and "when the people cry out against them the railroads will say that all the trouble was caused by the Interstate Commerce law." Like the problems of landholding and debt, the railroad question appeared to have no solution. Farmers needed capital and access to reliable and affordable transportation because agriculture, like other segments of the American economy, had entered a new age of commercialization and specialization.[77]

Farmers who produced for the market experienced greater fluctuations in income, but even at its worst in 1890, the average value of farm products (cotton and tobacco) in districts nine and ten were 28–48 percent higher than the more self-sufficient first and second districts. McKenzie's 1879 comparison of per capita income between market-oriented farm households and subsistence households found market incomes in West Tennessee more than fourteen times greater than those for nonmarket households. Such evidence led him to conclude that cotton "reward[ed] western farmers . . . at a level that far exceeded that of farmers anywhere else in the state." Little wonder that farmers annually risked everything to plant for commercial production. As farmers surveyed their barnyards, they assessed the greater potential for profits and their own needs for cash to pay taxes and debts. Wherever soil, climate, and access to transportation permitted, farmers planted for the market and prayed for higher prices.[78]

By 1890, many farmers believed that all avenues that had once held promise of an expanding, or even a modest, agricultural income were closed. Prices dropped, and the debts that had been contracted when cotton was high had to be paid with additional work. Farm size shrank, and with it opportunities to improve land, diversify crop production, and invest in agricultural technology declined, even as farm ownership rose. An increasing agricultural population meant that tenancy rates also rose, even as the likelihood of climbing the agricultural ladder declined. Individual farm families heeded the advice of agricultural leaders and their own experience to reduce expenses and return to the practice of self-sufficiency. Nevertheless, the "solutions" offered by political figures, agricultural leaders, and experience held out little hope for ordinary farmers caught in the web of debt and prices.

In Tennessee, census data and anecdotal evidence indicate that although no area escaped the problems of the age, the farmers experiencing the worst difficulties were those who produced for the market, especially the farmers in the ninth district. These middling-level farmers returned from the war, rebuilt their farms, invested in agricultural improvements in order to produce for the market, and now faced economic disaster. By every measure—farm size, tenancy rates, and mortgage indebtedness—farmers in the ninth congressional district were at greater risk than any others in the state. They became the first to organize in the 1880s and led other Tennessee farmers into what became an agrarian revolt. As they surveyed their barnyards, they saw no possible relief without organizing to protect their interests. To accomplish that, they would build on a long history of agrarian societies and political activism.

Farmers are learning to magnify their office, and are everywhere organizing societies for mutual instruction, and for the protection of their own interests.

—JOSEPH B. KILLEBREW, *Rural Sun*, 1873

CHAPTER TWO ORGANIZING FOR THE GRAND WORK

In the years following the Civil War, agricultural organizations sprang up to address the needs of farmers and their families. Ranging in scope and focus from agricultural societies designed to reinforce the political and social hegemony of planter elites to associations for rural uplift, institutes to advance scientific agriculture, and cooperatives to control commodity prices, these organizations reflected the desires of their members to establish order in the open country, where the impact of modern industrial capitalism undermined established social, political, and economic networks.

The most important organizations—the Patrons of Husbandry (the Grange), the Agricultural Wheel, and the Farmers' Alliance—attracted thousands of followers. These organizations built on extant communal ties of friendship and kinship, religious congregations, and military service and promised to establish a producer-driven cooperative commonwealth that would simultaneously uphold rural ethics and integrate the countryside into the emerging industrial economy.

Although local unions formed around traditional networks of support, they challenged popular notions of agrarian isolation from the dynamics of modernity in several important ways. Migration both before and after the Civil War created long-distance ties among churches, families, and towns across the state and facilitated the "education" of Tennessee farmers in the principles of the agrarian movement. Collectively, if not individually, farmers demonstrated an awareness of larger national and international events, formulated organizational methodologies, and articulated social, economic, and political objectives that suggested they gave thought to the changing dynamics of American life and acted in a rational, systematic way to protect their own interests. Moreover, the cooperative organizations they formed opened the doors to new social constructs as they admitted women to their unions and created parallel associations for African Americans. While never overtly challenging either male dominance or white supremacy, farmers nevertheless shifted the ground under these monoliths of southern life.

Finally, Tennessee farmers inherited a complex history of political division that forced them to tread carefully as they invoked the camaraderie of military experience. Rent by the Civil War, Tennessee was populated by both former Confederates and former Unionists. Agrarian leaders, therefore, championed the memory of Confederate service and promoted Americanism as they sought a middle ground. Under the circumstances, internal conflict was not surprising. Indeed, more remarkable was the degree to which Tennesseans made common cause.

Early Farmers' Associations

Tennessee farmers formed agricultural societies early in the history of the state. The first such association was the Cumberland Agricultural Society, organized in 1819 to exchange agricultural information and promote new technology. Subsequent farming associations spread across the state, and most counties boasted agricultural organizations by the eve of the Civil War.

In 1854, the general assembly acted upon the petitions of farmer organizations and created the Tennessee State Agricultural Bureau. The bureau's functions included certification for the incorporation of county agricultural societies, distribution of state subsidies, and sponsorship of agricultural fairs in each of the three grand divisions and an annual state fair. Biennially, the bureau reported on agricultural conditions and developments to the general assembly.[1]

Agricultural fairs were central in disseminating improved agricultural methods. Drawn to the entertainment provided at fair time, farmers and their families stopped to admire the newest equipment displayed by manufacturers or watched fellow farmers receive awards for samples of their crops or prize livestock. Country farmers came to recognize those among them whose efforts deserved special merit and whose advice could be sought and trusted.

In his study of Tennessee's antebellum agriculture, Donald Winters suggests that the county and regional agricultural leaders served as vital sources of information and promoted advanced farming methods. According to Winters, those with the financial means took the risks "to experiment with new crop varieties and cultivation techniques, to import blooded stock from abroad, and to test new farm machinery." As planters publicized their findings in farm journals and in speeches before county societies and fairs, poorer farmers were educated on the suitability of crop varieties, livestock breeds, and mechanical innovations for their own region. For reasons ranging from a lack of motivation to lack of financial resources, not all farmers took advantage of the opportunities offered by the fairs and the efforts of agricultural innovators. Nevertheless, thanks in part to the efforts of local supporters of scientific agriculture and fair organizers, Winters concludes that by 1860, "Tennessee agriculture had reached maturity . . . virtually all the land that could be adapted to farming had passed into private hands . . . [and] Tennessee had developed a diversified agriculture."[2]

Postwar organization began with the establishment of Farmers' Clubs, encouraged by a variety of commercial interests. In order to advance their agricultural views and sell newspapers, the editors of the Nashville *Union and American* hired Joseph B. Killebrew to travel the midstate to organize Farmers' Clubs and persuade farmers and rural merchants to subscribe to the newspaper. A New South advocate, Killebrew was a splendid choice for the position and came to exercise a considerable influence in Tennessee (and national) agriculture.

An antebellum Whig, the well-educated and well-married Killebrew spent the war years at home in Montgomery County. He apparently suffered few losses during the period of federal occupation and reaped an immediate reward with the postwar demand for tobacco. By 1870, his farm could be left safely in the hands of his wife and other farm managers while he pursued a career in agricultural journalism, public service, and New South boosterism. Killebrew launched his career as agricultural editor of the *Union and American* and quickly distinguished himself as an early champion of New South principles: improvement and extension of the system of public education, encouragement of immigration to the South, development of natural resources for industrialization, and scientific agriculture.[3]

In his speeches to the farmers' clubs, Killebrew urged cultivators to put aside antiquated farm methods and sectional prejudice in order to "take those steps for advancement that experience has shown to be necessary." From his perspective, a "well-organized farmers' club" offered the most powerful vehicle for change. In addition to his primary program for the advancement of scientific agriculture, Killebrew touched upon themes that characterized later, more radical, rural reform efforts. "We must . . . organize thoroughly and efficiently against the powerful combination of railroad interests . . . and against the machinations of banking institutions and speculators," he told the Montgomery County Farmers' Club in 1870. Denying that farmer organizations pitted class against class, he assured his listeners that he only wanted to see "each class in the community wield an influence proportioned to its importance."[4]

Whatever Killebrew meant by this, he never intended to overturn the hierarchy of power that had characterized rural communities of the past. As his actions and words over a lifetime made clear, he expected yeomen farmers to concern themselves with the improvement of their farms while voicing their support for the efforts made on their behalf by men of economic and political power. Killebrew never anticipated that ordinary farmers would step forward to exert political control. Indeed, Killebrew's biographer asserts that his "Farmers' Clubs appealed more to owners of large estates and plantations" than to the yeomen farmers. Nevertheless, Tennessee farmers saw him as one of their own and clamored to hear his speeches. As one admirer wrote, "Col. Killebrew is a farmer himself and knows how to talk to farmers, and although a man of great erudition and high mental culture, is not a mere theorist . . . but a clear headed, strong minded, plain, intelligent and thoroughly informed agriculturist."[5]

Killebrew's writings in support of scientific agriculture and his skills in organizing farmers attracted the attention of well-placed planters and powerful political figures, eventually leading to his appointment as Tennessee's first commissioner of agriculture, immigration, statistics, and mines, a position that provided him with every advantage for publicizing the state's potential for development. Like many New South advocates, Killebrew did not separate the interests of agriculture and industry, arguing that the development of local markets, the promotion of social order, and the establishment of good government—attributes fostered by industrialization—would benefit agriculture. To take advantage of the opportunities that industrial development provided, he promoted a four-part program that proposed breaking up large landholdings into smaller, independent farms; diversifying crops to reduce southern agricultural dependence on tobacco and cotton; enriching the soil through subsoiling, crop rotation, and use of commercial fertilizers; and attracting European immigrants to cultivate unoccupied land.[6]

With the backing of a number of prominent farmers of Middle Tennessee, Killebrew published in 1872 the first issue of a new, weekly agricultural newspaper, the *Rural Sun*. Although he identified the newspaper with "the farmers' interests," the views of ordinary farmers seldom found their way into print. Espousing the policies of the Tennessee Bureau of Agriculture and its New South supporters, Killebrew waged war on the cotton and tobacco economy of the region, urging crop diversification and home production.[7]

Killebrew's agricultural interests extended beyond Tennessee's borders and, with the backing of the Farmers' Clubs and others of like mind, he moved toward the establishment of a national forum for agrarian interests. In October 1871, delegates from eleven states and forty agricultural and mechanical associations assembled in Nashville for the first meeting of the National Agricultural Congress. William G. Jackson, assistant manager of Davidson County's Belle Meade plantation, welcomed the delegates and outlined the purpose of the organization, which he saw as providing accurate information on markets, prices, and crop estimates; dissemination of practical information; and the breakdown of sectional prejudice—common goals of New South agriculture. The delegates elected Jackson to the office of president of the congress, a position he continued to hold throughout the decade, and installed Killebrew in the position of secretary, an office he held until 1873.[8]

When the National Agricultural Congress met in St. Louis in 1872, delegates clearly hoped to expand their range of influence and voted to invite Oliver H. Kelley of the National Patrons of Husbandry and Samuel Bliss of the Vermont Dairyman's Association to attend the next meeting in Indianapolis, scheduled for May 1873. However, in neither meeting did the congress demonstrate a willingness to confront the problem of primary concern to most of the nation's farmers—railroad rates—an issue that Killebrew had previously raised in Tennessee. Rather than tackle the issue of railroad regulation, the congress supported resolutions for improvements in water transportation to meet critical transportation needs in the South. That unwillingness to antagonize railroad interests—in which many planters had a financial stake—and the tendency to concentrate on the less contentious issue of waterway improvement characterized the organization throughout the remainder of the decade.

Largely an association of southern planters, the Agricultural Congress never exerted as much influence regionally or nationally as its Tennessee organizers hoped. The congress had reached a critical level of decline when Robert Beverly of Virginia became president in 1880, a position he held until 1888. Although he claimed that the "Congress was the forerunner of the [Farmers'] Alliance, and prepared the way for the grand work it is now doing," there seems little in the Tennessee experience to justify this view.[9]

Patrons of Husbandry

Concurrent with the organization of the Agricultural Congress, the Patrons of Husbandry, better known as the Grange, appeared in Tennessee. Founded by Minnesota reforming agriculturalist Oliver H. Kelley in 1867, the organization emerged first as an effort by like-minded bureaucrats in the U.S. Department of Agriculture to restore southern agriculture to its prewar economic position and heal the breach between southern and northern farmers. Similar in organization to Masonry, the Grange emerged as an advocate for rural uplift and economic development, bringing to the forefront the problems of rural socialization, railroad regulation, and farm credit—issues with far more appeal for middling-level farmers than planters and industrialists.[10]

After a slow start, the Tennessee Grange membership peaked at 1,050 local granges, claiming 37,000 members in 1875. Granges attracted middling and poor farmers with the prospect of ending the crop lien debt cycle through a program of cooperative buying and selling. Likewise, Grange advocacy of public education that included coursework in practical farming methods drew families anxious for rural vocational schooling for their children. Grange cooperative stores briefly flourished before succumbing to outside commercial pressure and inadequate financial support, and tobacco farmers organized a successful tobacco warehouse in Clarksville. Unlike Grangers in other states, Tennessee Grangers never succeeded in gaining control over the state's land-grant college or in establishing local agricultural schools. As already suggested, they engaged in a complex critique of railroad development, generally adopting a more ambiguous stance that supported the construction of new lines, accepted whatever benefits railroads offered, and advocated modest regulation. West Tennessee Grangers marshaled support for a campaign to complete the Texas and Pacific railroad in order to establish a southern transcontinental route—a position they could defend as enhancing southern access to markets and encouraging competition to provide better transportation rates.[11]

The Grange also attracted a few prominent and wealthy farmers. Both Killebrew and Jackson as well as a number of state politicians associated themselves with the Grange, perhaps recognizing both the potential to shape grassroots agrarian reform to reflect their own interests and garner votes as the farmers' friend. Indeed, one historian remarked that these men did not "enter politics as Grangers, but joined the Grange as politicians." Like similar rural organizations, the Grange supported self-sufficiency through the reduction of expenses and increased production through the use of scientific farming methods, although it advanced a step further than the Agricultural Congress when it proposed the reorganization of business relations "to bring producers and consumers, farmers and manufacturers into the most direct and friendly relations possible." But like other agrarian organizations, the Grange disavowed political actions in the pursuit of its aims and included in its Declaration of Principles the unequivocal statement that "no grange, if true to its obligations, can discuss political or religious questions, nor call political conventions, nor nominate candidates, nor discuss their merits in its meetings."[12]

Such claims to the contrary, the Tennessee Grange nevertheless quickly incited the concerns of agricultural leaders who worried that ordinary farmers might organize politically and undermine the status quo. Governmental and educational institutions proved particularly effective in stemming grassroots reform efforts. Annual Farmers' Conventions and fairs brought together cultivators, livestockmen, and businessmen to discuss the state of agriculture and promote improvements in livestock breeding, the use of chemical fertilizers, and the latest technical advances. As a rule, the papers presented at these meetings reflected the vision and ambitions of New South proponents, who touted diversification and home markets as the most advantageous position for yeomen farmers.

Oliver Perry Temple, an East Tennessee attorney and long-term trustee of the University of Tennessee, was an active sponsor of scientific agriculture and the first president of the East Tennessee Farmers' Convention. As Temple's correspondence reveals, sponsors of the East Tennessee Convention feared control of the organization by the Grangers, who might subvert the convention's "original designs and principles" and provide "the extra ordinary spectacle of farmers convened for their own good thrusting their darts into the heart of their own prosperity." Indeed, one correspondent was so concerned that Grangers would overthrow the intentions of the founders, he asserted "if we have only lawyers, doctors, and merchants, I am for holding the convention." The perception that ordinary farmers had nothing to offer proponents of scientific agriculture permeated the thinking of agrarian leaders in the state agricultural bureau and the land-grant university, as well as among the "better class" of farmers and planters.[13]

The estrangement between Grangers and the state's foremost agriculturalists was more a matter of who said what rather than what was said. Like their wealthier counterparts, Grange leaders preached a deflationary program of hard work, self-sufficiency, and a pay-as-you-go plan. They recognized the potential benefits of home markets in an industrializing state. But for the majority of Grangers who lacked the means to abandon cotton, corn, and tobacco farming, the advice of men such as Killebrew and Temple was cruel. Historian Theodore Saloutos rejected such schemes in scathing terms, saying, "More unrealistic advice for the mass of sharecroppers and impoverished landowners could hardly have been given." His assessment is borne out in the diary of the previously introduced West Tennessee Granger, Archelaus M. Hughes.[14]

Hughes recorded a depressing litany of struggle and failure in his efforts to hold on to his farm during the period 1874–1877. Education and effort proved ineffectual against debt and bad weather, and Hughes eventually lost his land in a forced sale. Self-sufficiency provided no relief for him, as he recorded: "I have lived closer, and the whole family have worn old clothes and shoes longer than we ever did before. For myself I have bought one pair of pants and 3 yards of domestic in the last six months, and not much more for Hattie and the children." Six months later, he recorded no improvement in the family fortunes and used the occasion of the death of a prominent, though impoverished, acquaintance to comment on the problem of debt. "Debt is a grim monster," he wrote, "that drags down a man from his high position." Grange activities apparently brought little relief, and his final cursory reference to the organization included a telling observation. "Was at the grange yesterday," he wrote, "for the first time in five months. Nothing of importance was done."[15]

If the Grange failed in its political and economic efforts, it successfully challenged social mores, with important implications for the future. The Grange built upon familiar organizational networks of religious and social groups common to every county to create a sense of community among farmers and their neighbors. Indeed, historian Donald B. Marti characterized the Grange as a "family organization, where men and women regularly crossed the boundary separating their spheres to share work and social enjoyments." The Patrons of Husbandry insisted upon the inclusion of women in the organization, and women played a prominent role in the Tennessee granges, though their activities remained traditional and supportive. Nevertheless, Grangers themselves recognized the significance of their sexually integrated organization. As one Mississippi Granger wrote in the tristate newspaper *Patrons of Husbandry*, "The Grange is the only organization outside the church into which woman has been admitted."[16]

In the Wilson Grange No. 148 at Clarksville in Montgomery County, six men and six women enrolled as charter members when the organization was established in 1873. By the end of the year, membership stood at twenty males and twelve females. Ten men and seven women presented themselves for membership at the organizational meeting of the Mill Creek Valley Grange No. 182. By the end of the following year, sixty-one men and twenty-seven women were members. Grange women advanced through a separate ritual within the organization, held offices within their ritual sphere, presented essays before county meetings, and wrote letters to the *Rural Sun* and other Tennessee Grange newspapers.[17]

Membership in the Tennessee Grange reached a high point of 19,780 families in 1875. Thereafter, it began a steep and steady decline that ended in complete dissolution by 1880. In 1876, the Tennessee Grange reported a membership of 10,216 families. By 1877, it had dropped to 1,474. Perhaps the cause of the organization's demise rested in its view of agriculture. As the effects of the economic depression of 1873 ate away whatever advances farmers had made after the war, the emphasis on the value of scientific agriculture failed to break the debilitating cycle of credit and debt faced by the majority of the state's farmers.[18]

Nevertheless, the Grange laid important foundations for the future. Many Grange ideas—rural uplift, cooperativism, the role of women in agrarian reform—found new life in successive farm organizations. Later agrarian leaders also learned from the failures of the Grange and organized more effectively to exert political pressure and elect sympathetic legislators. Finally, to a remarkable degree, the Grange provided the leadership for the future. Thomas B. Harwell of Giles County, a country doctor and four-term member of the Tennessee General Assembly, was the 1878 Master of the Tennessee Grange. His advocacy for rural reform remained strong, and Harwell became an active member of both the Agricultural Wheel and the Farmers' Alliance. J. R. Miles, early organizer and state president of the Agricultural Wheel, "was one of the first in his neighborhood to join the Grange." Likewise, John H. McDowell, editor of the *Weekly Toiler* and second president of the Tennessee Farmers' Alliance, reportedly joined the Grange in 1871 as a "strong advocate of the necessity of organization among farmers." Leaders of the succeeding organizations—the Agricultural Wheel and the Farmers' Alliance—incorporated Grange ideology, even as they challenged old assumptions and formulated new demands.[19]

The Agricultural Wheel and the Farmers' Alliance

When Tennessee farmers organized again in the 1880s, they followed a familiar path as they joined forces with agrarian associations from other states—the Agricultural Wheel in Arkansas and the Farmers' Alliance in Texas—and built on earlier efforts. This time, farmers from the Volunteer State played a more active role in shaping the movements' ideologies and in directing their efforts regionally: they shaped agrarian agendas, held national offices, and organized economic boycotts.

Unlike the Agricultural Congress, the farmers' conventions, and the Grange, the new organizations limited membership to working farmers and rural ministers, doctors, and teachers. This allowed them to focus on working for fundamental economic and political changes for smaller, landowning farmers. Tapping into long-standing antiparty sentiment, these organizations blamed partisan politics for the rise of monopolies and the impoverishment of agriculture. Their remedies included agrarian control of agricultural markets through cooperativism and favorable legislation for the protection of farm interests.

The Wheel and the Alliance simultaneously represented connections to the past and the future. Steeped in the political rhetoric of Thomas Jefferson and Andrew Jackson and in the economics of the physiocrats, they upheld the centrality of agriculture to the political and economic health of the republic and promoted a producerist view that the proper determinant of farm prices was the value of labor's contribution to the finished product.

At the same time, they recognized a new world. "Monopolist powers" dominated access to capital, transportation, and the marketing of farm commodities and thereby threatened the existence of independent producers. Agrarian leaders adapted the rhetoric of the early republic that upheld the "common good" and applied it in new and radical ways. Banding together in cooperative associations, farmers justified their organizations in language similar to that of other professional groups, staking their own claim to legitimacy in the emerging order of industrial America. Pointing to the favorable laws, such as the tariff, that shielded American industries, farmers advocated protective legislation for themselves as well as the abolition of institutions they believed detrimental to their interests. Thus, unlike their forefathers, these farmers expected government to take an active role in mediating between competing interests. Farmers never demanded a return to the past, but they refused to become servants of the future.

The Tennessee reorganization of farmers, unlike that of most other southern states, began with the Agricultural Wheel, which merged with the Farmers' Alliance in 1889. Seven farmers in DesArc (Prairie County), Arkansas, organized the first Wheel in February 1882. They chose the name "Agricultural Wheel" for two reasons. First, in recognition of the industrial age in which they lived, they proposed that "no machinery can be run without a great drive wheel, and as that wheel moves and governs the entire machinery, however complex, so agriculture is the great wheel or power that controls the entire machinery of the world's industries." The second reason was political. Arguing that "the root of evil lay in the laws," they organized for the purpose of addressing economic inequities through political and legal means. Following in the footsteps of the Jacksonian antimonopoly ideology, they believed themselves to be surrounded by political "rings" and turned to the biblical symbolism of Ezekiel's wheel to characterize their own position.[20]

The Wheel quickly attracted members. After local incorporation in August 1882, the Arkansas State Wheel organized in April 1883. The State Wheel united with another state agricultural society, the Brothers of Freedom, headed by founder Isaac McCracken, in 1885. When the two organizations merged, and the Brothers of Freedom consolidated their efforts with those of the Agricultural Wheel, McCracken claimed 643 local units with a combined membership of

43,000. The following July, Tennessee sent five delegates to the Wheel convention at Litchfield, Arkansas, where they organized the National Agricultural Wheel, installing McCracken as the first president of the National Wheel. A. E. Gardner of Dresden, Tennessee, was elected as the national secretary, a position he would hold throughout the life of the Wheel, becoming the first secretary of the combined Wheel and Alliance in 1889. In November 1886 Tennessee Wheelers hosted the organization's first national meeting at McKenzie, Tennessee. The Wheel remained largely a southern association, spreading throughout the Mid-South and into Missouri. Membership in the Wheel rose from the initial seven to a reported 5,000 in 1884; 50,000 in 1886; and a half million in 1887.[21]

J. R. Miles organized the first Tennessee Wheel at Public Wells, in Weakley County, on February 1, 1884. Overcoming considerable opposition from crossroads merchants and disbelieving planters, he claimed five additional wheels by the end of March. From those modest beginnings, the movement spread across the northwestern counties and moved into the rest of the state. In July 1885, when the Tennessee State Wheel was chartered, it recorded 154 local wheels in the counties of Weakley, Henry, Obion, Carroll, and Gibson. Two years later, the State Wheel represented local organizations in twenty-nine counties. At the last meeting of the State Wheel in Clarksville in 1888, organizers claimed to represent over 42,000 members in forty-two counties, in addition to nine county organizations of the Colored Wheel, which represented 177 subordinate wheels. As the organizational report indicated, there still remained "a large scope of country in Tennessee where the wheel [was] not known." In the heart of the midstate, Davidson, Sumner, and Trousdale counties had not been organized. The same was true for DeKalb, White, and many counties on the southern Cumberland Plateau. In the Great Valley of East Tennessee, the Wheel claimed not a single unit.[22]

Although many leaders in the Tennessee Wheel traced their reform history to earlier membership in the Grange, they differed from the Patrons of Husbandry in important ways. W. Scott Morgan, the contemporary Wheel historian, clearly recognized the limitations of the Grange program and noted that it "failed because it didn't go far enough," adding that while it "sought to correct abuses by abolishing middle-men [it] did not allow political discussion." *Weekly Toiler* editor McDowell echoed Morgan's criticism and further asserted that the Grange "sprang into existence on the sole idea of knocking out the middle man." Its policy of "gathering its membership from most all avocations [sic]" hampered reform efforts. Indeed, he claimed, "the very men it proposed to knock out got in[,] and the order went into wild speculation and soon came to grief."[23]

McDowell also condemned the farmers who "went in with the expectation of getting rich in a year." Attempting to distance the new movement from past failures, the editor assured his readers that "the situation is different now." He cited the problems of "unequal laws and class legislation" that created millionaires and monopolies and "made slaves of the producers" and claimed that the hard times "created a necessity for organization to break up combinations and rescue the country from impending anarchy." He concluded that "the ballot box is the only resort and only hope for reform."[24]

In order to protect themselves against laws that favored the interests of bankers, lawyers, manufacturers, and railroads, Wheelers demanded the right "to manage their own affairs." They pointed out that "other professionals organize for improvement and do not ask the leadership of people outside the profession" in defense of their own exclusion of town leadership. They declared their independence from the influence of the established political and economic powers, arguing that farmers had "made up their minds to discuss matters relating to their interests among themselves, free from interference by men engaged in other pursuits."[25]

Like others trying to restore order in the wake of industrialization, farmers sought to define and protect their interests. But unlike bankers, lawyers, or industrialists, farmers in American society represented more than an economic interest group. With the United States on the brink of becoming an urban industrial society, farmers still made up the majority of the population in 1880. If adequately marshaled, they embodied a formidable voting bloc. Furthermore, in the collective American mind, farmers represented the soul of the American mythology. Their own awareness of their voting strength and their exalted place in the American heritage as the builders of the nation and the embodiment of the Jeffersonian republic added to their power. If farmers succeeded in wresting political and economic control from proponents of the New South and their industrial colleagues in the North and Midwest, Morgan's prediction of an "impending revolution" seemed prophetic.

Wheel ideology held that "the system . . . is at fault more than the men." From the organization's perspective, farmers shared no blame for the disorder experienced by late-nineteenth-century society; the problem evolved as the consequence of the violation of "natural law," particularly the subversion of the law of supply and demand by monopoly power. To Morgan, the growth of industrial society violated the elementary principles of labor. In his view, "man is naturally disposed to take pleasure in remunerative employment," and therefore, "he is justly entitled to the fruits of his own labor." He warned that "any violation of this natural law will breed social disorder," and "a universal violation will bring national calamity." In Wheel ideology, the problems of industrialization were solvable, but only if Americans rationally returned to the principles of natural law and abandoned their quest for riches through the exercise of monopolistic power.[26]

The Agricultural Wheel defined its ideology and adopted a list of demands at its first national convention in McKenzie, Tennessee, in 1887. John H. McDowell chaired the committee that formulated the demands and presumably influenced the final product. The fourteen demands incorporated familiar Gilded Age reforms and added new ones. The final product included a list of economic demands, beginning with calls for assurances that public lands would be "reserved for settlers only," and prevent aliens from acquiring titles to American land. Wheelers demanded the rapid payment of the public debt by the minting of gold and silver at full capacity. They pressed for the abolition of national banks and the substitution of legal tender notes issued by the national treasury in sufficient amounts to establish a cash economy. They called for congressional action to prevent dealing in futures and advocated a graduated income tax. They called for legislation to prevent the importation of foreign labor and an end to the practice of convict leasing. To overcome the threat of public utility monopolies, they demanded regulation or, if that did not work, public ownership of telegraph, telephone, and railroads.

Their demands for the election of all national government offices and for enforcement of election laws supported their desire to return government to the "people." Wheelers advocated the repeal of laws that "do not bear equally upon capital and labor," including unequal taxation, tariffs, support for the gold standard against silver, and legislation that permitted the development of monopolies and trusts. They also demanded protection of the Chickasaws and Choctaws against encroachments by railroads and wealthy syndicates. To prepare their children for the future, they supported "a well-regulated system of free schools." Finally, they called for an end to patent renewal. The list of demands concluded with a resolution that "this body will not support any man for Congress, of any political party, who will not pledge himself in writing, to use all his influence for the formation of these demands into laws."[27]

The list of demands compiled at the McKenzie meeting built on the reforms espoused by labor unions, Greenbackers, and nationalists and summarized the position of rural radicals for the next decade: an expansion of democracy, government control of the currency, government regulation of the economy, and public education. The platform articulated at McKenzie defined the demands of Tennessee's farmers and provided the yardstick by which the Wheel (and Alliance) would measure political candidates.[28]

The Wheel constructed parallel arguments to those being formulated concurrently by the Texas Farmers' Alliance. The Farmers' Alliance began in September 1877 when a group of farmers who described themselves as "comparatively poor," met at the Lampasas County farm of J. R. Allen and organized the Knights of Reliance (later the name was changed to Farmers' Alliance). The Alliance entered into what historians call "the culture of protest," a culture that included industrial workers as well as farmers, whose ideology rested on producerism and social relationships to the land. In Texas the initial organization pitted the small landowners and tenants against the advocates of scientific farming, fencing of the range, and the railroads. As historian Robert McMath argued, the poorer western farmers understood that the enclosure movement promoted by their opponents centered on the issue of control, no matter how nicely packaged in campaigns to upgrade herds and protect both cattle and crops. The first organization, after spreading into the surrounding counties, collapsed in its efforts to lead the Alliance into the Greenback Party.[29]

In 1883, the Alliance was saved from oblivion when S. O. Daws, who Lawrence Goodwyn calls "the first Populist," was named to the newly created position of traveling lecturer. A compelling speaker, the thirty-six-year-old Daws energized the dormant alliances with antimonopoly speeches that called for the development of buying and selling cooperatives. Daws's enthusiasm drew men with similar perceptions and abilities into the Alliance, including William Lamb, a self-taught frontier farmer from Tennessee. Lamb's indignation over the condition of southern agriculture and his ability to articulate the need for action made him ideal as state lecturer (1884–1885). Under his guidance, the Texas Alliance grew phenomenally. More important, he saw the value of a coalition between like-minded organizations, including the Knights of Labor.[30]

In 1886 the Texas Alliance met in Cleburne in an atmosphere of tension brought about by the rapid expansion of the membership, the strain of establishing cooperative ventures across the state, and the agricultural problems associated with a crippling drought in West Texas. However, the real issue facing the delegates was politics. When insurgents drafted a list of "demands" that were adopted after hours of debate, the Alliance split between a conservative minority that eschewed political action and a more radical majority that hoped to use politics without becoming partisan. The organization's president and treasurer and the conservatives bolted the convention. In the face of what appeared to be a repeat of the fate of the original Lampasas Alliance, the organization made the newly elected chairman of the executive committee, Charles William Macune, acting president.[31]

Over the next year, Macune revitalized the organization by focusing efforts on three objectives: the unification of the local cooperative efforts into a Texas state cooperative, the spread of Alliance beliefs into surrounding states, and the development of a national organization. As Macune explained, all farmers could unite under a universal principle of cooperation "in its broadest sense . . . that we will assist one another, that we will stand shoulder to shoulder in bearing the crosses and burdens of life, that we will intelligently pull together in everything; in buying and selling, in producing and consuming." But, as Macune argued, the solution to the farmers' dilemma extended beyond the borders of his own state and demanded the development of a national organization to exert pressure for change. Macune believed that "as the interests of the cotton-producers were identical, and the evils from which they were suffering general, the greatest good could not be affected without uniting the whole cotton belt." Local dissension would disappear under the grand march to organize and unify the cotton belt into an agricultural business alliance based on cooperative effort rather than self-aggrandizement and greed.[32]

In the winter of 1887, J. F. Alsup, a national organizer, met with Tennessee farmers and in March organized the first suballiance in Wilson County. A year later, the Tennessee Farmers' Alliance convened under the leadership of John P. Buchanan, a Rutherford County farmer and member of the general assembly. After the Tennessee Alliance merged with the Agricultural Wheel in 1889 as the Farmers' and Laborers' Union of America, the expanded state organization claimed 99,440 white members in 2,436 subordinate unions, covering ninety-three of the ninety-six counties of Tennessee. The Colored Farmers' Alliance added another 421 unions.[33]

Not surprisingly, before the merger, tensions arose between the Wheel and Alliance as organizers competed for members. The official Alliance histories only hinted at the strained relations between the two, but the Tipton County Wheel provided stronger evidence when the secretary described a recent "harmonious meeting," an apparent contrast to earlier ones. He continued: "the alliance was in this county before the wheel, and when the wheel came some jealousy arose, but I find that is now dead."[34]

Even as strife between the two organizations dissipated, tensions remained. Farmers from around the state testified to the need for unity and harmony while suggesting the sources of ongoing conflict. Sam Young, a Dyer County farmer and member of the Tennessee General Assembly, worried that "the greatest difficulty the Wheel has had to contend with is a want of unity of action." Suspended between hope and doubt, he observed that "the consolidation of the

Wheel and Alliance seems to have infused new life in most of the brotherhood, but we have a few men in our ranks yet who want to use the order to advance and build up their own interests."[35]

Across the state, the Knox County Alliance qualified its assertion that the members were "getting along nicely" by adding that they were "getting straightened out, at least." The East Tennessee farmers praised the efforts of their leaders, venturing to predict that they would one day "be recognized as public benefactors." The report concluded with an invitation to promote a more amicable unit by using care in the admission of new members, exhibiting a commitment to "be generous enough to act in harmony," and avoiding becoming "weary in well-doing."[36]

As the reports suggest, discord between organizations posed fewer problems than internal strife. In an apparent effort to foster harmony, Wheel and Alliance leaders moved to create a newspaper to provide readers with a unified voice on organizational matters and control dissent. Their efforts were rewarded with the establishment of the *Weekly Toiler* as the official organ, but this became one of the most tumultuous episodes of Tennessee Wheel and Alliance history.[37]

Two newspapers claimed to be the official organ of the Tennessee agrarian movement: the *State Wheel,* published by T. B. Ruff at Jackson, and the *Weekly Toiler*, published by John H. McDowell and M. V. Ingram at Nashville. No copies of the *State Wheel* survive, but contemporary newspapers, including the *Toiler*, reprinted its articles. Those articles and the debate that surrounded efforts to unify behind a single paper suggest that the *State Wheel* pursued a more inflammatory editorial policy than did the *Toiler*.

Ingram had been publishing the paper for three years by the time McDowell joined as editor in the spring of 1888. A seasoned newspaperman and farmer from West Tennessee, McDowell soon controlled the paper's editorial policies and instituted changes that improved the appearance of the paper and elicited praise from rival journals. The *Clarksville Chronicle* noted the improvements and assured its readers that while the *Toiler* "represents the farmer of the country," they should not "conclude from this that there is anything rural in its makeup." Somewhat less enthusiastically, the *Nashville Banner* reported that "the Toiler . . . shows every evidence of increasing support and success." Evan Jones, the first president of the National Farmers' and Laborers' Union, spared no praise when he labeled the journal "the best of its kind in the South."[38]

By early summer 1888, it had become clear to the rural readership that conflict between the two papers threatened the state's agrarian movement. Pursuing different editorial policies and promoting opposing positions on the same issues, both the *State Wheel* and the *Toiler* claimed official sanction. The problem of settling disputes between the two papers became so acute that State Wheel

President J. R. Miles intervened to demand that the editors behave like brothers, put an end to strife, and focus on the interests of the order.

According to Miles, McDowell complied, but Ruff continued the feud, and the matter occupied three meetings of the state leaders. Although McDowell had earned a reputation for editorial venom, he took the more conciliatory position, perhaps confident of his ultimate victory in the newspaper wars. Editorials in the *Toiler* focused on the need for an official source of Wheel information and noted that Tennessee was the only state in which the Wheel and Alliance had not designated a single official organ, a failure that compromised the work of the order.[39]

As Wheelers assembled for their last state convention in July 1888, everyone knew the newspaper crisis would be addressed by the delegates. In the final hours of the meeting, a resolution came before the convention to provide the state organization with an official journal, owned by the Tennessee State Wheel and operated according to the organization's principles. The resolution included a mechanism for funding the purchase of an existing newspaper or establishing a new one, by assessing each member fifty cents, redeemable as a six-month subscription. The resolution passed easily and, in a subsequent vote of 288 to 143, the delegates designated the *Toiler* as the official organ.[40]

Ruff declared the vote fraudulent, claiming illegal manipulation of the ballots. His unrelenting campaign through the columns of the *State Wheel* embarrassed the organization as antiwheel editors in Middle Tennessee predicted that the organization "would soon fall to pieces on account of corruption in the order." When the state leadership initially reacted to Ruff's charges in silence, believing that the accusations would fall of their own weight, the work in many subordinate wheels came to a standstill as members tried to determine the truth.[41]

Ruff maintained that the vote had been left until the final hours to rid the convention of West Tennessee delegates who might otherwise have supported his newspaper. Moreover, his description of the proceedings painted a sinister picture of election fraud, with Miles and his henchmen counting the ballots in a tiny, darkened room, far from the scrutiny of less partisan delegates. For weeks the *Toiler* printed letters from voting delegates and participants in the counting procedure refuting Ruff's allegations; no letters supporting the West Tennessee paper appeared in the now-official Wheel organ. From the one-sided presentation, it is impossible to determine the validity of Ruff's protests, but the furor quickly died down as the members accepted the explanations of state leaders.[42]

The episode reveals the thinking of the Tennessee agrarian leaders as they attempted to fashion a unified, responsive organization to protect their interests. Not unrealistically, they recognized the need to speak with a single voice, and they accepted the directives of the elected leadership. Perhaps the newspaper rift reflected local divisions between the Wheel and the Alliance (the *Toiler* represented both organizations, while the *State Wheel* seemingly spoke only for the Agricultural Wheel). Ruff called his newspaper the *State Wheel*, but he apparently pursued his own editorial policy, or at least one that had fallen out of favor. His pronouncements caused much internal turmoil and justified the criticism of outsiders who claimed the Wheel was the work of unprincipled men. When reined in by the organization, Ruff refused to make peace or follow the principles of the order. Both outsiders and members needed the assurance that the newspaper claiming the Wheel mantle spoke with the Wheel voice. Nevertheless, the episode left the Wheel and the Alliance open to charges of corruption and manipulation of information.[43]

Urban newspapers provided a statewide platform for the views of Bourbon and New South Democrats and the Republican Party, but the establishment of the *Toiler* gave the state's farmers a statewide forum and organ for expressing their position. The advent of the newspaper announced the independence of the farmers from the social and political dominance of the "best men," as upper-class men were known.

As editor, McDowell guided (some said manipulated) the direction of the agrarian protest movement from within the ranks of leadership (he was also state secretary of the Wheel and later of the combined organizations) rather than through contentious editorials. The editorial policy he outlined for readers depicted the *Toiler* as "conservative in tone, but firm in the advocacy of what we believe to be the interest of the great brotherhood." He promised that the newspaper would remain "non-partisan . . . [to] promote peace and harmony in the ranks" and asserted that "the ambition of my life is to see the tillers of the soil freed from the oppression of combined capital."[44]

Although McDowell's editorial style appeared moderate, even conservative, the journal generally maintained an open forum, publishing diverging views on matters ranging from the role of women in the organization to the efficacy of the subtreasury scheme. The pages of the *Toiler* therefore provide important clues to the problems organizers faced as they attempted to create a brotherhood in a state so strongly divided by class, history, gender, and race.

Class as a Dividing Wedge

The issue of class cut several ways: planter versus yeoman and tenant, farmer versus laborer. Michael Schwartz suggested that the Farmers' Alliance brought together the producerism of the yeoman farmer and the political activism of the more substantial planter. As he interpreted the development of agricultural reform in North Carolina, the advent of the Alliance placed wealthier planters in positions of state and local leadership. From their positions of power, these men led the reform effort into partisan politics, but deflected the more radical proposals. Roger Hart's analysis of Tennessee agrarian reform found a similar pattern of control of leadership positions by the more substantial farmers, but, as he found little evidence of the implementation of Alliance demands by members elected to the state legislature, he interpreted the presence of economically secure men as a sign of demagogic control by political outsiders who used the Alliance organization as a mechanism for obtaining political office. The Tennessee Wheel and Alliance, like the agrarian experience in North Carolina, followed a more conservative path to reform, but a closer look at the development of the state's agrarian organizations reveals a more complex problem than a delineation of wealth can explain.[45]

In Tennessee, planter-dominated agricultural organizations remained aloof from the Wheel and Alliance; local farmers' clubs and conventions continued to meet and read papers in apparent isolation from their scrappier neighbors. The Giles County Farmers' Association, for example, included a number of prominent planters who apparently found no need for cooperation with the Wheels and Alliances that met regularly in that county. In other parts of the state, local agricultural elites undertook initiatives in defiance of the efforts of poorer farmers. Madison County's rural leaders determined to build a cooperative creamery in what seems to have been an attempt to counter the attention gained by local Wheel and Alliance cooperative efforts.[46]

Both the Wheel and Alliance attracted substantial commercially minded farmers to positions of leadership, though neither apparently enticed the great planters or the most prominent New South agriculturalists. John P. Buchanan, the president of the State Alliance, raised blooded cattle in Rutherford County, where his landholdings placed him among the wealthier men of the area. Oakland House, the Weakley County home of Alfred E. Gardner, the state and national secretary of the Wheel and the combined Wheel and Alliance, stood as a model of Victorian architecture and a symbol of rural wealth. When John H. McDowell made a tour of the wheels in the Upper Cumberland counties, he stopped at the home of the Smith County Wheel president, William Ross Shaver. The former state senator (1885–1887) showed McDowell over "a magnificent blue grass farm of 500 acres" that encompassed the once-noted Clinton College. Shaver raised Durham cattle, and at the time of McDowell's visit had over one hundred head in his herd, among which was "the 20th Duchess of Crethmore, a pure Bates, [that] cost him $500.00 at three months old."[47]

Levi Thompson of Giles County praised the efforts of J. F. Black, an affluent farmer. "When the alliance tocsin was first sounded for our section," Thompson wrote, "Bro. Black was among the first to rally to the call. While other men of prosperity stood aloof and scoffed the institution, he paid no heed to criticisms of the wheel, but threw his shoulder to it side by side with his poorer neighbors." Wheel and Alliance members made no secret of the presence of affluent farmers in their organization. Indeed, in speeches and editorials, they frequently upheld these men, and their successes, as examples of the efficacy of good farm methods and the respectability of their organizations.[48]

The discriminating placement of prosperous farmers in leadership positions built on traditional southern deference to the hierarchy of wealth, but it also represented a realistic approach to the limited financial underpinnings of agrarian organizations and a valid assessment of the sometimes superior experience and education of agrarian elites. Frequently unable to meet even the minor expenses of officers who spent several days a week traveling to rallies and organizing events across the state, the Wheel and Alliance relied on men with the means to purchase their own railroad tickets and pick up their own hotel tabs. Officers and lecturers were farmers themselves whose traveling schedules necessitated long absences from the fields. A number of these men were engaged in raising blooded livestock rather than the more labor-intensive crop production, which could only be accomplished by renting out large tracts of land to tenant farmers.

Tax records and anecdotal evidence indicate that local and county offices also went to men who seemingly had nothing more to recommend them than the esteem of their neighbors. J. M. Glasgow of Obion County figured prominently in the affairs of the Wheel and Alliance; he was president of his county alliance in 1890, and served with J. H. Bittick and John McDowell as a delegate to the Populist Congressional Nominating Convention from the tenth district of Obion County in 1896. The county tax records for 1890 shows that Glasgow, a neighbor of Bittick, paid taxes on ten acres valued at $150. For the thirty-two Obion County property taxpayers who can be identified from the 1896 convention, the average land holding was 130.4 acres, with a range of 10 to 519 acres. Nineteen of the thirty-two held property of 100 acres or less. Only four owned more than 250 acres.[49]

The president of the State Wheel, J. R. Miles, described his impoverished background in his Confederate Veterans' questionnaire. His father appeared to be making some economic progress; his property included a three-room frame house and two or three slaves. But at his father's death, when Miles was twelve, his mother sold the slaves and moved Miles and his siblings to Weakley County. There the mother and children farmed 98 acres and lived in a two-room log house with a split-slab floor and "cracks you could run your hand through." Miles had only one year of schooling, explaining that his mother could not afford the $1.50 monthly charge to provide extensive education for her eight children. After the Civil War, he returned to farming and twice served as magistrate of his county.[50]

Among the first to join the agrarian cause, Miles paid a high price for his efforts. In October 1888, McDowell called reader attention to the plight of the state president. In order to promote the interests of the state's farmers, "the father of Wheelerism in Tennessee" left the management of his farm to sharecroppers, "who by bad management [and] sickness . . . have nearly ruined Brother Miles financially." Miles apparently had no other source of income, and McDowell, without Miles's knowledge, appealed to readers for contributions to relieve the president's financial difficulties, noting that "the enemies of the order would rejoice to see our noble old President go down financially." Miles seemed embarrassed by the attention and declined the assistance of the membership, but his situation was not unusual: absent from home for long periods and forced to assume the economic burdens associated with travel, agrarian leaders sometimes placed their farms, and thus their financial well-being, in the hands of less capable tenants or sharecroppers.[51]

Nor was Miles the only one to experience adverse economic consequences as a result of participation in leadership roles in the Wheel and Alliance. In 1889, as the State Wheel closed out its books in anticipation of its merger with the Farmers' Alliance, it became apparent that W. T. Davis, secretary of the State Wheel, had embezzled $2,100. His five bondsmen eventually prosecuted their former "brother," who was sentenced to five years in the state penitentiary. The bondsmen, of whom Miles was one, struggled to reimburse the organization for the stolen funds. Again McDowell called upon the members of the Alliance to help. Alliance women became especially interested in this cause and urged one another to take up penny collections, much as they did for church missionary work. The episode and the plight of the bondsmen remained news for the *Toiler* for several weeks, and sporadic reports demonstrated a concern and a desire to help, but generated only small donations to the cause of the beleaguered bondsmen.[52]

The problems of finances extended across the organizational spectrum. The state business agent in Memphis, R. H. White, finally resigned after several months without pay after West Tennessee county and local unions failed to collect the one-dollar assessments for the support of the cooperative work. McDowell experienced repeated difficulties in meeting the expenses of the *Toiler* and made weekly appeals for subscriptions, offering prizes and free subscriptions to members who brought in new, paying readers. He occasionally reminded readers to enclose stamps if they expected a reply to their letters of inquiry about the work of the organization: the *Toiler* could not afford the postage to answer their questions.

The financially strapped editor identified other "hidden" economic costs assumed by officers in the organization. He received many invitations from local and county Wheels and Alliances to speak at meetings, rallies, and "jollifications." His enthusiasm for the cause kept him traveling most of the time, crisscrossing the state by horse, buggy, wagon, and train. While many, and perhaps most, Wheels and Alliances provided meals and a bed for him in the home of a local member when the activities kept him overnight, some did nothing. In exasperation, McDowell lectured readers on the costs associated with travel. "When you send for speakers to educate you in the good work that benefits all," he wrote, "remember that the speaker has to pay hotel bills and railroad fare besides loosing [sic] his time, and you should always pay his actual expenses if no more. Don't ask a man to benefit you at his expense." In closing, he referred to the principles of the organization, reminding members that "the laborer is worthy of his hire, be it in the corn field or lecture room." McDowell pointedly thanked Smith County for the round-trip ticket that accompanied its invitation to speak. R. W. Tucker provides another example of the costs of leadership. A young teacher and farmer from Wayne County, Tucker rose to the position of State Alliance Lecturer and became the subject of a biographical sketch in the *Toiler*. He served as the Wayne County business agent for fifteen months from 1887 to 1889. He reportedly "rode horseback 3,000 miles and delivered over 100 lectures; during that time, he also paid out three hundred dollars more in expenses than he received as wages."[53]

The problems of funding hampered not only the Tennessee reformers, but the national officers as well. When the Farmers' Alliance and Laborers' Union of America met in St. Louis in December 1889 following the merger of the Wheel and Alliance, President Evan Jones reported on the state of the organization. After assuring the delegates of the nobility of their purpose and the surety of success, he confessed to his own misgivings. The dissolution of the two national agrarian organizations and the emergence of the Farmers' and Laborers' Union placed Jones in what he described as "a very awkward and embarrassing situation." With "demands coming from the various States for organizers, new rituals, secret work, and other printed matter, and having no funds in the treasury for defraying expenses," he was compelled "to draw upon my own private funds for the defraying of all my office and official expenses." The financial shortfall, combined with the "considerable division and dissension in some of the States," left Jones with "great forebodings." He expressed his gratitude to those who came to his assistance, including J. H. McDowell, who printed 50,000 rituals with the new secret work, and had them "ready for distribution to State secretaries within thirty days from the issuing of our official proclamation." Given McDowell's chronic need for money, one can only speculate that the cost was added to his accumulating debt.[54]

Despite the need to elect state officers who were capable of paying their own expenses, Tennessee farmers remained reluctant to permit domination by "the best men." Several skirmishes provided ample evidence of attempts by wealthier members, or even outsiders, to control the Alliance. Tennessee distinguished itself by becoming the first state to ratify the merger of the Wheel and Alliance in a joint state meeting in July 1889. Almost immediately, an effort emerged to petition the national organization at the St. Louis meeting in December 1889, for the removal of "Laborers" from the name of the organization (Farmers' and Laborers' Union). Supporters of the demand justified their position by arguing that farmers would be better served by a farmers' union focusing only on their interests. Local unions quickly condemned the action in language that suggested that the plight of the farmer was not at the center of the demand.

The Coffee County Alliance clarified the class conflict inherent in the dispute: "We are organizing for the mutual protection of the laborer as well as the farmers; for the tenant as well as the land holder; for the men who teach our children to read; for the men who operate machinery or draw water or hew wood for their 'daily bread'; as well as the man who bosses numbers of hands on his large plantation," they wrote. The Poplar Hollow Alliance more pointedly asserted that "labor is, and always has been the bed rock of civil society . . . and it is clearly self-evident that the fight now on, when simmered down is simply a fight between capital and labor." The Poplar Hollow Alliance men considered it "necessary for the salvation of the country—for the preservation of liberty—and the perpetuation of a genuine republican government, for the United States of America, that the laborers in all the legitimate occupations of the civilized world should join in heartily together in one grand, yea in one gigantic effort to free ourselves from this hydra headed monster, monopoly." At least for these farmers, the attempt to exclude laborers from the union represented an effort by the well-to-do planters to control the organization and eliminate the potentially radical element. In a repudiation of planter domination, farmers declared their solidarity with laborers.[55]

Nevertheless, the union of the Southern Farmers' Alliance and the Knights of Labor (KOL) remained tenuous. Nowhere was the potential for contested ground greater than in the affiliation of class and race represented by the alliance of farmers and workers. Simultaneous with the rise of the agrarian movement, the KOL expanded nationwide, moving into the South with the railroads and the coal and iron industries. Under the leadership of Terence Powderly, the Knights organized both skilled and unskilled workers and admitted women and blacks to union ranks, positions not permitted by more conservative and restrictive trade unions. Like the Farmers' Alliance and the Wheel, the ideology of the KOL centered on producerism and cooperativism. Although most Americans would have associated the Knights with a successful strike against Jay Gould's railroad and the violence of the Haymarket Riot, the organization officially denounced the use of both methods.[56]

Knights of Labor organizing efforts in Tennessee centered on the coal and iron fields in Middle and East Tennessee, especially those at Tracy City in Grundy County and Coal Creek in Anderson County. The Knights also moved into the state's industrial cities and eventually organized eleven urban locals. Aside from affirmations of official policy positions, Tennessee's farmers provided few indications that they either sought or wanted the support of miners and workers. Thomas F. Carrick, miner and leader of the Tracy City Knights of Labor, occasionally wrote to the *Toiler* endorsing a closer affiliation between miners and farmers, but without apparent success.[57]

Although the Knights of Labor represented an important opportunity for organizing East Tennessee, farmers failed to develop this potential for at least three reasons. First, racism played a prominent role as farmers, no doubt, shied away from the biracialism attributed to the KOL. As Karin Shapiro's study suggests, some of the state's black and white coal miners lived and worked together more closely than whites elsewhere could accept. Part of the miners' acceptance of blacks as workers and union members rested on the role of African Americans as voters with the Republican majority in East Tennessee. The KOL's close alliance with labor was therefore called into question, as was the agrarian commitment to the Democratic Party and white supremacy for the Wheel and Alliance. Second, farmers repeatedly raised concerns that labor's views on property and taxation were incompatible with their own. Finally, many farmers were also employers and feared that strong support of labor would raise expectations and encourage organization among sharecroppers and pickers. Together these issues limited the Alliance commitment to labor and thereby affected the organization's influence in East Tennessee.[58]

The "Lost Cause" and Agrarian Organization

Exacerbating the divisions between farmers and laborers was the pervasive force of the "Lost Cause." In Tennessee, the industrializing regions fell almost exclusively in East Tennessee, where Civil War Union support and Republican

Party sentiment were strongest. Middle and West Tennessee, the strongest agricultural regions, had embraced the Confederacy in 1861 and now showed the strongest support for Lost Cause sentiment.

The romanticizing of the military feats of the Confederate army began with the death of Robert E. Lee in 1870 and gained strength with each passing year. Veneration of southern soldiers and reverence for the imagined harmony and gallantry of the antebellum South captured even the most committed of the New South advocates. Seemingly unmindful of the paradox, Henry Grady, whose most famous speech announced the death of the "South of slavery and secession," praised the "Imperishable knighthood" of the earlier age as well as the "exquisite culture" of the Old South. Indeed, historian W. J. Cash proclaimed that "far from representing a deliberate break with the past, the turn to progress clearly flowed straight out of that past and constituted in a real sense an emanation from the will to maintain the South in its essential integrity."[59]

The Lost Cause permeated public life in the South. Ministers imbued the Lost Cause ideology with religious significance as they tied morality and ethics to the fallen heroes of the Confederacy. In churches across the region, congregations demanded particular hymns because they evoked the antebellum or Civil War experience. Public rituals in the form of commemorative celebrations offered the occasions for speeches and sermons in praise of southern virtues. As the century came to a close, Confederate veteran groups and women's organizations, particularly the United Daughters of the Confederacy, collected contributions to raise statues in honor of the southern war dead. Histories rolled from the presses, and forgotten skirmishes assumed new life in the journals and biographical accounts that vied for public attention. No facet of southern life escaped the imprint of the Lost Cause.[60]

By 1890, on the twenty-fifth anniversary of the war's end, the number of Confederate veterans residing in Tennessee had dwindled to 31,727. The decrease in the number of living veterans spurred the romantics to greater efforts to record, justify, and enshrine the Confederate history in an unassailable regional culture of memory. As a result, the veterans themselves, particularly those with war wounds, became the objects of public philanthropy. In no small measure, the sudden spurt of charity on behalf of ill and destitute Confederate veterans arose from the fury southerners stifled as they watched the well-pensioned Grand Army of the Republic accept public accolades and tax dollars. Robert Cartmell, a Madison County farmer, complained bitterly about the "paternalistic" government of the Republican Party that voted pensions for what appeared to be a growing army of northern veterans. Condemning the pension system with the assertion that "no such fraud and outrage was ever perpetuated by any

Party or on any People before," Cartmell's observations reflected the expected bitterness of a failed cause as his complaints also unintentionally pointed to the economic advantages of Republican paternalism with its infusion of cash into northern communities, a benefit southerners could only envy.[61]

During the 1880s the Tennessee state legislature enacted two modest pension programs for disabled Confederate veterans. The first act provided a lifelong monthly payment of ten dollars for Tennessee veterans who had lost their sight as the result of battle wounds. The second act raised the payment to twenty-five dollars for sightless veterans and extended the pension to include amputees. Finally, in 1889, James M. Crews, a former Confederate soldier and member of the Farmers' Alliance from Shelby County, introduced a bill to "provide and maintain homes for needy ex-Confederates of good character, with their widows and their orphan children." The resulting establishment of a Confederate veterans' home on the grounds of the Hermitage, Andrew Jackson's plantation in Nashville, physically placed the old warriors alongside the state's most cherished historical figure.[62]

The marshaling of public opinion on behalf of the Lost Cause and in support of impoverished veterans opened avenues to recruitment of like-minded farmers who shared a Confederate past. However, evoking the Lost Cause could be counterproductive for Wheel and Alliance reformers: romanticizing the past potentially isolated East Tennessee farmers and laborers from their brothers in Middle and West Tennessee, who did not share a common past—and the break point occurred at the Civil War. Equally troubling, if not as obvious in its implications, the inculcation of Lost Cause ideology bound together planter and yeoman or (white) tenant farmer—men who otherwise championed opposing solutions to the problems of the Gilded Age political economy.

Like James Crews, a substantial number of county and state agrarian leaders claimed military service. Most, because of age and economic background, remained in the enlisted ranks, although there was a sprinkling of junior officers, few above the rank of captain. Both the president of the State Agricultural Wheel, J. R. Miles, and the president of the State Farmers' Alliance, John P. Buchanan, served in the Confederate ranks. Miles, a self-proclaimed subsistence farmer, enlisted in Company I of the Twentieth Tennessee Infantry in November 1863. A man of modest means with only one year of education, Miles did not advance in the ranks, but his prior service apparently earned him the respect of his neighbors, and he was elected to the position of magistrate twice during the postwar years. Before the war, Buchanan's father had owned substantial holdings in Williamson County. In 1864, sixteen-year-old John joined Roddy's Escort and served until the end of the war. Like many others, Buchanan returned to a ruined farm and was forced to start over in nearby Rutherford County. John H. McDowell, the editor of the *Weekly Toiler*, first attempted to enlist in the Confederate army at the age of sixteen. In 1861, he signed up with the Tennessee Seventh Infantry and began training before his age became known. He was sent home but successfully enlisted at a later date, served at Vicksburg and Meridian, and remained a lifelong supporter of veterans' causes.[63]

Obion County, a hotbed of Alliance and Populist sentiment, had also been the scene of considerable wartime skirmishing and the site of bitter partisan unrest after the war. Bonds forged in battle and during the postwar disruptions remained decades later. Indeed, eighteen of the fifty-two known members of the Obion County Farmers' Alliance who later joined the People's Party served in the Ninth Tennessee Infantry. J. H. Bittick held minor offices in the county Alliance and became a trusted member of the inner circle of men who engineered the election of Buchanan to the governor's office. As a reward, Bittick received an appointment as supervisor of the state capitol, a highly regarded patronage position. Bittick had joined the Ninth Tennessee in May 1861 and saw his first action at Shiloh, where he was "desperately wounded." He described his recovery as "a long serious struggle," though he was back with his unit in time for the Chattanooga campaign. Lingering weakness compelled him to request a transfer to the cavalry, and he finished his service in Forrest's Cavalry, where he advanced to first lieutenant.[64]

Other Obion County veterans claimed similar histories. Bittick's brother-in-law, Thomas Henry Latimer, joined the Thirty-third Tennessee Infantry. After his capture at Missionary Ridge, he walked to the federal prison at Rock Island, Illinois. In 1896, Latimer was an active member of the People's Party, serving as a delegate to the state nominating convention. T. P. (Pled) Callicott, a Wheeler and Allianceman, as well as a member of the Tennessee General Assembly, shared the experience of capture and imprisonment. Captured at Collierville, Tennessee, in February 1864, he was released from the prison at Johnson's Island, Ohio, in June 1865. Of the twenty-two Obion County Confederate veterans who attended the 1909 dedication ceremony for the second of Union City's two Civil War monuments, ten (45 percent) could claim past membership in the Wheel and Alliance. Six of the ten had actively participated and held offices in the Wheel, Alliance, and People's Party.[65]

Like other communities, Obion County veterans and supportive women's organizations collected funds to erect monuments to the Confederate dead. Union City erected two monuments. The first one, dedicated in 1859, was a relatively plain obelisk that honored the twenty-nine unknown Confederate soldiers buried nearby. In 1909, Union City residents erected a second

Confederate monument depicting a uniformed soldier. This memorial extolled the Anglo-Saxon heritage of those who suffered and died in words evocative of the Lost Cause:

To the Confederate Soldier of Obion County
Who was Killed in Battle
Who Was Starved in Federal Prison
And Who Has Preserved Anglo-Saxon Civilization in the South[66]

Confederate service, federal imprisonment, and the lingering effects of war wounds characterized many of those who rose to positions of leadership in the suballiances and county organizations, raising the question of how common their wartime experience had been and what effect this had on recruiting efforts for agrarian reform. Fred Bailey's analysis of the Tennessee Civil War questionnaires suggested that among the 1,253 veterans who responded to the 1915 inquiries about their military service, the percentage reporting war wounds was equally distributed across economic and class lines. Bailey reported 20.6 percent wounded among the poor yeomen veterans, 24.4 percent among nonslaveholding yeomen, 23.8 percent among slaveowning yeomen, and 21.3 percent among elites. The consequences of such injuries were anything but equal, however. Men with little education, no wealth, and few skills faced lives of poverty and pain. Scattered sources indicate that communities may have provided for a few disabled men by electing them to minor political offices as constables and magistrates—positions threatened by the urban-based progressive reformers of the 1890s, who demanded educated bureaucrats. Bailey reported that former prisoners of war "tended in later years to be more antagonistic toward the North than other former Confederates." If the Wheel and Alliance held out potential for alleviating the social and economic crises of the period, its affiliation with like-minded northern farmers perhaps turned away many potential recruits.[67]

McDowell personified the paradox of remembrance and reunion. A lifelong advocate of reconciliation, he used the pages of the *Toiler* to smooth over sectional antagonism by providing weekly column space for news from state and local alliances in Kansas and Missouri. He reinforced the common ties between the farmers of Tennessee and Missouri when he presented a biographical sketch of the president of the Missouri State Wheel, H. W. Hickman. Born in Obion County, Tennessee, Hickman served in Company A of the Thirty-third Tennessee Volunteers. Like many other soldiers from the area, he was captured by federal troops and incarcerated at Johnson's Island.[68]

Antebellum nostalgia threatened agrarian organization as surely as did antagonisms lingering from the war. For reformers, an uncritical ennobling of the past spelled disaster: if the past was not open to critical analysis, change was out of the question. As Fred Bailey pointed out in his assessment of the Tennessee Veterans' Questionnaires, most men professed little class consciousness in the antebellum epoch, a sizable minority expressed discontent based on social and economic status. Intensity of class animosity increased from east to west and was most severe in the cotton southwest. Likewise, the poor were more likely to express awareness of class-based antagonism; the wealthy were least likely to do so. From the perspective of the success of agrarian reform efforts, the most troubling answers to questions about upper-class patronage asserted that elites had done more before the war to assist the lower classes in their struggles for economic independence than they did after the war. The widespread acceptance of this southern truism left reformers vulnerable to the suppression of their views in the name of southern patriotism. If the men crying for relief became convinced that contemporary elites, like their predecessors, could be counted on to protect the interests of "lesser" men, the movement for agrarian reform would collapse. The avenue for making that claim led through the nostalgic rituals of the defeated Confederacy.[69]

In 1892, when the Democratic Party of Tennessee sought an opponent to the Alliance Governor's claim for a second term in office, they selected the feeble State Supreme Court Justice Peter Turney, who enjoyed impeccable credentials as an unreconstructed southern patriot. The old fire-eater made his way through West Tennessee accompanied by his platform partner, Mississippi Congressman John Allen, the "only Confederate Private in the U.S. Congress." In the war of words and ideas, upper-class political brokers used every opportunity to remind veterans of the sacrifices made on their behalf by officers who accepted the responsibility of caring for their men. Reunion speeches and published war stories enshrined even minor skirmishes and attached great significance to the small gestures of Confederate officers, imbuing them with evidence of an egalitarianism that did not exist. If veterans could be organized by appeals to comradeship and mutual suffering, all but the most committed to the agrarian cause were equally vulnerable to the Lost Cause nostalgia for a return of the paternalistic rule of the elites under the guise of equality.[70]

While Tennessee members of the Wheel and Alliance grappled with the lingering emotions of Confederate service, they also participated in activities that emphasized American unity and an emerging nationalism. In the name of agricultural interests, many former Confederate soldiers laid aside their prejudices and hatreds to make common cause with Union veterans.

Twenty-five years after the war, the mythology of the Lost Cause found an active competitor in Americanism. On July 4, 1890, Chattanooga held a highly publicized Independence Day parade in which Confederate veterans marched under a banner that proudly proclaimed, "No More Loyal Citizens Today Than Are the Boys That Wore the Gray." Wheel and Alliance leaders encouraged the use of Independence Day as a showcase for the organization. A few weeks before the 1889 celebration, the *Toiler* suggested that "the brethren of the wheel and alliance have big Fourth of July Rallies in every county . . . and try to stir up . . . that spirit of freedom, manhood and resistance to oppression that characterized our forefathers who handed down to us that proud fabric of American liberty and freedom." As editor of the *Toiler* and as an agrarian leader, McDowell consistently promoted reunification between the North and South. When wartime animosities threatened the national organization at its Memphis meeting in November 1892, McDowell led a veterans' "love fest" to tamp down sectionalism. As late as 1912, he was still staunchly advocating an end to wartime sectionalism. On the occasion of his seconding speech for the nomination of Theodore Roosevelt as presidential candidate for the Progressive Party, McDowell focused on the need to bridge the chasm between North and South and proclaimed the old Rough Rider the only man capable of reuniting the nation across the sectional divide.[71]

McDowell's treatment of the most emotion-laden event of the Confederate South—the death of Jefferson Davis—is especially illustrative of the editor's sentiments and the agrarian effort to unify the nation's farmers in one grand cause. Buried in the interior of the *Toiler* between a poem titled "The Incubator Chicken" and testimonials for the cure of blood poison, catarrh, scrofula, and rheumatism, the account used the death of the Confederate president as a vehicle for upholding the national unity of the Farmers' Alliance. The first paragraph noted that Davis, "the hero of the South," had died in New Orleans. The report depicted a South "draped in mourning" with "evidence of great love and respect for the Southern chieftain shown by the mass meetings in every section of the South." After listing Davis's accomplishments, ending with his service as president of the Confederate States, McDowell shifted tone. "We will not undertake to argue the question of whether secession was right or wrong; the cause did not succeed," he wrote matter-of-factly. The result of secession and war was a lingering sectional hatred, "blinded by prejudice." He took the occasion to lecture the two political parties for their willingness to promote the campaigns of demagogues who "kept alive these prejudices to the injury of the whole country . . . and the destruction of the free institutions of our government." In contrast, the editor noted the significance of another event that occurred the same day as Davis's death—the unification of northern and southern farmers at the Alliance's St. Louis convention. McDowell interpreted this demonstration of "brotherly love and fraternal union" as the beginning of a movement that would "wip[e] out in so far as the farmers of our nation are concerned sectional lines."[72]

In his depiction of agrarian unity, McDowell pursued a theme also expressed by the president of Tennessee's State Alliance six months earlier on the occasion of the union of the State Wheel and Alliance. John P. Buchanan hoped for a future where farmers would be "so fully imbued with the fundamental principles of our organization" that every call to resist and overthrow monopoly would be as fully obeyed as "the organized, well-disciplined, brave patriotic soldiers obeyed the order of Lee, or Grant." Clearly, Wheel and Alliance leaders sought to capture the emotions associated with military service and the Lost Cause to support their own agenda while simultaneously replacing sectionalism with Americanism and agrarian unity. Such a precarious position left the movement vulnerable to attack from both southern and northern partisans.[73]

Women in the Wheel and Alliance

Like the problems of class and sectionalism, the issue of gender presented the Wheel and Alliance with a potential source of conflict as well as a mechanism for establishing harmony. As they presented themselves for membership in the Wheel and Alliance, women confronted a variety of personal, social, and organizational ambiguities. Unlike the Patrons of Husbandry, which insisted on female membership and provided women with a separate and complementary role, the newer reform movements encouraged female membership without clearly defining the status or activities of women. While some states, such as North Carolina, offered few obstacles to the admission of women or their inclusion in public activities and "stressed [female membership] forcefully," the best that Tennessee women could hope for were occasional resolutions favoring their admission and the opportunity to defend the case for their inclusion in the Ladies Column in the *Toiler*.[74]

Tennessee farm women apparently did not play as large a part in Alliance affairs as women from other regions, but they did expand their activities beyond the traditional roles that southern womanhood allowed. While more vocal and notorious women in the prairie states enjoyed a more expansive role in public and political affairs, historian MaryJo Wagner concludes that "joining the Alliance and supporting the Populist Party did not involve a change in ideology concerning women's roles" in defense of farm and family. Studies of the far western states also reveal a pattern of utilizing gender-specific roles to further political ends. Marilyn P. Watkins suggests that the maintenance of community ties, a traditional female role, was vital to agrarian reform movements while it furthered women's political awareness and demands. Alliance activities broadened the political outlook of rural women, but the degree to which they could, or did, avail themselves of the opportunity for expanded roles and further demands for equality and suffrage appears to have been dependent upon local and regional expectations and definitions of the proper sphere for women.[75]

Despite its commitment to "equal rights to all, special privileges to none," not all women had ready admission to the Wheel and Alliance, and the pages of the *Toiler* frequently carried letters voicing their complaints against their exclusion. Some women expressed their desire for membership, though a lifetime of inequality made them reluctant to be the first to step forward. Mrs. C. A. Mitchell of Cheatham County supported the idea of women in the Wheel and Alliance, yet she was not a member because "we have no lady members in Oak Lawn, and I am only waiting for others to join." Others decided to wait to apply for membership until the most prominent woman (often the wife of the local or county president) joined. Even that strategy did not always work, as a Weakley County woman complained: "some of the best ladies of our community petitioned for membership, but were blackballed." A number of women thought they knew who stymied their entry into the order and pointed accusing fingers at the "old bachelors" and widowers in the local organizations. Women in Sullivan County named the culprit, the first president of the Stony Branch Alliance, "Uncle Jimmie," who openly stated that "he had been both a widower and a bachelor and he thought it best to have no lady members."[76]

Some women cried "How Cruel Uncle Jimmie!" while others supported a ban on female membership. One Sullivan County woman boasted she was "not a member of the Farmers' and Laborers' Union—for this reason: I am a lady and argue in the order is not woman's proper place." Across the state, a Lauderdale County "Wheeler's Wife" wrote that "the wheel is a good organization for men but [I] do not think it is for the ladies." She encouraged women to "stay at home and help their husband, father, or brother get ready to go to their meeting." She feared that Wheel activities drew men and women from more important religious duties and that they would "turn out to their wheeler meetings at any time, rain or shine, [but] when it was prayer meeting at their church they would say it is too cold or too rainy or dark." Clearly, even where Wheel and Alliance leaders welcomed women and encouraged their membership, many found it difficult to gain admittance to the local unions. Notions of propriety, along with the public and political nature of the organizations, operated against their active participation. If women wanted a place in the order, they would have to justify their inclusion. The attempt to provide that justification resulted in a public debate on the pages of the *Toiler* that offered a rare look into gender issues and a glimpse into the conflicts raging within the local unions.[77]

Women joined the Wheel and the Alliance because they agreed with the organizations' objectives. "I shall ever be ready to perform any duty assigned me that is calculated to promote the cause of the laborer," promised Cora McLean in a welcoming address to the Carroll County Wheel. Her friend Maggie Gwin followed her to the podium and read an essay praising women's faithfulness to the Wheel, describing the Wheel woman as always "ready to help in a cause when there is so much at stake—the prosperity of her people and the liberty and freedom of her citizens." Letter writers displayed the same commitment to Wheel and Alliance principles. Characterizing herself as "a sister who loves the principles of our great and grand organization," Nancy Tarpley of Warren County urged women to join the Alliance, "for we should all work together in love and harmony and help to down old monopoly [and] combined trusts." As a former Granger, Tarpley's dedication to Alliance programs displayed a zeal common to many *Toiler* correspondents. "Let us throw off our yoke of oppression," urged a Jefferson County woman, "and fight like men and women and work while 'tis called to-day." It was an enthusiasm born of experience, for women, no less than men, had a stake in the success of the Wheel and Alliance.

Where women were admitted, they held minor offices, read essays, and, in at least one case, contributed to the establishment of a cooperative store. Martha M. Johnson, the corresponding secretary of the Decatur County Wheel, reported that the organization had amassed $1,300 for the construction of a cooperative store. Eight men pledged $100 each, several donated $30 and $40, and "one Lady subscribed $10." Johnson declared of the lone female contributor, "She is a wheeler of the right kind." As working partners women shared the toil, and as petty producers they encountered the risks associated with agricultural production just as their husbands did.[78]

Traditionally, women's agricultural work was an extension of their domestic duties. The production of poultry, eggs, butter, jams, and jellies contributed crucial cash to the family farm without upsetting the gender roles dictated by tradition and society. Women took pride in their contributions, and when they met resistance to their participation in the activities of the Wheel and Alliance, they asserted their rights to membership and full participation rested on the recognition of their labors as agricultural partners with their husbands.

Letters from women to the editor of the *Toiler* routinely recorded their work in the fields as they claimed a share in the work of reform. As one woman wrote:

"It would be better, methinks if the men would say, 'come join us in the fight against your enemy' with as bold a front as he says 'come Betsy, help me hang the meat and drop the corn and potatoes. Or, while you are driving home the cows, just step over to the next meadow and salt the horses and let down the gap and drive the sheep through.'" The demands of farm life reduced the barriers that separated the work of men from that of women and roused agrarian women's interest in reform. Fannie Davis, a frequent contributor to the Ladies Column, defiantly asserted the rights of Alliance women, reminding her readers that "we help to produce as well as to consume." Pressing her point further, she asked: "how many bales of cotton do you suppose the fair daughters of your honest farmers help to prepare for the manufacturers and speculators? Then allow us room among you." Women demanded the right to participate in a fight that affected their interests, as well.[79]

Minnie Herrell of Anderson County sarcastically asserted that "if the union is such a disgraceful organization that ladies can't join they should disband and let monopoly rule." Biblical reference to women as helpmeets and reminders that the value of a good woman exceeded that of rubies filled the Ladies Column for weeks. A man from McLemoresville contended that a woman's "place is anywhere in God's universe that she can be instrumental in elevating and reforming any of God's creation." Furthermore, he argued that "her interest is our interest, and no one has more at stake in the upbuilding of the home than the wife and mother." Since, he added, "the Wheel in the beginning only aimed to elevate the conditions of home and to strengthen the moral atmosphere of society, it is woman's friend and women will stick with it with all the ardor of their nature."[80]

Women seemed to agree with this assessment of their role. Fannie Davis observed that "not only are the girls calculated to help do up the [farm] work, but they can cheer and assist society by their presence, for if love and peace does not dwell where ladies are, it's not to be found anywhere."[81]

Wheelers and Alliance men found merit in the assertions that women provided a moral atmosphere conducive to harmony and unity. After a particularly vigorous election campaign in 1888, the Wayne County Wheel resolved to exert a special effort to attract women to join the organization in the hope of developing a more harmonious association of farmers. John McDowell favored the admission of women because "all sensible men admit the refining, elevating influence of association with the good women of earth." Furthermore, he believed "the presence of ladies in our lodges prevents rude or uncouth language, one towards the other, and often produces harmony where otherwise strife and dissension would prevail." A man from Giles County agreed with McDowell, noting that "their presence [has] a tendency to keep order and their counsel of wisdom certainly [can] be felt and realized." Betty Gay, a contributor to N. A. Dunning's contemporary history of the Alliance, claimed that the soothing and refining presence of women had the effect of controlling strong tempers and placed "a premium upon politeness and gentility." Without the presence of women in the organization, she declared, "it would be a failure."[82]

The presence of women in the Wheel and Alliance introduced conflict within those local organizations that resisted female membership by calling into question the proper role of women in late-nineteenth-century American society. Women welcomed into local unions expected to be sources of strength and harmony, not to change the sphere in which they and their sisters lived and worked. Like their husbands and brothers, women participated in secret work, made speeches, wrote letters to the *Toiler*, served in minor offices, and promoted cooperative efforts. The boldness with which women claimed the rights of membership belied their modest demands. Nevertheless, the presence of women in an organization already suspected of radicalism worried opponents and provided grist for the antireform mills.

Whites and Blacks Working Together—Separately

The race record of the Wheel and Alliance has produced various interpretations, from praise for the progress toward unification of class interests to condemnation for consent to disfranchising legislation and Jim Crow laws. In fact, whites in the Wheel and Alliance could and did exemplify both behaviors, often simultaneously. The white Tennessee Wheels and Alliances managed to project the appearance of encouraging black brotherhood while actively supporting registration and poll tax legislation. At best, Tennessee Wheelers and Alliancemen acted paternalistically toward their counterparts in the Colored Wheel and Alliance and largely avoided the overt race-baiting and violence common to the period.[83]

Both blacks and whites made claims for hundreds of subunions and thousands of members in the Tennessee Colored Wheel and Colored Farmers' Alliance. Without accurate membership rolls, it is impossible to refute their claims completely, but available evidence supports the view that the organizations never attained the strength reported by the leadership. In his study of black Populism, Gerald Gaither suggests that inflated membership claims resulted from the practice of never purging the rolls of inactive members; attendance at a single meeting with no further activity may have resulted in "membership" in some instances.[84]

Even allowing for inflated membership claims, the Colored Wheel and Alliance engaged in active recruitment, overcoming tremendous odds to establish a viable organization in the western and north central counties of the state. The Tennessee Colored Wheel met at the same time and place as its white counterpart, with joint meetings in the opening ceremonies and an exchange of official representatives during the course of the proceedings. Both the Colored Wheel and the Colored Farmers' Alliance designated the *Weekly Toiler* as their official organ, and the newspaper periodically carried announcements and news of county meetings, plus occasional letters from members. When the Shelby County Colored Wheel experienced internal conflict that threatened the future of the local organization, black Wheelers invited the white county Wheel president to hear their complaints and settle the dispute. African American Wheelers and Alliancemen paid the one-dollar assessment and traded through the state agency "on the same conditions as enjoyed by the whites." Blacks expressed gratitude for the opportunity to use the agency and concluded that the "sympathy among our white members in Tennessee for the poor, oppressed, colored farmer is unexcelled in any other state, north or south."[85]

Whites also praised the efforts of black Wheelers. In July 1888, McDowell traveled to Covington in Tipton County to address the Colored Wheel on the cooperative system. Several local members of the white Alliance attended, including the county business agent and county president. In the course of his two-hour speech, McDowell reported that "the colored people showed an eagerness for information and enthusiasm for our cause and principles, which was unsurpassed by any audience that I ever addressed." Warming to his subject, McDowell concluded that "the prejudices growing out of their former condition are rapidly disappearing and if white brethren will only treat them kindly and show them that we are their friends, and that our fight for the elevation and prosperity of agriculture is a common fight for their interest as well as our own—they will join in with us and stand by as unflinchingly in every movement where unity is necessary to success." He came away convinced that "the colored farmers of Tennessee, if properly educated will make active working members of the order."[86]

McDowell especially praised the efforts of Tipton County Colored Wheel President James Y. Bernard, saying he "makes a good presiding officer and shows a knowledge of parliamentary ruling that would astonish many of the white brethren." Seven months later, his view of Bernard reflected increasing tension between the white and black unions. McDowell reported inquiries "from many parts of the State" from white members who wanted to organize Colored Wheels at the insistence of local blacks. The editor referred the requests to Bernard, who was also state president of the Colored Wheel. However, McDowell expressed little hope that the requests would be acted upon. "To be plain," he wrote, "the president don't [sic] want any white man to know their secret work, and therefore won't commission them [whites]." Since the editor believed it was "impossible to find a sufficient number of colored men in the State who are able and willing to enter the field and organize the colored people over the State," it seemed obvious that "the poor, colored people who need help even worse than the whites" would be "deprived of the benefits of the wheel." Although McDowell claimed that he favored "only colored organizers organizing colored wheels," the situation under the leadership of Bernard had satisfied the editor that "if the colored people in forty or fifty counties, where not a wheel exists, get the benefits of the order, their white brethren will have to assist them." McDowell attributed the unhappy situation to Bernard's "short-sighted policy of race prejudice."[87]

McDowell's editorial tirade against Bernard illuminates the complexity of the relationship between white and black Wheels. As McDowell certainly knew, the bylaws of the two organizations forbade whites from joining Colored Wheels, and blacks from white unions. Common signs and symbols permitted recognition between white and black Wheelers, but their secret work remained separate. No doubt whites would have refused to divulge their secrets to black Wheelers, no matter how noble the request. Yet, blacks, who refused a similar demand from whites, exhibited "short-sightedness" and "racial prejudice." As Bernard clearly understood, black Wheels, operating in secret and outside the supervision of whites, threatened the hierarchy of white supremacy that supported southern society. Throughout the history of the organizations, an uneasy coalition bound the black and white Wheels and Alliances together. Blacks needed the relatively stronger white organization to establish and maintain the institutions of cooperative buying and selling, but they resented the paternalistic concern for their success that only lightly masked a desire to control the Colored Wheel and Alliance. On the other side, whites needed the support and votes of black farmers. More to the point, since white membership rested on the landowning class of farmers, they wanted a labor force organized in support of their own views toward reform. However, in organizing blacks, whites may have worried that perhaps their opponents were right, and that they were undermining southern solidarity.[88]

The Tennessee Wheel and Farmers' Alliance organized under the shadow of past agrarian failures and a divisive history, as well as more contemporary divisions of class, race, and gender. The weaknesses of the Grange, the elitism of New

South initiatives, and the dominance of Bourbon planters soured many farmers on the possibility of Wheel and Alliance success. For the Tennessee agrarian movement, problems of history and Civil War service were not occasional issues confronted at national meetings but everyday problems in a state strongly divided between former Unionists and Confederates. Lack of financial support limited agrarian efforts, both in organizing and in the establishment of cooperative agencies. Wealthier men rose to positions of leadership, in part because their money provided the means for expanding the membership through lectures and publications. Nevertheless, their constituents kept a careful eye on their activities and pronouncements and voiced loud objections whenever they believed Wheel and Alliance leaders failed to defend the grassroots interests.

Defining the membership produced divisions within the organization and initiated some of the strongest external criticism. After excluding bankers, lawyers, and merchants as incompatible with agrarian interests, many found the alliance with industrial labor incongruous. The admission of women to the organization also created pockets of local turmoil. Wheels and Alliances that readily admitted women defended their presence as an aid to harmony; Alliances that refused admission sometimes found themselves under fire from women angered by their exclusion. Likewise, the Wheel and Alliance walked a tightrope in their relations with African American farmers. Wheel and Alliance leaders maintained their commitment to white supremacy and adopted a paternalistic attitude toward Colored Wheels and Alliances. Extending their hands in help, whites nevertheless expected the usual deference accorded to their race. When African Americans protected their Wheels and Alliances from outside interference and insisted on controlling the establishment of local unions, whites reacted in shock and dismay. The veneer of cooperation between the two races proved to be thin indeed. In order to foster unity between the disparate and destabilizing forces within the Wheel and Alliance, Tennessee agrarian reformers followed a conservative plan of cooperativism and antiparty rhetoric.

Brethren, we must meet trust with trust, organization with organization, and agree to abstain from using an article that has been cornered, no matter what it is.

—S. F. NEEDHAM, Secretary, Gibson County Wheel, 1888

CHAPTER THREE
BUILDING COOPERATIVISM

Gilded Age agrarian reformers built their organization on the existing rural community. The establishment of local Wheels and Alliances drew together friends and kin, farmers with similar experiences and histories, and extant community institutions that included country churches and other quasi-religious organizations such as the Masonic lodges. Tapping into existing community bonds strengthened the Wheel and Alliance even as agrarian leaders introduced a new ideology that undermined traditional politics, social relations, and agricultural practices. Paradoxically, dependence on country ties to kin and family also weakened agrarian reform as localism prevailed, allowing prejudices ranging from racism to sectionalism to dampen more progressive ideas.

Farmers organized to overcome the economic disabilities inherent in industrialization and production for the market economy and to give themselves a voice in the shaping of modern America. Incapable of resisting or limiting market intrusion, farmers eagerly sought the financial and material benefits of commercial agriculture while searching for ways to minimize the risks and maintain existing communal networks.

Divided as they were by geography, politics, and history, Tennessee reformers found traditional rural ideologies and paradigms effective organizing principles. Thus, organizers focused first on local networks of communal life that centered around rural congregations. Second, they addressed economic constraints through cooperative buying and selling agencies. Finally, they marshaled agrarian anger to stage a successful boycott against the jute trust. In all three endeavors, Tennessee farm organizers emphasized the conservative nature of the state's efforts and avoided stances that might point up the deep divisions between competing agricultural interests.

Religious Foundations of the Cooperative Community

The growth of national markets fostered conformity that purposefully undermined localism. The national progressive critique that developed in the early twentieth century suggested that the perceived backwardness of rural families stemmed from strong attachments to country churches and familial bonds that stymied change. Modernity, as historian Robert Wiebe explained, demanded an end to America's earlier "island communities." The Wheel and Alliance critiques of American industrial society defied the tenets of modernity and relied on the local bonds of friendship and kinship to effect social and economic change beneficial to farmers and their families. Thus, the Wheel and Alliance built their organizations on the extant community and closely identified with rural congregations. Church buildings offered convenient meeting places for local wheels and suballiances, and congregations provided ready-made organization. Furthermore, the moral language of the rural reform movement tapped common beliefs and ideas regarding the right ordering of society.[1]

Scattered and often isolated families in the rural South formed communities that did not conform to urban concepts of discrete neighborhoods where friends and kin were more immediately available. The boundaries of agrarian communities were less predictable, frequently ranging over miles of open country, bypassing some households and embracing others. As historian Jean Friedman observed, community consisted of "complex ties of kinship" that bound rural people together, making "community less a village of proximate neighbors than an understanding of the heart among distant kinfolk and neighbors." Visiting between families and friends separated by miles of alternately dusty and muddy roads was a staple of rural life. Farm families traveled to celebrate special occasions, bury loved ones, or assist in the rituals of work.[2]

Men and women in the Wheel and Alliance called one another "brother" and "sister," as they did in traditional country churches. Nor were these titles always simply social constructs, since many Wheel and Alliance members could say with Mattie Mahaffy, "I have a good portion of friends and relations who are members of the union." Churches provided accessible and familiar meeting places, reinforced communal ties, and confirmed an ideology of cooperativism and agrarian life.[3]

Obion County, in northwestern Tennessee, offered clear evidence of the thick web of kinship, friendship, and religious life that characterized local Wheels and

Alliances. The Bitticks, Catrons, Latimers, Glovers, Joyners, and Cloars formed a close network of agrarianism and shared family histories that overlapped with church membership. Fate Glover was John E. Cloar's brother-in-law; they served together in the Civil War and supported one another in the Wheel, the Alliance, and the People's Party. Likewise, John H. Bittick, the county Alliance steward, married Harriet Latimer, whose brother Thomas joined the Alliance and later was a delegate to the 1896 Populist Gubernatorial Convention. Family connections beyond the county also helped create an effective state organization. John McDowell's brother, William Wallace McDowell, was a Shelby County attorney, ineligible for Wheel or Alliance membership. Nevertheless, during William's 1889–1891 term in the Tennessee General Assembly, he was a strong advocate for agricultural reform.[4]

Neighbors and fellow church members also influenced one another to join the Wheel and Alliance. In 1890 J. M. Glasgow served as president of the Obion County Alliance and later became a delegate to the Populist Congressional Nominating Convention. County tax records show that Glasgow, a neighbor of John Bittick, paid taxes on ten acres of land valued at $150. This seemingly inconsequential farmer's access to an important countywide Alliance position perhaps resulted from his proximity to the more influential Bittick, as well as his association with the Beech Cumberland Presbyterian Church, where he had served as clerk since 1888. Fate Glover, John Cloar, and J. D. Killion were also neighbors who supported the Wheel, Alliance, and People's Party throughout the 1880s and 1890s. Of the thirty-seven Obion County agrarians known to have traversed the organizational path from Wheel to Alliance to Populism, and whose landowning can be confirmed, nine farmed adjoining land with one or more Wheelers and Alliancemen.[5]

County histories confirm the religious affiliations of fourteen Obion County agrarian leaders: five Cumberland Presbyterians, four Baptists, three Methodists, and two members of the Christian Church. Perhaps more interesting than denominational affiliations were the roles played by these men in their churches: particularly among the Cumberland Presbyterians and Baptists, agrarian reform leaders were sons of founding members of their local churches. Moreover, their close association with their family congregation and their activism in the voting districts where they worshipped suggest that the agrarian organizations coalesced around particular churches and around men with deep ties to the local church. Certainly the Beech Cumberland Presbyterian Church represented a hotbed of Wheel and Alliance activity. The Latimers, Cloars, and Bitticks were among the founding members and the most active participants in the Beech Church. Having migrated from Sumner County in Middle Tennessee, the founding families named their new church in honor of the one they left behind. This close association of settlement pattern and religious affiliation suggests a possible avenue for introduction of Wheel and Alliance ideas as agrarian reform took hold across the state. As the reform movement spread east, Cumberland Presbyterians in Sumner County were more likely influenced by organizers tied to them by history, family, and religious belief than they would have been by men of unknown affiliation.[6]

Likewise, Obion County's early and strong advocacy of Wheel and Alliance principles and its familial ties to Middle Tennessee supports a similar assessment of cross-county relations as a facilitator in agrarian organization. Although settlement began in Obion County in the 1820s, the largest migration to the region occurred after the Civil War, when the population rose from 15,500 in 1860 to 27,273 in 1890. The identifiable agrarian leaders were almost evenly divided between long-term residents and more recent arrivals. The majority in both groups migrated from Henry, Hickman, Humphreys, Perry, Lincoln, Marshall, or Giles counties—all actively involved in the Wheel and Alliance movements. If rural Americans lived in "island communities," the Obion County experience suggests that those communities were not as isolated as the designation implies. Visits among family members and exchanges of letters spread information and solidified ties beyond the immediate circle of acquaintances. While Middle and West Tennessee quickly organized in the mid-1880s, East Tennessee remained more resistant to agrarian rhetoric. Fewer familial ties bound the mountainous region to the rest of the state and, no doubt, inhibited the spread of the agrarian movement.[7]

Holding Wheel and Alliance meetings in local church buildings implicitly imbued the organizations with moral purpose and both shaped and sanctioned union activities. Men and women called upon God to bless their efforts in speeches that allied agricultural reform with God's will. Meetings opened with prayer and singing. Women frequently moved into more public roles in these meetings and rallies, suggesting similarities between reform speechmaking and religious "testifying." Thus, when Cora McLean exhorted Alliancemen to remember that "man's first duty is to God" and encouraged her listeners to "grasp each other by the hand and resolve by the help of God that our condition shall be bettered and the right of farmers and laborers shall be respected," her public proclamation made no new demands for women but rested firmly on the traditional female claim to moral and religious speech. Women prayed and exhorted within the communal response to perceived economic injustice. When the editor of the *Toiler* raged against trusts and unjust laws, he likewise couched his anger in biblical terms. "God is a just God," he wrote, "and to say that he endorses the burdens of oppression that rest upon us as a class on account of unjust laws . . . is certainly contrary to the teachings of the Holy Bible." The Williamson County Alliance forecast the construction of a "noble agricultural and mechanical commonwealth" founded "on the side of truth and right worthy [of] the highest type of God's created being." By expressing their reform in religious rhetoric, Wheelers and Alliancemen both expanded and limited the terms of reform. On the one hand, they claimed sanctification for their political and economic programs—programs many critics believed to be radical departures from past American constructs. On the other hand, they embedded their organizations in rural traditionalism that stopped short of proposing revolutionary change. Thus, Tennessee reformers offered no objections to female speechmaking when it occurred within the accepted domestic sphere of moral uplift or religious testimonial. And men took heart that their efforts to rein in industrial capitalism enjoyed God's blessings.[8]

Wheel and Alliance procedures often mimicked the celebrations and rituals of rural churches. As farmers confronted internal quarrels, they drew on church procedures to reinforce traditional mechanisms for settling disputes. Likewise, Wheel and Alliance mass meetings resembled revivals, brush arbor meetings, and dinners-on-the-ground, which were staples of rural religious life.[9]

When quarrels between Wheelers and Alliancemen arose, local unions upheld personal and customary usage in preference to legalism and court-ordered action. Reformers worried that the intrusion of courts and tort law in the wake of rising commercialism would destroy local cooperative networks and the foundations of rural community. To counter the growth in lawsuits that accompanied the cash nexus, Wheel and Alliance leaders urged farmers to settle their disputes among themselves, using the union as arbiter. The plan would have seemed familiar to men and women reared in country churches where family disputes, quarrels between neighbors, and disruptive behavior found resolution in church meetings and councils.[10]

W. P. Morris's 1888 letter to the *Toiler* demonstrated the effectiveness of local Wheels as arbiters and promoters of organizational unity. Under the headline "How Wheelerism Works," Morris described an Obion County experience that illustrated the efficacy of arbitration. According to the writer, two Wheelers, one a member of Wheel 445 and the other a member of Wheel 112, came into disagreement and filed lawsuits against one another. Concerned members of their respective Wheels intervened and arranged for them to present their views informally to the combined membership, effecting a settlement acceptable to both parties. In Morris's opinion, arbitration by fellow union members was preferable to any alternative because "the trial was private and no meddlesome mischief-makers [were] present to aggravate matters." Clearly, men of good will, acting as "brothers" in the agrarian "faith," could settle matters in the time-honored tradition of rural congregations without involving lawyers and courts. Using the forms and rhetoric of religion, the Wheel and Alliance claimed the middle ground between tradition and modernity: union men could engage in modern commercial agriculture and maintain the benefits inherent in rural exchange economies.[11]

The county rallies and "jollifications" that brought together farm families for a day of excitement and edification built upon the revivalism expressed in camp meetings and the socialization of dinners-on-the-ground. James D. Vaughan reported a meeting at Bodenham in Giles County, where farmers listened to speeches outlining the principles of the Wheel and Alliance. He assured readers that "if there were any doubting Thomases" in the crowd, "they cried 'we believe,' and will henceforth talk and preach Alliance and Wheel doctrine to rescue our down trodden farmers from financial destruction." He then compared American farmers to the "children of Israel . . . groaning under Egyptian bondage," whose "yoke of oppression" was thrown off by Moses. Farmers could take heart, knowing that "He is the same God, yesterday, to-day and tomorrow, and will listen to the cries of distress just the same now as he did then. "If God be for us," one correspondent to the *Toiler* wrote, "who can be against us." Wheel and Alliance lecturers self-consciously cloaked themselves and their program in the blessings of God, crafting edification speeches that resembled religious revivalism in rhetoric and style.[12]

Reformers carried their claims of sanctification further with their preparation of rally sites and their choice of supporting organizations. Masons and Wheelers held a joint picnic at Price's Chapel in McNairy County in 1888. Under a brush arbor, Baptist preachers, Masons, and reform leaders regaled a crowd of two thousand people with speeches and sermons. The Farmers' Alliance of DeKalb County held a Fourth of July celebration in conjunction with the "various Sunday Schools" of the Antioch, Jefferson, and Indian Mound communities. The Farmers' Alliance clearly initiated and directed the events, which included speeches on patriotism, "the importance of being true" to the Alliance, and Sunday School attendance. The Collins River Band performed for the event, which ended with a reading of the Declaration of Independence. Estimated at "something like three thousand people," the crowd shared a meal that left everyone "bountifully supplied." When the leftovers were placed in boxes and baskets, the correspondent to the *Toiler* was reminded of the parable of the loaves and fishes.[13]

The DeKalb County reporter was not alone in his comparison between meals shared at Alliance rallies and the biblical story. McDowell's reports on visits to county meetings fell into a standard pattern: when the speeches and organizational business ended at noon, Alliance women "spread out before us a bountiful supply of life's essentials and the inner man was truly refreshed." J. R. Miles, president of the State Wheel, was appreciative of the skills of women when he complimented their "baskets and boxes 'chuck full' of as good 'grub' as ever went into a union man's mouth." For men, the dinners offered a respite from long-winded speeches and provided them with the energy to undertake a long afternoon devoted, again, to "opening the eyes" of farmers weighed down by the effects of courthouse rings and monopolies.[14]

The dinners also symbolized community for the women who prepared the meals, though they spoke of it in different terms. A Marion County woman reported the activities of Sweden's Cove Union, which included a picnic under a grove of trees. Judging it "one of the best dinners I ever saw," she noted approvingly that "everyone ate until they were perfectly satisfied, and there was much of it left to be taken home." The occasion made such an impression on the writer that she declared it "will be a bright spot in my memory for the rest of my life." Allowing for the overblown rhetoric of the age, the impression remains that, like the report of the DeKalb County farmer, these communal meals offered more than filled stomachs. Sharing food with men and women of distant communities bound them together in a joint effort. The comparison of their shared meals with the biblical meal of loaves and fishes provided a living example of the benefits of cooperation that sustained Alliance efforts as readily as long-winded speeches.[15]

The identification of the Wheel and Alliance with Tennessee's rural churches strengthened the reform movement by providing a highly symbolic space in which to hold meetings and an accepted language for framing action. Claiming divine sanction, reformers fashioned a critique of American economic development that resonated among men and women who embraced the example of the first-century Christian church as the model for their own religious lives. Stark and antihierarchical, these Christians maintained a prickly sense of autonomy that supported their sense of indignation over the perceived oppression that kept the poor in bondage.

Tennessee claimed a rich and complex religious history. As the scene of the first manifestations of the religious fervor associated with the Cane Ridge revivalism and the Second Great Awakening, Tennessee churches divided and splintered in a dizzying array of sects and congregations, with some claiming "sole validity and unbroken succession" to the early Christian churches. During the nineteenth century, Cumberland Presbyterians broke with the Presbyterian Church; the Church of Christ moved away from the Disciples of Christ; and Landmarkism threatened the association of Southern Baptists.

As the Pentecostal and Holiness movements gathered momentum in the early twentieth century, Tennessee's churches stood on the cusp of a new era of factionalism and splintering. In each of the religious breakaway movements, class and the perception of economic and social oppression played a role. The expectation of a social and democratic revolution that would set the world right fed into the emerging agrarian critique of modern industrial capitalism and bridged the chasm between white and black Christian farmers.[16]

Paradoxically, historians have also associated the agrarian movement with the Social Gospel, a movement generally identified with urban Progressive reform. Focusing less on personal salvation and more on social change, the Social Gospelers likewise crossed racial and ethnic lines to champion education, improved housing, labor rights, and Prohibition. Among Tennessee's agrarian leaders, evidence of a nascent Social Gospel is not hard to find.[17]

When John P. Buchanan assumed the office of president of the Tennessee State Alliance in 1889, he tied his official duties and those of the membership to religious obligations, saying, "There are no three greater duties of practical Christianity, than to strive for the interests of one's own household, and for the interests of one's fellow laborers, and for the interests of one's fellow citizens." The message was clear: rural Christians belonged in the Alliance and in the service of man. Like Buchanan, "Little Hatchet," a frequent contributor to the *Toiler*, professed his belief that "helping people who cannot help themselves, that is what the meek and lowly Jesus taught." Tying his Social Gospel to political action, he continued, "our rulers have reversed it . . . the toilers are the oppressed people and they need protection." Bailey Basford of Oak Lawn Wheel submitted a poem to the *Toiler* titled "The Toiling Men," in which he outlined the plight of the oppressed and prophesied a new day:

Ye toiling men, ye toiling men,
  Your burden will grow lighter;
The night at last is giving way,
  The day is growing brighter,
Oh, may the light of justice shine,
  And men of every station
Love and obey the golden rule
  Throughout our lovely nation.[18]

Like Christian social reformers from Thomas More to Washington Gladden and Josiah Strong, the poet viewed social justice and reform within the context of the "Golden Rule." The problem was not with the toilers themselves as individuals, but with the subversion of justice. Even more to the point was the anonymous poem, "Am I My Brother's Keeper?"

My days are filled with comfort,
  And nights are passed in ease;
'Tis presumption to annoy me
  With questions such as these.

They tell me of the paupers,
  The sick and the oppressed,
The worn and weary workers,
  The sorrowing and depressed.

The young, whose noble longing
  Perish in the sordid fight for gain
Who fall into the tempter's toils
  And those whose struggles are in vain.

The waifs, half clothed and starving,
  The orphan and the widow's wail—
Why haunt my downy pillow
  With their specters gaunt and pale?

Yes, we are our brother's keepers;
  Who lives unto himself,
To gain, to trade, to barter,
  To gather gold and pelf.

To ignore the pleading glances
  Of poverty and woe,
Shall have small capital to start with
  In that land to which we go.

With hand outstretched and empty
  Like the beggar at our door,
Shall we reach those shining mansions
  Where our loved have gone before.

Oh, Lord! shamefaced and contrite,
  Let us not waiting stand;
By love and suffering sanctified
  Let us enter in Thy land.[19]

The best-known Tennessee Social Gospeler, Mark Allison Matthews, imbibed the radical politics of the Farmers' Alliance as he preached the Christian gospel in the Presbyterian churches of North Georgia. A frequent speaker at farmers' meetings in the 1880s and an admirer of Tom Watson, Matthews moved to Jackson, Tennessee, in 1896 and launched a crusade for social reform that confirmed the city had been touched by the Social Gospel. In his six-year term as pastor of the First Presbyterian Church, Matthews started night classes for workers, built a public library and a hospital, persuaded the railroads to provide coal for the poor and to support a health insurance program for railroad workers, fought for the regulation of cocaine and morphine, established a Ladies Bible Training Class to work with the poor, organized a humane society, and brought the YMCA to Jackson. His energy and accomplishments exceeded those of the Alliance rank and file, but he touched a widely felt belief that the solution to the problems of industrial society rested in a reestablishment of social justice and a return to the Golden Rule. It was an idea expressed by John Buchanan in his inaugural address as Governor of Tennessee. "Progress does not mean a mushroom growth of 'booms' and wild speculation," he lectured, "in which a few are enriched to the detriment of the many." As Michael Kazin noted in his study of popular movements, the unification of moral (religious) goals and economic and political objectives remained a defining characteristic of late-nineteenth-century populism.[20]

As elites among New South men and the Bourbons understood, tapping into the religious volcano beneath the surface of every denomination threatened the hierarchical order that ruled the South. Southern preachers would rather have sermonized on sin and salvation and avoided the issues of oppression in whatever form, but the energized rural communities were asking questions about social justice. The potential unification of agrarian demands and religious discontent heralded a revolutionary combination when extended to the more politically pious African American Christians. Southern elites hoped that the seeds of disunity inherent in the localism of the rural congregations would break the momentum of the reforming spirit.

Indeed, if communal meals and jollifications facilitated harmony, it was sorely needed. Tennessee's rural communities were no exception to the pettiness, backbiting, and suspicion of change that characterized localism and communal networks. As State President J. R. Miles prepared for the meeting of the State Wheel in July 1888, he attributed the "lack of fraternizing within the ranks" to several sources: "the rapid organization, the lack of thorough education in the principles of the order, the power of hostile combinations which confront the organization, and the various demoralizing agencies at work." Others offered more specific assessments of the cause of trouble afflicting the unions and warned, "our organization, like all others have bad men who enter it for personal gain and care nothing for the principles we advocate."[21]

To prevent the domination of the organization by such self-interested men, Wheelers and Alliance leaders advised members to "keep an eye open for the 'wolf in sheep's clothing,' [and] do not put too much confidence in new converts with an axe to grind." The Obion County Alliance acknowledged that they had "many weak-kneed brethren in our ranks who are ready to cry out that our order is almost a failure; that they have never benefited by it, and are becoming stupid and very slow to act." A member from Madison County reported that "a few strife producers and bad management on the part of officials have in a number of counties discouraged those who entered the order with pure motives and not for gain, but as these individuals drop out confidence is restored and the grand work moves on." One observer attributed "the want of unity of action" to "too many solomans [sic] in the order, who if they can't work in the lead, they kick out of traces." Another county union traced its problems to "a tattler and strife producer who has nearly destroyed the peace and harmony of two unions." Though the Wheels and Alliances made no secret of their problems, they were careful, as Miles indicated in his 1888 speech to the State Wheel, not to provide their opponents with the means to defeat their efforts.[22]

Solutions to the problem of disunity took several forms. Words of encouragement filled the weekly issues of the *Toiler*. Calls for "stickability" and jokes about "weak-kneed, cotton-stringed backbone fellows" bolstered sagging spirits. Organizers encouraged members to "strike straight from the shoulder" and "make no half-hearted 'milk and cider' work." Women especially ridiculed the lax and motivated the reluctant. Unity and harmony also resulted from at least three Wheel and Alliance activities: the education of farmers in the political economy of the nation, the organization of a state cooperative agency, and the marshaling of an incipient antiparty spirit to correct the wrongs promulgated by trusts and monopolies.[23]

Educating Farmers in the Political Economy

One of the least discussed but most important efforts of the Wheel and Alliance focused on the education of farmers in the political economy of the nation. So important were these initiatives, particularly in the largely unschooled and underschooled South, that Alliance organizers claimed to have "done more to educate the great mass of people in the principles of government than all the schools and colleges have in the past century." In speeches, letters, and editorials

Wheelers and Alliancemen repeatedly voiced the necessity of "opening the eyes" of farmers and laborers to the need for reform. Before men and women joined the cause, they had to be made aware of the source of their problems. Afterward they needed to be educated in "the sciences of social, financial, and political economy," an endeavor for which the Alliance molded "its every effort, with the firm belief that every citizen in the U.S. who loves freedom, justice, and law will rally to its support." It fell on locally elected lecturers to catechize new members in the secrets of the order and promote the latest agricultural methods promulgated in farm journals. Meanwhile, state lecturers fanned out to reinvigorate sagging enthusiasm and establish new wheels and suballiances.[24]

From local Wheels and Alliances came praise for the efforts of lecturers and calls for more education in the principles of the organization and the political economy. The Wilson County Alliance passed a resolution in favor of "broader education of the farmer." Three months later, the same Alliance expanded its support and became more specific in its endorsement as the union resolved to support "the education of our sons and daughters more fully in the science of government in order that our sons may vote more intelligently and that our daughters may understand more thoroughly the laws to which they are subjected." Tipton County correspondents assured readers of the *Toiler* that "the farmers of Tipton have begun to read and think for themselves, and keep posted on the various matters of the day" with the result that they "are not quite the blockheads they used to be." J. A. Scott of Maury County believed that participation in the activities of the local union prepared farmers for understanding the issues as well as developing future leaders. The education on pending legislation, the experience in the use of parliamentary procedures, and the public presentation of views developed skills largely absent among farmers. As a result of these grassroots "schools," Scott predicted future legislatures would be more representative of the agricultural element.[25]

Speeches and lectures inspired members to greater activity, but the crucial elements of maintaining an active organization centered on regular attendance at union meetings, where farmers were encouraged to speak and to hone their rhetorical skills. Union members reinforced what they learned at local meetings by reading farm and economic news in the *Toiler*. Both the Wheel and the Alliance took the position that to be effective as farmers and as voters, members must be informed, it being "utterly impossible to make a good wheeler of a man who neither reads nor takes any interest in his lodge meetings." Wheel leaders believed that "mental indolence [was] far worse than common laziness, and when the two [were] combined in one man, he [was] very near a hopeless case."[26]

Advocates for farm journal subscriptions argued that hard work and diligence alone could not produce prosperity and supported their contention with the example of a Robertson County farmer, who was as "moral and clever as men get to be, but never read a newspaper in his life." His farm presented the picture of good management: fat cows, good fences, and clean fields. Nevertheless, "if he made anything some speculator would get it for less than the cost of production." The lesson was clear: being uninformed left the farmer at the mercy of sharper men, a fact that his hard work could not overcome.[27]

A sharp criticism of the presumed benefits of scientific agriculture rested at the heart of this educational effort. Reform leaders recognized the advantages of New South agricultural ideology—incorporation of advanced farming methodologies, crop diversification, home production—and they frequently encouraged Wheel and Alliance members to follow these practices. But they also noted the limits of hard work, thrift, and improved technology in the battle against low commodity prices and the inadequacy of farm credit. Like industrialists, farmers needed access to capital and control over costs and commodity prices, but were stymied in their efforts by the banking laws, the crop lien, and the transportation and storage systems that favored urban, industrial development. Men on both sides of the issue understood that producer control over marketing of agricultural commodities represented more than prices: it wrested social and political control from "the best men" and handed an economic advantage to smaller commercial farmers. Neither southern industrialists nor old-line planters were willing to step aside for such a radical transformation.

Cooperative Buying and Selling

Both the Wheel and the Alliance advanced cooperativism as the mechanism for reducing costs and raising farm prices. Basing their cooperatives on the British Rochdale plan and drawing on the experiences of the Grange, state leaders established a purchasing system of county and state agencies to buy plows, seeds, sugar, and salt at bulk rates and resell these necessary farm inputs at discounted prices to union members. Some county unions also constructed warehouses to store local agricultural commodities while state leaders negotiated to sell cotton, tobacco, and peanuts in bulk and thereby exert greater control over farm prices. In their most grandiose schemes, farmers anticipated a day when they could control corn and wheat sales to cotton farmers in the Deep South and handle cotton on its way to New York. In many ways, farmers plotted to create an agricultural cartel that would out-trust the industrial trusts in much the same way that New South promoters hoped to "out-Yankee the Yankees."

Achieving their goals demanded a commitment that many were unwilling or unable to make, and complaints abounded among the more dedicated members of the Wheel and Alliance against the "converts" whose only aim was "cheap goods." Many Wheelers and Alliancemen thought these do-nothing members were the most critical opponents when the agrarian organizations failed to meet their elevated expectations for ending the immediate economic problems of southern farmers. Nevertheless, poor farmers desperately needed the cheap goods as an incentive to develop the perseverance necessary for the long-term challenges of confrontations against trusts and monopolies. The Tennessee Alliance president anticipated the argument of historian Lawrence Goodwyn when, in an address to the state convention, he asserted that "we had but little co-operation in our organization in Tennessee, until we adopted the State Agency system; this, though still in its infancy, has worked wonders for the general interest of those engaged in agricultural pursuits." "Co-operation," he enthused, "is the key that unlocks our future."[28]

Tennessee's cooperative system got underway in the spring of 1888 and met immediate opposition. The Nashville-based state agency under the direction of the State Business Agent George A. Gowan encouraged members to pay a one-dollar assessment in order to trade through the store, which sold farm equipment and bulk items such as flour, sugar, and salt at wholesale prices. Organizers expected the assessment to provide the fiscal foundation for demanding reduced prices from manufacturers and suppliers and establishing lines of credit to purchase items for future sales. Operating expenses and salaries came from the interest accumulating on the banked fees. Agency directors opened an account in the Safe Deposit Trust and Banking Company in Nashville, where Wheel and Alliance funds received 4 percent interest.[29]

Nashville merchants acted quickly. In late spring, as soon as the venture "began to show signs of success," the state agent was arrested for failure to pay the merchant tax required by state law for all retailers. The business community ridiculed agrarian claims that the agent was not a merchant but acted only on behalf of the farmers' organizations. When the case came before the Nashville city court, the judge ruled against the Wheel and fined the state agent twenty-five dollars.[30]

After an unnamed Springfield judge apprised agency directors of a different ruling in a similar case, they appealed to the Circuit Court. Fighting the judgment provided two potential benefits: unification of disparate farmers in a fight for justice, and the setting aside of a ruling that threatened the economic viability of the cooperative effort. Letters soon arrived at the *Toiler* office urging unity in order "to fight the battle with brave hearts and a determination that will gain victory for free men." Local Wheels and Alliances passed resolutions in support of the state agent and collected nickels and dimes to meet the expense of court action. McDowell warned that if the court ruled against the state agent, the cooperative system "would be closed out at once" and leave farmers "at the mercy of combined capital." His assessment of the situation rested on the realization that if Nashville merchants successfully forced the payment of the tax, other cities and counties would follow suit. In incorporated cities, the cost to the Wheel would be $606 per agent, while in crossroads towns and counties, the cost would be $404—a prohibitive tax that would effectively end agrarian efforts to control their costs and market their crops. The refusal to capitulate paid off when Judge W. K. McAlister ruled in favor of the state agency, and Nashville merchants declined to press the case further.[31]

Fearful the suit might still produce similar action in counties across the state, the Wheel and Alliance agitated for permanent legal protection. After a fierce campaign notable for fraud and violence, the fall elections elevated forty-six Wheelers and Alliancemen to the general assembly with the demand that they enact legislation to exempt the state agency from merchant taxes. When the state legislature met in January 1889, Wheelers J. M. Crews in the senate and C. C. Adams in the house introduced the State Agency bill. Business interests attempted to block the intent of the legislation by tying it to the state funding bill, a ruse to force a biennial defense of the agency's protection as each new funding bill came under consideration. When the agency bill stalled in the house, Representative John P. Buchanan, the State Alliance president, brought the farmers' legislation to a successful vote. In April 1889, Governor Robert L. Taylor signed the act, and farmers proudly boasted that Tennessee was "the only State in the Union that ha[d] such a law."[32]

Despite its obvious success in protecting the agency against legal harassment, the Wheel and Alliance still faced serious obstacles in providing adequate funding for the cooperative system. State leaders followed a conservative approach and developed the agency on a cash basis, relying on the accumulating assessment fees to provide the financial support for buying in bulk. Wheelers offered no apologies for their decisions and, indeed, "rejoiced to know our agent is steering clear of indebtedness," as this offered the "plain and sure road to success." As a result of this decision, Tennessee's cooperative system developed more slowly than agencies in other states. Supporters of the Tennessee agency acknowledged that "we are not boring with as big an auger as Texas and perhaps some other states," but offered the reassurance that "we are doing wonders for our limited means and steering clear of credit snags, while some other states have got in water too deep to swim, and are struggling hard to keep their heads above water."[33]

The state agency operated on a contingency basis, purchasing goods only after receiving orders from county business agents. Those county cooperatives followed one of two methods for providing low-cost tools and furnishing to local farmers. In some counties, Wheels and Alliances organized an independent cooperative using paid subscriptions to finance the endeavor. Other counties with fewer resources or less support named a local mercantile establishment to act on behalf of the union. Generally these merchants granted discounts to Wheelers and Alliancemen on all purchases made through the store. Creating a cooperative store from scratch demanded strong local support in money and commitment. Few communities sustained such endeavors for long. Although the second option seemingly made cooperative buying easier, it was discouraged by state and national leaders and carried its own risks. Competing merchants resented what they viewed as underhanded practices by the cooperative merchant and pressured him to relinquish his position. Moreover, merchants themselves had little incentive to continue to offer discount prices after the initial flurry of excitement.

On the first anniversary of the state agency's founding, McDowell issued an upbeat report on the progress of the cooperative effort. He claimed that the "permanent fund has reached nearly ten thousand dollars," with monthly agency transactions of $15,000 to $25,000 in plows, sorghum mills, and other agricultural implements. He anticipated an expansion of agency offerings and noted that the state agent had negotiated arrangements to buy sugar directly from Louisiana as soon as the new crop became available. He further encouraged farmers to buy through the agency in order to receive the 25 to 50 percent reduction in costs over similar purchases bought on credit through local merchants (a 15 percent savings for farmers who usually paid cash).[34]

McDowell also reported a bustling sale of crops through the agency, but lamented, "complete arrangements could not be made for want of warehouses over the country to store in until buyers could be found." Still he enthused, "the brethren see the necessity of these houses and in some counties are preparing to build them for the coming crop," a claim borne out by numerous county resolutions throughout the following spring and summer.[35]

County organizations vigorously joined McDowell in supporting the cooperative effort, though none matched the enthusiasm and audacity of the Wayne County Wheel. Convening in a self-described "harmonious and business-like" meeting, Wheelers pledged "to buy at least $15,000 worth of goods within the next sixty days." The county organization required each subordinate Wheel president to call a meeting for the purpose of taking purchase orders. The county business agent was ordered to keep regular hours at an office located in Waynesboro, the county seat. On Mondays and Fridays from 10 a.m. until

3 p.m., members could place orders and pick up goods. The assembly demanded immediate payment from recalcitrant members who had not paid the one-dollar assessment. The meeting closed with appeals "to the honor and stability of the brotherhood to build up and sustain the agencies they have established." Perhaps symbolic of larger organizational limitations and problems, the Wayne County Wheel issued no further reports on the successes or failures of this get-tough business approach to building support for the state agency.[36]

McDowell's enthusiastic anniversary report avoided the thornier subject of the Memphis Agency. Wheel and Alliance leaders bowed to the geographic and agricultural differences in the state and established three agencies, one in each of the grand divisions. A Memphis agency followed the establishment of the Nashville agency by a few months. East Tennessee, which was still in the early stages of organization, did not receive an agency until 1889. While the Nashville agency experienced legal problems, the Memphis agency suffered from indifferent funding.

By November 1888, the failure of the West Tennessee members to pay the one-dollar assessment jeopardized the future of the Memphis agency. R. H. White, the Memphis agent, received no salary for three months and paid operating costs out of his own pocket. In September, the State Executive Board attended a joint meeting of the Tipton county Wheel and Alliance, where they complained that funding was "insufficient to establish a trade agent at Memphis" and warned members they could not sustain the effort without the assurance of the agent's salary. Tipton County farmers praised White as "the right man in the right place," and recommended an additional 25¢ fee for each member of the West Tennessee unions in order to secure White's salary. Resolutions in support of the assessment sailed through Wheel and Alliance meetings in other West Tennessee counties. But the expected flow of money was not forthcoming.[37]

At its October meeting, Clopton Wheel in Tipton County proudly reported that twenty-two of the twenty-seven members had paid the $1 assessment, while eighteen or twenty had paid the additional 25¢ fee. Clopton Wheelers proved exceptional in their diligence. The State Executive Board openly chastised the cotton-producing counties of Chester, Crockett, Dyer, Fayette, Gibson, Hardeman, Haywood, Henderson, Lake, Lauderdale, Madison, McNairy, Shelby, and Tipton for their failure to collect the fee and ordered them to "have the matter up at once and arrange to pay it direct to Brother White," if their members expected to trade at Memphis. Neither the publication of praises for compliant Wheels, nor the public dunning of the recalcitrant ones raised the needed funds. As late as April 1889, the Gibson County Wheel led all the other West Tennessee counties in collecting the $1 assessment, with a total collection of $900, a figure that represented payment by only 60 percent of more than 1,700 members.[38]

Members wrote letters to the *Toiler* in support of the assessments, offering encouragement to laggers, demanding accountability for the promises made, and predicting doom if the agency effort failed. J. S. Aycock called on White at the Memphis agency and found him "doing a great deal of work . . . undoubtedly saving the counties adjacent to Memphis a vast amount of money." Despite the adversities associated with his position, White appeared "not greatly discouraged [and] has faithfully discharged his duty." J. C. Castles reflected on the failure to follow through with pledges of support and offered the view that "some of our order would be better if they would talk less, pass less iron-clad resolutions, and carry some they have already passed into effect." He expressed amazement that "people claiming to be conscientious, honest, upright, Christian men can take a solemn obligation such as the alliance demands and then willfully disregard it."[39]

Both Aycock and Castles predicted dire consequences if the Memphis Agency failed. Proclaiming the farmers "utterly lost" without the agency, they feared the farming community would become "nothing more than monopoly slaves." After castigating those he believed had failed in their duty, Castles pledged a loan of fifty dollars to the agency if two hundred other men would do the same, or one hundred dollars if fifty men would match his pledge. Despite their efforts, money did not arrive soon enough to aid White. News of his resignation appeared in the November 28 issue of the *Toiler* alongside the usual exhortations to pay the assessments.[40]

Although discussions about the West Tennessee agent's salary quickly disappeared from the columns of the *Toiler* with White's resignation, money remained a serious problem, and county Alliances repeatedly resolved to collect assessments. In 1889, J. C. Castles again wrote to the *Toiler* to suggest that when delegates gathered to write a new national constitution after the proposed merger of the Wheel and Alliance, they should make it "compulsory on the members to pay all assessments." Echoing the concerns of the Executive Board, he warned that failure to address the problem of overdue assessments would place the organization in jeopardy, and "I fear we will never be able to raise money enough to give our state agents a good financial standing." Castles attributed the inability of Tennessee to fund the agency adequately to the prevailing wait-and-see attitude of those farmers demanding success before committing to the enterprise. "This talk about being too poor is all bosh," he asserted. "We have as poor men in our subordinate alliances as are in the state and all of them have paid the one dollar assessment and the twenty-five cents assessment . . . and every man in the state could do so if he wished." Rather than poverty, he opined, "we have too many weak-kneed members that are an encumbrance to the order anyway, and the sooner we adopt a sifting process to get rid of such the better." In his criticism of those who failed to pay their assessments, Castles touched on a deeper question that plagued the organization throughout its existence: should agrarian leaders organize broadly or limit membership to a committed few?[41]

Castles found little support for his view that a smaller, more committed group of men could accomplish more than a hundred thousand "weak-kneed, cotton-string backbone" followers. One of the strongest objections to limiting membership came from the editor of the *Gibson County Herald*, a rural newspaper that frequently carried Wheel and Alliance news. Concurrent with Castles' complaint, the *Herald* noted that "in some of the wheels, the knife has been applied (rather vigorously) in lopping off some of the dead timber." The editor doubted the wisdom of the move but postulated that it might be justified in consideration of "the light burden of dues" demanded of each member. The excision of "dead timber" included a ruling by the executive board that prohibited representation in the county or state Wheel for any subordinate Wheels with delinquent assessments. This seemed unduly harsh to the editor, who provided additional reasons for the failure to meet Wheel obligations, noting, "there are very few blessed with surplus funds to enable them to do this, the assessments being too insignificant to meet present demands."[42]

The question of whether or not farmers could pay remained a source of conflict among union members. Those who paid viewed their delinquent neighbors as lacking in conviction for the reform cause. No doubt their constant cajoling provided the necessary excuse to avoid monthly meetings and eventually abandon the organization altogether. A survey of the published county reports makes it clear that many must have felt as if they were being nickel-and-dimed to death, with new assessments appearing at every meeting. As the Gibson County editor noted, the small amount demanded of each member probably burdened few, but likewise, it required a substantial membership to support the cooperative effort.

Tennessee's Wheel and Alliance, unlike their counterparts in other states, accepted little or no support from nonagricultural sources, a source of pride for the agrarian leadership. Indeed, Tennessee's cooperative agency limited its activity to consignment purchases, buying farm inputs on order from county agents. This fiscal conservatism hindered the development of cooperative buying in two connected ways. The purchasing strength of the agency lay in the payment of individual assessments. Poverty prohibited payment of even a one-dollar assessment for some farmers, while others refused to risk any hard-earned cash on unproved schemes. The organization of the agency around cash payments rendered the scheme useless to many sharecroppers and farmers mired in the crop lien system. With little access to cash, sharecroppers and poor farmers needed credit, which the state agency could not provide. For Tennessee's African American farmers, trading through the cooperative agency meant defiance of local merchants and planters, a barrier many found insurmountable. Although the state agency operated to process orders from both black and white Wheels and Alliances, cooperativism proved an insufficient weapon against the economic and racial barriers that characterized the southern credit system.

As historian Robert McMath noted, "the most effective cooperatives in Tennessee were local operations." But county business agents encountered their share of problems as well. From across the state, Wheelers and Alliancemen reported the difficulties of establishing local cooperative stores. Under pressure from merchants, wholesale houses refused to sell to county business agents. Murray, Dibrell & Co., a Nashville boot and shoe wholesaler, "positively refused to sell any boots or shoes to any alliance merchant in White and Warren counties." Farmers in those counties furiously called for action to avenge "the slur cast upon the good name of our order," encouraging farmers in the midstate to inquire about the wholesalers patronized by local merchants. If boots and shoes sold locally came from Murray, Dibrell & Co., union supporters believed buyers should "have the courage and self-respect to just quietly refuse to buy or wear them."[43]

In East Tennessee's Washington County, the Alliance resolved to establish a cooperative store after learning that the retail merchants who supplied their needs in the previous year now refused to do so again under pressure from Knoxville wholesale merchants. As one writer noted: "clearly the decree has gone forth. The Knoxville gods declare that we can have supplies if we will come as plain John Smith, but if we come as John Smith, Pete Jones, and Sam Brown, members of the farmers' alliance, not a pound of coffee, ounce of Soda, or yard of calico can you have." Although not always as obvious as the actions of the Knoxville wholesalers, the power of wholesalers and retailers to circumvent the cooperative plans of local unions rendered farmers' schemes impotent across the South.[44]

A Cocke County merchant's experience demonstrated more disturbing intimidation. At their second county meeting in January 1889, the Cocke County Alliance appointed a committee to "make contact with some merchant or merchants to sell goods to the Farmers' Alliance." All the local merchants refused to make such arrangements, and in April, the Alliance resolved to "organize a capital stock for the purpose of establishing a store and warehouse at Newport." The organization finally negotiated an agreement with two merchants, identified only as Boyd and Clevenger, who agreed to open an Alliance store. Enthusiastic reports followed until July 1890, when the county Alliance formed a committee to negotiate with Valley Home and Fire Insurance company to determine "why they had failed to pay the insurance to the Clevenger firm of Newport, Tenn." The county minute book makes no further mention of the store. It seems reasonable to assume that the firm met the fate of Alliance stores in other states and was burned by men who felt threatened by the organization of farmers.[45]

A letter from "Claudius" outlined the ease with which cooperative efforts could be destroyed by playing to the needs of poor farmers. In Bradyville, an enthusiastic member of the Alliance identified only as Bro. Shacklett, "who owns one of the finest mills both for wheat and corn on Stone [sic] River," offered to grind wheat for alliance members at the considerably reduced rate of 1/10, the usual rate being 1/8. Shacklett tendered the offer in a secret session and intended it to stay within the brotherhood. A few "talkative" members soon made the proposition common knowledge, and other millers immediately dropped their fees to 1/12. Farmers reported that some millers even stopped wagons on the road to redirect them away from Shacklett's mill. As a result, Shacklett was "nearly ruined," and the farmers gained nothing, having fallen for what Claudius regarded as "an old dodge." Unfortunately for the cause of reform, Shacklett's experience proved to be a common occurrence, and Alliance leaders were hard pressed to hold the ranks in boycotts against local merchants or national cartels. The poverty of many farmers led them to value the few cents saved in any deal rather than hold out for long-term savings.[46]

Bedford County farmers reported a more positive experience. In 1888, the county Wheel took action to control the sale of the wheat crop, eliciting the "jeers and scoffs of contempt and ridicule" from the merchants, who had heretofore controlled the local crop and the prices for wheat. The merchants reminded the farmers that it "took brains as well as money to handle produce and merchandise" suggesting that the wheelers had neither. Nevertheless, Victor Flouring Mills at Shelbyville agreed "to pay the farmers the highest price for wheat that either the market for wheat or flour would bear . . . [and] furnish sacks and a suitable warehouse for storing grain free." In return the farmers agreed to "give the mill preference of buying the grain when [it was] ready to sell." The merchants met the news with condescending nods and reassured themselves that "there [was] nothing in this contract with the mills," and that "these wheelers are a set of darnphools [sic]."[47]

Their attitudes sobered somewhat when they learned that the Alliance county agent, W. F. McAdams, would receive, grade, and superintend the sale of wheat. "Uncle Bill" McAdams enjoyed a distinguished reputation among wheat growers. Indeed, farmers and merchants alike considered him a "regular gilt-edged, diamond breast-pin wheat man from the soles of his feet to the crown of his head." The so-called "darnphools" reportedly handled 250,000 bushels of wheat at ten cents a bushel. The writer, known only as Plow Boy, concluded that "the organization in this county is solid, and it owes its present state of solidity, in great measure to its solid, level-headed County Agent." Perhaps in recognition of the economic benefits available through their local organizations, Bedford County consistently elected Wheel and Alliancemen to the state legislature, and even gave the name "Wheel" to one rural community.[48]

In one area, Tennessee cooperativism enjoyed almost universal support—the substitution of the Chattanooga Chilled Plow for the better known and more widely used Oliver Chilled Plow. When the state agency first organized, Chattanooga industrialist Newell Sanders agreed to sell his plow to the Wheel and Alliance on the same terms he provided to wholesalers, a savings to farmers of 40 percent. Approached by the state agent for a similar deal, the manufacturers of the popular Oliver plow refused to offer the same terms, standing by their regular salesmen and against the cooperative stores. Almost every issue of the *Toiler* carried exhortations to buy the Chattanooga plow and boycott the Oliver company. Local and county Wheels and Alliances routinely passed resolutions and memorials in praise of the Tennessee-manufactured implement and promised to purchase plows only from the Chattanooga manufacturer. Indeed, McDowell's exhortations to buy the plow sounded at times like those of New South advocates as he called upon farmers to buy the Tennessee product in order to save money and support local industry.[49]

The issue proved to be especially popular in East Tennessee. Here Sanders's well-known activities in the Republican Party perhaps added to the support for the union and the plow, bridging the political gap between state organizers linked to the Democratic Party and potential Wheel and Alliance members. On a trip to upper East Tennessee, McDowell reported a Sullivan County rally in which Alliance men and women marched to the meeting ground under a banner depicting the Chattanooga Chilled Plow.[50]

As historians suggest, the success of Wheel and Alliance cooperative efforts always depended on the individual decisions of thousands of farmers. Rich or poor, they weighed the value of immediate needs and potential long-term gains. In every part of the state, farmers' efforts to establish financial independence through cooperative buying met solid resistance from local merchants and distant wholesalers. The fact that the cotton counties experienced greater difficulty in sustaining their regional agency than other parts of the state points to the possible influence of the crop lien system and its ability to inhibit autonomous actions. Even independent farmers faced difficult economic choices, and in the poor market of the 1880s and 1890s they saw no leeway for error. They remembered the enthusiasm that met Grange cooperative efforts, an enthusiasm that quickly dissipated as buying, processing, and marketing schemes fell to the hard reality of insufficient funding and poor planning. A Rutherford County Wheeler reported that opponents often "contended that it [Wheel] will fall through as did the grange." Naysayers characterized the Wheel as "nothing but a speculating scheme, as was the grange . . . all excitement, and will soon die out."[51]

Uniting Farmers behind the Jute Boycott

Wheel and Alliance leaders distanced themselves from the Grange failures by pointing to what they considered profound differences between the organizations. In their view, the recent establishment of trusts created a new set of circumstances that made their organization necessary. "Twenty years ago we had no such thing in existence as a pool, trust, or corner to raise and lower values, contrary to the law of supply and demand," went the standard refrain. To bolster their argument, reformers also reminded farmers that "twenty years ago there was [sic] not two dozen millionaires in the nation; farmers were prosperous. It was impossible when the Grange went down to bring together five hundred million dollars to buy up all the coffee, sugar, etc., on the market and tax you by raising the price, or to corner the wheat, corn, pork, or tobacco market, and force the price down just as these products of the soil are ready for market." In 1889, agrarian leaders claimed the nation harbored over three thousand millionaires, "some of whom are worth over two hundred millions each, making possible any kind of combination or the bringing together a thousand millions if necessary to raise and lower values." In other words, reformers argued against basing actions on past experiences and encouraged farmers to recognize the reality of the new business world, even as they used familiar and time-honored language to build a cooperative trust to fight the trusts forged by commerce and industry.[52]

Changes in business practices validated the politics of antimonopoly, which held an exalted position in southern history. From Thomas Jefferson to Andrew Jackson to Andrew Johnson, the fear of the loss of individual freedom to the power of money and monopoly resonated through the social, political, and economic language like the strains of a familiar hymn. Powerful planters expressed hatred of monopoly as vigorously as the lowliest tenant farmers. The Democratic Party utilized the term as a pejorative for every form of rival activity and legislative position. The familiarity of the antimonopoly refrain drew men into the agrarian movement and provided a potent weapon in the Wheel and Alliance rhetoric. Agrarian leaders staked out new positions in a fight against structural changes in the nation's financial markets and industrial business practices through the use of familiar political language. The antimonopoly rhetoric at once described the problem of trusts and market domination, while evoking political biases that needed no explanation in the minds of rural voters. The Wheel and Alliance utilized the antimonopoly tool most effectively in its 1888–1889 boycott of jute.[53]

In 1888, the jute manufacturers formed a cartel that doubled the cost of preparing cotton for market by raising the price of the bagging that covered every bale of cotton. This decision by the bagging industry gave the Wheel and Alliance an instant villain against whom farmers could unite without fear of division. Although historians of the Alliance seldom fail to acknowledge the jute boycott that followed the news of the formation of the cartel, it generally receives only cursory attention in their monographs. The exception has been the work of William F. Holmes, whose research demonstrated "the clash between agriculture and business at a pivotal point in the development of the two interest groups." The actions of Tennessee's farmers in the jute boycott offer a window onto the continuing efforts to find a rallying point for the state's diverse agricultural producers as well as an opportunity to assess the influence of State Wheel and Alliance leaders in the regional organization.[54]

By the time Tennessee farmers became aware of the jute trust in August 1888, the cartel had already made all the necessary arrangements to carry its plans into action. Jute manufacturers facing the prospect of higher prices for raw materials under a pending congressional bill that called for the removal of jute from tariff protection quickly replaced their initial panic over the price hike with a plan to "coordinate the distribution and sale of jute bagging for the entire nation." Warren, Jones, and Gratz of St. Louis, a firm that controlled approximately 25 percent of the baling in the United States, organized the cartel. In July 1888, the price of jute bagging jumped from 6.5¢ per yard to 12.5¢ per yard.[55]

The St. Louis *Dispatch* announcement of the cartel in jute bagging caught the Wheel and Alliance unawares and scrambling to meet the sudden threat. Like their counterparts across the South, Tennessee Wheels and Alliances, meeting in the late summer and early fall, vowed to fight the trust by boycotting the use of jute bagging. When the Tipton County Wheel met on August 30, it "earnestly protest[ed] against the unjust trust or pool on jute bagging by Warren, Jones and Granty [sic] of St. Louis" and recommended that farmers of the Wheel and Alliance in all the cotton states "use no jute bagging." In addition, they urged the state and national organization to take "prompt action . . . against this iniquitous trust swindle." W. H. Gouch, president of the Crockett County Wheel, exhorted "the poverty stricken farmers of the county . . . [to] rise in their might and shake off the chains of slavery that are so rapidly being forced on them, or else submit to the money power," which would leave them "in the condition Ireland is in." The Gibson County Wheel joined in the boycott and declared, "a monopoly is just as much an enemy to orderly society as an anarchist." To hard-strapped cotton farmers the claim was no exaggeration. The doubling of jute bagging prices when the price of cotton hovered at 8.55¢ per pound (down from 10.66¢ in 1880–1881) represented a considerable cost. The penny differences in costs meant the success or failure of independent farmers.[56]

In a state like Tennessee, where cotton farming did not dominate agricultural production, the jute boycott provided a mechanism for uniting all farmers: cotton farmers had an obvious stake in the outcome, and other farmers could join in the great fight without risk. Non-cotton–producing counties quickly passed resolutions in support of the jute boycott in an effort to establish their solidarity with the beleaguered cotton farmers. Hickman, Humphreys, and Benton County Wheels, areas already organized in the Peanut Association of Tennessee, loudly voiced their support. Peanut growers determined to avoid the use of burlap bagging "held by the trust," and they recommended the use of "domestic as a substitute." In colorful language, one county resolved to "roll the wheel over them [the cartel] at whatever cost, and we can soon land all the thieves on the other side of Jordan, where the sulphur trust reigns supreme, and where they will be treated to a feast of smoke and perpetual conflagration."[57]

As historian Robert McMath noted, the jute boycott also transcended the divisions between merchants, bankers, lawyers, and farmers, setting aside the usual economic differences in one grand alliance. Generally, townsmen "deflected some of the agrarian hostility aimed at them" by joining the fight on the side of the farmers. But, as the Tennessee experience shows, there was ample room to profit from such expressions of solidarity.[58]

The Hebron Alliance in Shelby County called a meeting of all Wheel and Alliance members, planters, and merchants "living in the trading bounds of Memphis." Approximately a hundred farmers from West Tennessee, North Mississippi, and Eastern Arkansas attended the late August meeting at the Memphis Cotton Exchange. No merchants met with the convention of farmers, "as was expected and desired." That the merchants were absent did not prevent the presentation of their views. As soon as the meeting got underway, one man rose to urge "co-operation between the merchant and planter as the surest means of success." A. M. Scarborough made a speech in behalf of the cotton buyers, in which he advised conservative action, linking the interests of the merchant and the planter. He cautioned the planters against decisions to withhold cotton from the market as a protest against jute bagging, saying that "they could not afford to hold cotton, as such a course might seriously embarrass commerce."[59]

Convention chairman J. M. Crews appointed a committee of six planters and three merchants to write resolutions for the body. When the committee retired to carry out their mission, they were joined by Pat Gatins, "a well-known cotton buyer," and J. M. K. Speed, the local representative of Warren, Jones, and Gratz. The resolutions committee listened to Speed's recitation of the jute industry's economic problems, but ultimately favored unlimited support for a boycott of jute bagging. A minority report of the committee suggested that the decision to boycott could be superseded by the action of the National Alliance at its Meridian, Mississippi, meeting scheduled for October 1888. The convention gave its approval to the majority report.[60]

The determination to boycott jute produced a few last-ditch efforts to forestall the action, as well as exasperated ravings from planters, who advocated stronger action. One man suggested a mechanism for reducing costs by increasing the size of cotton bales to 700 or 800 pounds to save on the amount of bagging needed to cover a single farmer's crop. Two planters insisted that farmers lost no money, even at the higher price for bagging, but conventiongoers dismissed them as "sockdologers." Finally, in the heat of the moment, N. B. Branch declared that he "favored burning cotton rather than sending it into market in 15 cent bagging." Cotton farmers adjourned the meeting with a sense of accomplishment; despite the pressures exerted by the cotton buyers and merchant apologists, and the ravings of more excitable or misinformed planters, the Wheel and Alliance had developed a reasoned and reasonable plan of action to deal with the bagging crisis.[61]

The decision to boycott jute proved easy compared to the search for a reliable substitute bagging in adequate quantities to cover the soon-to-be-harvested 1888 crop. Suggestions ranged from the practical to the ridiculous. News arrived that farmers in North Carolina wrapped bales in straw cloth similar to matting. In North Georgia, farmers covered cotton in pine boards. A Tennessee man offered a similar suggestion for using "thin, tough boards and cheap, light hoop iron." He pointed out that "mills and machinery in Memphis . . . can turn thin boards faster than any mill can manufacture gunny cloth." With an "inexhaustible supply of cottonwood on the Mississippi . . . that can be had for the cost of cutting and floating to the mill," he estimated the cost of wrapping cotton could be reduced to 25¢ or 30¢ per bale. A persistent alternative suggested using convict labor to produce jute bagging in Tennessee. Many farmers simply took matters into their own hands, like the Tipton County man who showed "Alliance Grit," when he brought his cotton into town "wrapped in carpeting."[62]

By far, the most popular substitute for jute bagging was osnaburg, a heavy, durable cotton cloth of a texture "sufficiently open to permit proper ventilation" if the cotton became wet. According to its promoters, osnaburg was cheap enough to be used profitably, and its production from low-grade cotton provided an added benefit; proponents estimated that cotton bagging in sufficient quantities to cover the entire crop would require approximately 150,000 bales of low-grade raw cotton. The Lane Mills at New Orleans took the lead in offering osnaburg for bagging.[63]

In Shelby County, the agitation against the jute cartel went beyond the boycott. As the largest inland cotton market in the country and the supply depot for five states, Memphis merchants soon felt the effects of the farmers' refusal to cover the cotton crop in jute as rolls of jute bagging sat unsold on the docks. Public indignation pressured the fall term of the Shelby County Grand Jury to "look into the operations of the bagging trust," to determine if the cartel operated in violation of state law. Attorney General George P. Peters "made a thorough examination of the law on the subject," calling a number of witnesses to testify. On October 26, 1888, the grand jury returned an indictment against the partners of the St. Louis firm, charging them under the Tennessee law with "combining, confederating and conspiring together for the purpose of oppressing, injuring and interfering with the course of commerce, trade and morals of Shelby County," a felonious offense. The attorney general immediately applied to Governor Robert L. Taylor for authority to issue arrest warrants against the directors of the St. Louis firm. While farmers waited in anticipation of the demise of the trust, the bagging dealers laughed off the indictment with the dismissive statement that "there was no telling what a country judge might do."[64]

Although Memphis newspapers warned that Judge DuBose was "engaged in a righteous cause" and was unlikely to "be silenced in this contemptuous manner," the trust members had little cause to worry. Governor Taylor refused to sign the necessary papers to make the indictments effective, and the case died. At its January 1889 meeting, the Gibson County Wheel prepared a facetious memorial to Taylor in which the farmers expressed appreciation for "the conscientious scruples which his Excellency, the Governor of Tennessee, may have had in refusing to grant the necessary requisition papers." Wheelers deplored his action and requested that he "reconsider the matter in the interests of the agricultural classes and not in that of the capitalists engaged in this combine."[65]

As farmers soon discovered, appearances of support for the boycott by public officials, merchants, and cotton ginners seldom led to substantive action. Indeed, public declarations of camaraderie often masked attempts to consolidate economic advantages at the expense of the farmers. In Gibson County, the cotton ginners met at the Trenton Opera House for the purpose of "taking action in regard to buying cotton bagging for the present season." The ginners condemned the jute bagging trust as "a hateful monopoly for the purpose of trampling the cotton industry of the South," and they recommended the use of "any substitute that may be agreed upon." After the passage of the resolution, "all who were not members of the wheel remained in private session." During this meeting the ginners formed a Cotton Association and selected agents to ship and receive cotton. Wheels soon charged that the exchange established by "merchants and grocery men of Trenton" operated at prohibitive costs to farmers and urged cotton producers to avoid the excessive charges and make other arrangements.[66]

Despite setbacks and attempts by the merchants to control the boycott to their own advantage, farmers ended the 1888 cotton season with a feeling of victory over the jute trust. But by April 1889, it was clear that they had celebrated too soon. The *Toiler* reported that the jute trust, which appeared defunct only a few weeks earlier, now "bounded to its feet again, full of life and vigor." Given the successes of the previous season and more time to deal with the problem of securing suitable alternative bagging, McDowell confidently expected the farmers to "deal the monster a blow that will be slow to heal." While the state agents once again scrambled to find enough bagging to cover the crop, writers churned out indignant letters and enthusiastic poems and songs to marshal the agrarian spirit for battle. One such poem, "Fare Thee Well," promised the jute cartel, "Cotton will bury you." With no hope of resurrection, the trust's tombstone would read, "Died of combine fever / In eighteen hundred and eighty-nine."[67]

After showing initial reluctance to act, National Alliance President Charles W. Macune agreed to meet with disgruntled members and formulate a national plan of action. In May 1889, Wheelers and Alliancemen met in Birmingham, Alabama, and discussed further maneuvers against the jute trust. Tennessee Alliance delegates included John P. Buchanan, Thomas O. Abernathy, John F. Black, and W. T. Grant. George A. Gowan, R. W. Tucker, Alfred E. Gardner, and J. W. Zelner represented the Wheel. Tennessee Agricultural Commissioner B. M. Hord also attended and became one of three men named to the Cotton

Executive Committee, which represented the interests of cotton planters across the South.[68]

Agents from the jute trust appeared before the bagging committee of the convention and proposed to supply nearly three million yards of jute bagging at reduced prices for Wheel and Alliance members. The proposal promised to have bagging delivered at Charleston and St. Louis in August, September, October, and November. Prices for the first shipment would range from 8.75¢ for 1.5-pound bagging to 10.75¢ for 2.25-pound bagging for the August delivery. The price would advance ⅛¢ per yard on each subsequent delivery. Sure of their victory in forcing the trust to negotiate and confident of further concessions, the bagging committee and convention delegates concluded that the proposal was no bargain and recommended "the permanent use of cotton bagging as a covering for cotton." In order to gain the best price for farmers using osnaburg wrapping, the delegates appointed a committee to negotiate with cotton buyers to adjust the tare to reflect the lighter weight of cotton bagging. In the event that buyers and manufacturers refused to accept an allowance of 8 pounds per bale on cotton-wrapped bales, the delegates advised the members of the Wheel and Alliance to withhold cotton until the allowances were made.[69]

On September 11, 1889, representatives of fourteen cotton exchanges, Wheel and Alliance leaders, and the agricultural commissioners of five states met at the invitation of the New Orleans Exchange to discuss the question of the tare. The exchange agents agreed to accept a 24-pound tare for jute-wrapped cotton and a 16-pound tare for cotton-wrapped bales after October 1. Tennessee farmers soon realized they had not achieved a victory, however. On October 2, the *Toiler* announced the Memphis Cotton Exchange's opposition to the tare allowances. According to John Zelner, a Shelby County Wheel leader and a delegate to the Birmingham Convention, the decision to abandon the exchange convention agreement came at the insistence of Memphis merchants, who were "loaded with jute," and unwilling to support cotton bagging until all the surplus jute had been sold. McDowell resigned himself to the fact that "Memphis cotton merchants are . . . as clever generally as merchants of other cities, but self-interests come in and dictate their action."[70]

In the meantime, negotiations opened with Liverpool, England, the ultimate destination of most cotton, and farmers withheld their crop from the market pending the outcome of those discussions. State Agent George A. Gowan made arrangements to "store, transfer, and compress cotton" assigned to the state agency. To assure union members that their cotton would be handled profitably, he reported the conclusion of contracts with New York and New Orleans agents to obtain the latest market information by wire and keep abreast of price fluctuations. In order to take advantage of this arrangement, Gowan encouraged farmers to allow him to use his own judgment in selling their cotton.[71]

For a while, the resolve of the farmers held. McDowell crowed that cotton sales in Liverpool "were smaller on Sept. 27 than on any other day since the war." Gowan reported that only three hundred bales of cotton could be found at the Atlanta exchanges, where usually thousands of bales stood ready for shipment. But in the end, when news of Liverpool's refusal to accept the agreements reached the United States, farmers had few choices but to sell their cotton. When jute prices dropped, cash-strapped farmers no longer bought the higher-priced cotton bagging despite encouragement from Alliance leaders to hold the line against the trusts.[72]

The jute boycott represented a high point in Wheel and Alliance history, despite its long-term failure. The ability of the organizations to present a reasoned approach to the problem and to keep the cartel at bay for two growing seasons surprised opponents. The prominent role played by Tennesseans in the events no doubt gave pause to politicians facing Wheel and Alliance opposition. With their lack of financial support and their inexperience, farmers could be defeated, but it was becoming increasingly evident that they would not be defeated without a fight.

Tobacco Farmers versus Buyers

While cotton farmers battled the jute trust, tobacco planters waged their own war against monopoly. Throughout the summer of 1888, the *Toiler* carried information about a forthcoming Tobacco Congress to organize growers and warehousemen, gain control over the dark-fired tobacco crop in the Clarksville District, and raise tobacco prices.[73]

When the tobacco men met on July 27, they organized the Western Tobacco Growers Association. As the organizers explained, having been "reduced to poverty by the outrageous robbery of trust combinations," farmers decided to "fight fire with fire," and resort to "the schemes of the tricksters to beat their trickery." What the tobacco men proposed, as would become clear within a few weeks, was the establishment of a growers' trust. Couched in the agrarian language of cooperativism, the organizers argued that the decade-long depression in the tobacco markets and the "continuous depression of our financial condition . . . warrant us as business men [to] organize in the cooperative action that has worked so well for other businesses."[74]

Even as they argued for the creation of a farmers' trust, planners recognized the formidable problems confronting them, not the least of which was the history of agrarian organizational failures. "It has been thought that nothing could be done in the way of organizing farmers to resist the tyranny of combined capital and break up the combinations," they acknowledged. "But, necessity is a mother of inventions and earnest investigation leads to many new things and opens up new ways and outlets to progress," ran the enthusiastic promotional mantra. In an attempt to offset early disappointment with the cooperative effort, the organizers warned that "it is not expected the plan will thoroughly accomplish all desired in a year or two years." Planners encouraged enlistment in the cause and promised much "may be accomplished with the present growing crop if all tobacco growers will promptly go into the organization." The congress extended an invitation to all tobacco farmers in the Clarksville, Upper Cumberland, and Western Kentucky districts to join the cooperative effort against the stemmers, speculators, manufacturers, and buyers.[75]

The congress plan presented a well-thought-out mechanism for achieving producer control over the crop. It seemingly included benefits for the large planter as well as the small farmer. Moreover, it built on the organizational structure already in place in the tobacco region. Organizers assured farmers that the scheme would cost them nothing. The association collected no dues, since it planned to operate through existing granges, wheels, and alliances. The plan called for the construction of sheds and warehouses but relied upon each farmer to sort, classify, and prize his own tobacco.

Success depended on the honest packing of tobacco, with "but one kind or quality in a hogshead—no mixing of the short with the long, or packing inferior tobacco in the corners as is the practice with rehandlers and speculators." Congress organizers attributed low prices to underhanded practices, since buyers expected deception and allowed for it in their pricing. Honesty promised to raise the price by drawing buyers to clean, well-packed tobacco and creating competition for their crop.

To encourage farmers to bring their tobacco to the association's warehouses, the organizers promised that members could draw half the value of their tobacco immediately upon deposit. This inducement allowed farmers to meet immediate expenses (taxes, mortgages, store credit, etc.) and gave the association the flexibility to market the crop at intervals or hold out for higher prices. Near the end of the association's presentation, the organizers touched on the keystone of the plan—controlled planting "to prevent over-production or great surpluses . . . by decreasing or increasing the acreage as the necessity may require—just as the whiskey trust manage their distilleries." As the congress ended, the association members confidently advised cotton planters to adopt a similar scheme.[76]

At the November meeting of the Tobacco Congress, the association adopted formal resolutions justifying the creation of the organization and establishing the hierarchy of officers. The resolutions blamed "unusually large crops in the barns throughout the country," and the "combination of monopolistic capital" for the tobacco prices "far below the cost of production." Condemnation for the situation also extended to tobacco farmers, who "have hitherto been (as a class) plodding single-handed along in the tobacco business as in all other farm pursuits without any sort of organization or co-operative action."[77]

From the defense of their organizing efforts, the resolutions once again moved to the heart of the association's program. In order to "avert the evils now pending," and achieve "immediate relief," the association recommended "the adoption of total abstention of tobacco planting for the year 1889." Defending the abstention proposal with a classic statement of New South agrarianism, the association leaders couched their plan in language that mimicked the high-riding Wheel and Alliance. If farmers "determine[d] not to even sow a bed for 1889, put their land in wheat and clover, make their own bread next year which some are not doing now," they argued, "besides resting and recuperating their over-taxed land, they will find that they are master of the situation." Apparently worried about opposition to the enforcement of the nonplanting policy and the inflammatory nature of ordinary farmers, the association went on record as favoring "all legitimate means of carrying out the object and intention of these resolutions."[78]

Wheels and Alliances in the tobacco district quickly fell into line with the plan. Hard-pressed farmers pointed to sales of tobacco at $217.50–$243.50 per hogshead and lamented that the price "won't half way pay expenses of raising and yet it [tobacco] was raised on land worth forty dollars per acre." In October, McDowell spoke to the Stewart County Wheel and visiting Wheelers from Kentucky at Big Rock Church, "crowded to its utmost and many could not get even standing room." He reported that the people listened for two and a half hours, even postponing dinner to satisfy their interest in the nonplanting policy.

McDowell found Stewart County "heart and soul for this or any other practical method by which a fair remuneration could be obtained for the present large crop." Among the farmers in the audience was E. P. Bozart, "one of the largest dealers of tobacco in Stewart County." After hearing the proposition of the association, Bozart pledged that "not an acre on any of his fourteen farms should be planted if the plan was agreed upon." Support for the policy extended into West Tennessee, where both Weakley County and Obion County resolved to abstain from planting tobacco.[79]

Dissenting voices soon rose to challenge the view that nonplanting benefited the smaller farmers. Members of the Colored Wheel of Montgomery County in the heart of the Tennessee tobacco district issued the first challenge to the proposal. At its October 13 meeting, the Colored Wheel questioned the justice of allowing "the tobacco men and speculators to dictate to us as to whether or not we shall raise tobacco." They encouraged Wheelers to "give the tobacco question fair consideration and do what they think will be best for the farmer." Perhaps in reference to the Montgomery County resolution, McDowell noted before the November meeting of the Tobacco Congress that "the idea that the Clarksville Tobacco Warehousemen want to use the Tobacco Congress to enrich themselves, regardless of the farmer's interest, seems to deter some good farmers from joining in the cessation policy."[80]

By late November, at least one Stewart County Wheel also voiced second thoughts. Oak Hill Wheel noted that of the "two crops in American hands—one of the meanest ever raised . . . belongs to the speculator;" the other crop "an extra good one [is] in the hands of the producer." The memory of past ill treatment by the Clarksville warehousemen added to the suspicions of the rank-and-file Wheelers. Stewart County farmers remembered that earlier requests for assistance met with derisive comparisons of the Wheelers to the Grangers, with assertions that like the Grangers, they were "snow-men, and would pass away with the morning's sun." No doubt some farmers had a hard time reconciling the abstention policies of the Clarksville group with the enthusiastic reports of the Fulton Tobacco Warehouse in Fulton, Kentucky.[81]

During the winter of 1887–1888, the Tobacco Congress and the Agricultural Wheel of Kentucky met in Fulton to establish a tobacco warehouse. Fulton citizens enthusiastically supported the decision, since the city had no other tobacco warehouses. The citizens secured a building to use temporarily and promised both to purchase a lot that the farmers wanted and to erect a warehouse if the farmers proved unable to build the structure. With the backing of the town and the banks, the warehouse prospered and "sold more tobacco than [the Wheel] or anyone else expected, and at a higher price." The congenial spirit evaporated, however, when the farmers selected their lot and asked the town to fulfill its promise. Although the town offered to purchase a different lot, the warehouse manager complained that the town acted in bad faith. Yet, despite the rocky beginning, the warehouse reportedly prospered and the manager didn't mention low prices or unsold tobacco from previous crops.[82]

In February 1889, the Fulton warehouse manager, D. C. Reams, issued the *Second Annual Circular of the Farmers' Tobacco Warehouse* in Fulton. He reported an increase in the size of the facility to handle 10,000 hogsheads of tobacco.

He claimed that "the growth of our market has created the necessity of the organization of a new bank at this place, which will open on the 25th." Stock in the warehouse also increased during the year from $5,000 to $50,000. Reams assessed the past year enthusiastically: "prosperity has smiled upon us; everything seems to have gone our way." Furthermore, he asserted the western markets "singularly clear of old tobacco," and he encouraged farmers to bring their tobacco to Fulton, where he expected to "close the season of 1889 with the best tobacco market in the west." Four months later, the Obion County Wheel confirmed Reams's assessment of the warehouse. The Wheel encouraged farmers to patronize the "oldest and most successful co-operative enterprise established by our organization." The West Tennessee farmers claimed the Fulton Warehouse successfully "forced the tobacco buyers to advance their bids on loose tobacco . . . advance[ed] money on tobacco in the hands of brethren . . . [and] stands the peer of any market in the west in prices . . . careful handling . . . honest weights and impartial inspection." Although it is impossible to validate the Fulton report, the contrast with the Clarksville claims certainly raised questions in the minds of many farmers. The Fulton Warehouse claimed to offer the same services as the Alliance of Clarksville warehousemen and showed a profit in what everyone regarded as a disastrous year, without resorting to a nonplant program.[83]

A Stewart County farmer, who identified himself only as "Alpha," posed a thorny question early in the discussions over the nonplanting policy. "If I am not badly mistaken," he wrote, "the mission of the wheel and Alliance is to protect and foster our interests as a laboring class and not to combine with monopolies for the purpose of breaking down and cheating each other out of a just reward for our labor to gratify monopolistic greed." He could not reconcile the farming interests with those of the warehousemen, and wondered, "if Clarksville's tobacco warehouse-men entertain such a great degree of sympathy and anxiety for the welfare and prosperity of the toiling tobacco grower why don't they reduce their profits some and give us the benefit of it." His concerns, first dismissed, proved prophetic.[84]

The association between farmers and warehousemen fell apart in the spring of 1889, when the Wheel decided to establish its own warehouse in Clarksville to handle the tobacco of Wheel and Alliance members. Located near the passenger depot, the warehouse was "entirely under the control of [the] order in the Clarksville tobacco district." C. P. Warfield, "an experienced tobacco raiser and a man of unimpeachable integrity," managed the warehouse. Farmers received assurance that Warfield was "well known to the brotherhood of his county," and supporters referred to local resolutions professing "perfect confidence in his judgment, intelligence and ability to manage the Wheel and Alliance Warehouse successfully."[85]

On May 1, 1889, the *Toiler* reported the disturbing news that M. V. Ingram, editor of the *Clarksville Cultivator* (and former editor of the *Toiler*), refused to support the Wheel warehouse. McDowell initially confessed that he "did not know the facts" and would therefore "withhold comment." Three weeks after the first announcement of trouble, McDowell charged that "a number of tobacco speculators" had established the *Cultivator* and placed Ingram in charge with the sole mission "to prove to the ignorant farmer of the Clarksville district that his interest and that of the speculator were identical." The reason for the efforts, just as critics had charged, centered on the abundant crop of the previous year. In the eyes of many farmers, the speculators overestimated their ignorance; some "insisted that the object of the Cultivator in advocating the abstention policy . . . was not a mission of mercy to the oppressed farmer, but merely a trick resorted to by the speculators to get rid of a large quantity of old stock that was dead weight on their hands." Having discovered the plot, McDowell generously hoped that the talents of his former partner would be "appreciated by the public in a better cause."[86]

At the end of May, the Wheel warehouse petitioned for membership in the Clarksville Tobacco Board of Trade. The rules of the board required that applications for membership lay over for thirty days. The warehousemen saw their chance for revenge, however, and with the knowledge that they represented a majority, moved for a suspension of the rules to act on the application immediately. Buyers voted for admission and warehousemen against. The Clarksville Board of Trade excluded the Wheel warehouse from its association and refused a resubmission of the application. Wheel leaders charged that "inspectors were threatened with expulsion if they dared to inspect a hogshead of Wheel-Alliance tobacco." An editorial in the Clarksville *Chronicle* warned that "buyers will not be allowed to bid on this tobacco, nor will inspectors be allowed to inspect it." Many buyers and inspectors rejected heavy-handed attempts to intimidate them, and inspectors "notified [the Wheel] superintendent that they would inspect our tobacco regardless of what the warehouse men might say." Eighteen buyers soon signed an agreement "to bid . . . without discrimination on all tobacco stored with the Wheel-Alliance house, as from any stored in this city."[87]

Organization in the tobacco districts involved more than control of the crop—it demanded a change in the political and social hierarchy. The small growers in the Wheel and Alliance continued their agitation for reform, but they failed to achieve the shift in power necessary to their success. Powerful tobacco planters like Joel Fort made no secret of their hatred for the Wheelers and Alliancemen; along with the warehousemen, these planters worked to achieve control over dark-fired tobacco production without regard for "popular sentiment." In 1888, the aroused farmers had their "eyes open" to manipulation, and the planters' schemes failed. But, likewise, the Wheel and Alliance warehouse plan came to naught. Clearly, organizing against monopolies and trusts had its limitations—more evident in the experiences of tobacco farmers than in the jute boycott.[88]

By the end of the 1880s, Tennessee farmers had created an effective organization that proceeded cautiously and pragmatically in its efforts to address the problems of agriculture in the age of industrial capitalism. Building on the existing structure of rural community and personal networks of kin and friends, agrarian leaders evoked the blessings of God in their work to educate farmers in the political economy of the nation. Forced to overcome a legacy of skepticism handed down from past agrarian failures and hobbled by inadequate financial support, the Wheel and Alliance adopted a conservative plan to establish a state agency that operated on the commission system. Stymied in their efforts to find a single rallying point for agrarian action in a state marked by agricultural diversity, state leaders nevertheless realized both successes and failures in the cotton and tobacco regions. On the plus side, they marshaled considerable support for the jute boycott and advanced their own claims for regional leadership. Less successfully, Wheelers and Alliancemen struggled to build a tobacco warehouse in Clarksville. Here they encountered the firm resistance of entrenched economic elites. Both their successes and failures convinced many farmers that traditional methods no longer sufficed. As a result, the drumbeat for political action sounded louder with each election. It remained to be seen if Tennessee farmers had the tenacity to sustain the union as Wheelers and Alliancemen gained political offices.

If a man loves his party better than the great welfare of the laboring masses, he had better stay with his party and not disturb the harmony of our order.

—ROBBIE RUSKINS, correspondent to the *Toiler*, 1888

CHAPTER FOUR
FARM ISSUES IN THE POLITICAL ECONOMY

Wheel and Alliance ideology always left the door open to political action. As McDowell so often noted, the solution to the agrarian crisis rested in the ballot box, and the problems encountered in establishing the state's cooperatives made politics more attractive. In addition, farmers needed state law to protect their business arrangements, and they wanted control over rural institutions: local schools, the land-grant university, and the Bureau of Agriculture. As important as state political action was, advancing rural interests nationally took precedence. With the merger of the Wheel and Alliance, the agrarian movement shifted from an emphasis on community action to a more focused effort to elect like-minded men to the state legislature and congress. Although the majority of rural voters cast their ballots for Democratic Party candidates, the nonpartisan policy advocated by the Wheel and Alliance forced candidates to measure up to the agrarian "yardstick" in their campaign promises.

The questions farmers raised framed the conflict of the age—the role of the individual and the community in the battle between capitalism and democracy. Agrarian leaders understood that they must act nationally to protect their local communities and their livelihoods as independent producers. On the national scene they demanded transportation and communication regulation or nationalization as necessary to the common good. Farmers advocated abolition of the national banking system, federal regulation of currency, and the subtreasury scheme to improve access to capital, provide farmer control over commodity prices, and eliminate the crop lien system. They promoted the popular election of U.S. senators and demanded a federal income tax as measures that would redistribute power downward and restore farmers and laborers to the mythical position they had claimed from the earliest days of the republic.

Antipartyism as a Political Tool

Wheel and Alliance leaders framed their political claims in the language of antipartyism, with words as familiar to the rural voter as the hymns he sang on Sunday. The growth of the agrarian cooperative organizations educated the state's farmers in the political economy and provided the momentum for political insurgency. The rallies, parades, and barbeques brought together men and women suffering from the constraints imposed by an industrializing and deflationary economy. Responding to familiar scenes of political and social celebration, farmers listened intently to the speeches that promoted and renewed Wheel and Alliance enthusiasm, supplied rural families with an explanation for their economic decline, and evoked a heritage of grassroots political action that reached back to the Jacksonian era, encouraging usually quiescent farmers to challenge powerful entrenched elites.

Antipartyism enjoyed a glorious reputation in Tennessee political history, evoking shades of Andrew Jackson and Andrew Johnson. More than a simple demand to "throw the bums out," antipartyism enshrined the role of the yeoman farmer and articulated a critique of American society that tapped into the perception of an inherent conflict between the political and economic ideologies that supported the nation and demanded a necessary balance between democracy and capitalism. In the late nineteenth century, antiparty rhetoric pointed to rapid industrialization with its concurrent rise in economic centralization and demanded an accounting from legislators and congressmen who were believed to have facilitated the accumulation of fortunes by the few through favorable legislation. Antiparty proponents faulted both Democrats and Republicans who served the interests of the elites and neglected the interests of farmers and laborers—the men and women whose work produced the agricultural and industrial commodities and made American democracy possible.

Farmers had participated in the state's political life since the return of two-party government in 1869 and the drafting of a new constitution in 1870, albeit generally as voters, seldom as legislators, and never as members of the executive branch of government. Candidates for the state's elective offices generally acknowledged the concerns of the so-called "wool-hat boys" in their biennial canvasses—a fair and equitable system of taxation, fiscally responsible state government, farmer-oriented agricultural institutions, defense of common rights, and business regulation—but they seldom acted on those concerns. Politically savvy office seekers exploited rural discontent to further their own aims while successfully avoiding substantive reform. By the early 1880s, however, national economic and social changes had so transformed the context in which farmers planted, harvested, and sold their crops that politics as usual no longer sufficed.

Pressure to enact favorable legislation for the benefit of Tennessee agriculture was not a new concept for Gilded Age politicians. In the years following the Civil War, the state increasingly acted to regulate Tennessee farms both to rationalize agricultural markets and institutionalize social order. Legislators passed lien laws, prohibited night sales of cotton, provided premiums for fairs, and debated the definition of legal fencing. These legislative actions generally benefited substantial planters by protecting the property rights of landholders and securing an adequate supply of agricultural labor.

Smaller-scale farmers looked to rural organizations such as the Grange for solutions to farm problems and sought answers through cooperativism, initially avoiding political activism. However, by the mid-1880s, many farm leaders reluctantly recognized the limitations inherent in under-funded cooperatives and anticipated a greater role for state governments to act as arbiters and regulators over regional, national, and international markets with their dense, overlapping networks of storage, transportation, processing, and long-distance sales. As historian William Cronon so eloquently described the changes in the Midwest, sales shifted from the marketing of sacks of wheat to the storage and processing of golden rivers of grain, a transformation in which the individual producer could no longer distinguish what he had grown from that of his neighbor or the farmer across the state. With prices determined by international supply and demand, farmers believed they needed new laws to protect their interests.[1]

With the merger of the Wheel and Alliance in 1888–1889, the agrarian political agenda assumed a more central role as farmers looked beyond locally directed cooperativism to government intervention at both the state and federal levels and regulation as the mechanism for stabilizing the agricultural economy. Rural voters gradually shifted from an antiparty rhetoric to more a sophisticated and more defined articulation of the agrarian political position. Initially operating as nonpartisan voters, farmers limited their support to candidates who measured up to the Alliance yardstick of demands, and they exercised increasing control over the selection of candidates and the election of legislators. A careful assessment of antiparty rhetoric, a delineation of the demands championed by Tennessee's farmers, and an outline of the political context in which agrarian insurgency operated provides a basis for analyzing the strength of rural activism and its political limits.[2]

Wheel and Alliance activities in the 1880s developed within an existing climate of antiparty protest that culminated in what Robert McMath called a new political culture. As we have already seen, the growth of the agrarian cooperative organizations educated the state's farmers in the political economy of their day and provided the momentum for political insurgency. The rallies, parades, and barbeques brought together men and women suffering from the constraints imposed by an industrializing and deflationary economy. Responding to familiar scenes of political and social celebration, farmers listened intently to the speeches that promoted and renewed Wheel and Alliance enthusiasm, supplied rural families with an explanation for their economic decline, and evoked a heritage of grassroots political action that reached back to the Jacksonian era, encouraging usually quiescent farmers to challenge powerful entrenched elites.[3]

Historians' attention to antiparty rhetoric has produced an interesting interpretive evolution from John Hicks to Richard Hofstadter to Lawrence Goodwyn. These historiographic frameworks variously characterized the agrarian critique as vital to the development of modern Democratic and Republican thought, a backward-looking example of political outsiders afflicted with status anxiety in a modernizing social and economic climate, or the last moment for the fulfillment of the democratic promise made by the founding fathers. More carefully nuanced recent scholarship suggests that antipartyism represented "a parallel framework of nonpartisan belief and experience [that] coexisted alongside, and interacted with, partisan politics . . . helping to shape third party eruptions and American political thought."[4]

The principal analysis of Tennessee's experience in agrarian antipartyism and third-party development in this period is Roger L. Hart's 1975 book, *Redeemers, Bourbons, and Populists.* Hart's work focused on contemporary political changes in the Tennessee General Assembly. Utilizing the computer-driven methodologies of voter analysis, Hart followed in the interpretive footsteps of Hofstadter and characterized the state's farmer-politicians as backward-looking political outsiders anxious to gain office and thereby restore their declining status as voters and community leaders. In general, Hart's dismissal of the agrarian movement has shaped successive state histories. But close attention to the voices of the farmers sheds new light on their antiparty position and the fate of third-party activity in the 1890s.[5]

Operating both within and outside the established parties, antiparty yeoman farmers claimed a near-mythical position in the political ideology of the nation.

The words of Thomas Jefferson supported farmers as "the true representatives of the great American interest," and asserted that they alone were "to be relied on for expressing the proper American sentiments." Although historians later dismissed these sentiments as agrarian fundamentalism, farmers were not alone in their belief that agriculture represented the true nature of the republic. National political leaders, especially Democrats, extolled agrarian life as the bedrock of American liberty and echoed the words of Andrew Jackson who praised the independent farmers as "the best part of [the] population . . . the basis of society and the true friends of liberty."[6]

Farmers and political elites anticipated different outcomes when they evoked the Jeffersonian ideal. Public speakers praised farmers and the agricultural life as a brake against the radicalism commonly assumed to be associated with industrial labor, but farmers interpreted political ideology as literally as they interpreted the scriptures and laid claim to the inheritance offered to them in the national mythology.

Warren Smith, a former Greenbacker, prolific writer and agricultural organizer, utilized the national mythology to recall a sense of pride in his readers when he wrote, "Before the brawny arm of the sturdy and iron-muscled pioneer, the forests disappeared and in their place were seen broad fields of golden grain prophesying of bounteous cheer for all." Smith attributed the garden of America, which "outshone the lands of Nimrod," to "the genius of labor [that] bestowed it all." Ordinary farmers took up the cry and proclaimed that "this glorious producing country of fertile fields was made by the farmer's strokes." By their work, farmers claimed, a great nation was established, and by their votes, that nation would continue into the future.[7]

The agrarians also claimed a share in the historic defense of liberty. While waging their own battles for equalization of the tax burdens, Wheelers and Alliancemen were reminded of "those humble farmers who were oppressed by the misrule, tyranny and over-taxation which was imposed by the royal government of England." When wool-hat boys encountered the hostility of entrenched political interests, they were buoyed by the memory that "the father of our country was a farmer, and he retired from the highest earthly position to enjoy the quiet rural life at Mt. Vernon." New South proponent Henry Grady suggested that farmers had not always acted as conservatively as elites hoped when he attributed the success of the French Revolution to the actions of the peasant farmers in the summer of 1789, and praised the farmers of Mecklenburg, North Carolina, who "made the first American declaration," as well as the "embattled farmers who fought at Lexington and Concord." In the same speech before the farmers of Elberton, Georgia, Grady quoted General Toombs, who reportedly noted that "the farmer, considered the most conservative type of citizenship, is really the most revolutionary."[8]

Agrarian leaders repeatedly urged farmers to "put their hands to the plow" and "never look back." Their cause was clear: "march onward until the great wheel plow has uprooted all such pestilent weeds as monopoly, unequal taxation and other glaring inequities." Rural organizers reminded their constituents that their mission encompassed more than a fight for their own interests. Once the farmers performed their political duty and "cleared the ground around the noble tree of freedom which [their] fore fathers planted for [them] of all nauseous and poisonous herbs," they could rest in the knowledge that their "children may partake of the fruits of the said tree to [their] hearts content."[9]

Women especially embraced antipartyism as the most viable mechanism for necessary weeding of the American political garden. As historian Sara Evans has noted, periods of political turmoil frequently provided opportunities for the disenfranchised to voice their demands. Empowered by Wheel and Alliance rhetoric, women readily stated their views on the political economy and on reform in the Ladies Column of the *Toiler* and in their addresses before organizational conventions and rallies, chastising reluctant farmers and excoriating political elites. Since John McDowell would have edited all letters before publication, it is impossible to know whether some women pressed for greater political changes. On the one occasion when he voiced an opinion, McDowell acknowledged that country people generally shied away from women's suffrage, believing "it will lower our respect and love for woman to mix her up in the rude political world, and that her proper sphere and influence for good can be better exerted by steering clear of politics." Whether McDowell's caution resulted from political pragmatism or genuine rejection of women's suffrage is hard to say; it is his one verifiable statement on the issue. But women could and did speak out in favor of rural reform, participating in the political insurgency that swept the countryside.[10]

Antipartyism focused rural anger on the "courthouse rings" that monopolized political power and stymied reform. Mrs. M. E. Hill of Jefferson County expressed the bitterness of many when she observed that "the men of this nation once voted like kings and were free, but today they look up to monied [sic] men and bosses." On the opposite side of the state, a Lake County farmer asserted, "we have been voting for cliques and rings, for professional politicians, dead beats and jack-leg lawyers, until we are almost in bondage." Others wanted party leaders to know that "we are sick and tired of old political frauds as representative men, generally in the person of some lawyer that loves the people well, but himself better." Farmers recognized that such a man "knows how to take care of his own interest and represents the people accordingly." A Tipton County farmer summed up the feeling of many when he complained that farmers had "listened to silver-tongued orators, court-house cliques, professional politicians, etc., and did their biddings, voted as they told them, and went on like a blind horse until, alas! They fell into the ditch."[11]

Antiparty rhetoric galvanized dispirited voters, but rhetoric alone could not transform the countryside. Nonpartisanship easily dovetailed with antipartyism as agrarian reformers tapped into the anger farmers readily vented against the "rings" that controlled political action and access to offices. Nonpartisan voting provided the means to translate that anger into action.

The official position of both the Wheel and the Alliance, like the earlier Grange, was one of nonpartisanship that was spelled out in the Alliance Constitution of 1887. In the Declaration of Purposes, farmers resolved "To labor for the education of the agricultural classes in the science of economic government, in a strictly non-partisan spirit." In the same declaration, they offered a critique of contemporary officeholders as they vowed to return to the principle of "letting the office seek the man, instead of the man seeking the office." Finally, they attacked common campaign tactics as they promised to "suppress personal, local, sectional, and national prejudices" in the furtherance of their aims. As the Agricultural Wheel and the Farmers' Alliance moved toward union, representatives of the two organizations met first in Shreveport, Louisiana. The Conference Committee, chaired by Tennessee's John McDowell, offered a resolution to "accept, as a basis of union, the secret work of the Alliance and the national constitution of the same." Wheelers and Alliancemen thus agreed to the policy of nonpartisanship (among other things), as they moved toward an expanded union. Again, at the Meridian, Mississippi, meeting (the second step in the merger process), National Alliance President Charles Macune charged farmers to "leave partisanship to the individual but discuss political economy as a class." Lest Alliance members misunderstand his charge, Macune advocated a rule that would strip any local Alliance of its charter if members engaged in partisan politics.[12]

Nonpartisanship suited the purposes of Tennessee's agrarian leaders. With the state's seemingly unbridgeable political, social, and geographic divisions, nonpartisan action represented the most viable mechanism for uniting farmers across the state, but Wheelers and Alliancemen harbored no illusions regarding the strength of party adherence among Tennessee voters. D. A. Vaughan warned farmers that "old party prejudices and ties are so strong that reason and even the general good of mankind are really lost sight of and trampled under foot to gratify this demon of party spirit." A month after Vaughan's appeal, the secretary of his county Wheel, J. H. Reed, gave assurances that nonpartisanship prevailed in Bedford County: "We have all, or nearly all, concluded to lay aside politics and vote for the man, let him be Democrat or Republican, from school commissioner to congress and governor, believing it to be the only way we can be successful in electing officers to represent us." To those who could not set aside party to bring about reform, Reed offered no remedy but to "scratch, brother, scratch."[13]

Through speeches and editorials, agrarian leaders hammered home the belief that "party spirit has robbed agriculture of its precious boom and lost all of its charms for the tillers of the soil," while simultaneously exhorting farmers to recognize that "the ballot box is the only resort and only hope for reform." Blending antiparty rhetoric with nonpartisanship, one Wheeler asserted that farmers had "determined to no longer blindly follow the lead of party spirit or flinch under party lash, but will go for the best thing and the best men in sight and teach parties a lesson." As the reform movement picked up steam, it was this determination to jettison partisan support that agrarians believed troubled "political tricksters."[14]

It proved easier to denounce partisanship than to remove it. Not unexpectedly, as local Wheels and Alliances became embroiled in election campaigns, partisan feelings rose to the surface, producing dissension and turmoil. At the meeting of the Wayne County Wheel in December 1888, the twenty-five Wheels represented took note of the recent election and the problems it had created for the organization by refocusing attention on cooperative efforts and the credit system. To facilitate harmonious meetings, they encouraged active female membership in the Wheels. The members prefaced their resolutions toward these objectives by noting that they acted with the purpose to "shake hands across the sea of political strife, burry [sic] all ill feelings and with a united effort to go out anew to win the prize." Likewise, when William Dixon called a meeting of the Colored Wheel of Henry County for January 1889, he encouraged attendance by reminding the members that "the election is over, now let us rally the old wheel onward."[15]

Clearly, despite assurances that the Wheel and Alliance could act to influence politics without being in politics, members felt the pressures of partisanship and found support of the party of their fathers difficult to reject. For many, the security of tradition that surrounded the established parties proved impossible to ignore, and nonpartisan voting on the basis of self-interest remained problematic. For not a few Wheelers, the seduction of political campaigning overwhelmed other considerations. Relishing the excitement of political campaigns, these farmers found postelection activities too tame. One enthusiastic grassroots campaigner compared the 1889 postelection activities of the Gibson

County Wheels to "the churches after Christmas." But, he assured his readers, in an obvious reference to future political races, "We are educating, fortifying, drilling for the conflicts of '90 and '92."[16]

The Dynamics of Tennessee Politics

Despite efforts to enforce nonpartisanship, the active presence of both major political parties seemed to offer Tennessee farmers a unique opportunity to achieve their demands. As Jeffrey Ostler suggested in his study of the prairie states' agrarian movement, two-party politics provided reformers an opportunity to pit one party against the other to gain concessions and facilitate reform. For Tennessee farmers, however, the racial and geographic distribution of power operated against such action.[17]

The political arena farmers entered was as divided and divisive as any in the country. Redemption of the state in 1869–1870 did not destroy the Republican Party but largely reduced its control to the mountainous East Tennessee. Sizable pockets of Republican support remained in the old plantation counties, where large African American populations voted for the party of Lincoln, and among white populations living near the Tennessee River in the western portion of the state, where Union support had been strong during the secessionist crisis. Republicans dominated political offices in the first and second congressional districts of East Tennessee and frequently won hard-fought elections in some counties of West Tennessee, but the real menace of Republicanism rested in its potential as the spoiler in any political battle that threatened to divide the Democratic Party. In 1880, when Democrats split into two factions over the question of the settlement of the state debt, the Republicans captured the governorship for the first time since Redemption and the last time until Prohibition offered a similar opportunity in 1912.[18]

Internal dissension and factionalism characterized the Democratic Party of the late nineteenth century, although, in the interests of party unity, candidates who lost the party nomination for office dutifully supported the final ticket and praised the political and personal qualifications of their recent rivals, even as they plotted the next factional battles. Not unexpectedly, the Democratic Party of the period would be better characterized as an amalgam of interests and views rather than a unified political whole. Political expediency joined the interests of the New South industrial heirs of antebellum Whiggery with the plantation-based defense of states rights offered by the Bourbons. The precarious balance of interests that supported the Democratic hegemony over state offices and the Republican regional control left little room for either party to initiate political or economic reform. As a result, Democrats focused attention on the familiar platform staples of low tariff and fiscal conservatism. Republicans responded with their well-known program of industrial protection and sound business practices. Since partisan responses remained fixed, personalities played a large role in the selection of candidates, and family history, war records, and material achievement figured prominently in every election. With neither party willing to tackle the problems associated with the economic and social transition, many voters might have agreed with the farmer who expressed his disgust by noting that "the most difference between the Democrats and the Republicans is a hustle to get there with both feet."[19]

By 1880, a powerful Republican bloc survived on the liberal distribution of patronage resulting from the party's apparent monopoly of the presidency. In the second congressional district, Congressman Leonidas Houk used his access to patronage to build a political machine comparable to that of northern urban "bosses." He kept detailed information on the voters of his district and claimed to be able to call out two thousand campaign workers at a moment's notice to rally support for a candidate. Houk's brand of politics centered on white supremacy and local issues. By 1894, within three years of his death, political control had shifted to H. Clay Evans, a third-district Republican with broader, national interests.[20]

Factionalism emerged as some in the party opted to sacrifice traditional reliance on black voters in West Tennessee in order to entice discontented whites and "Old Whigs" into a New South Republicanism. The national party led the retreat by offering patronage positions to white Democrats in preference to black Republicans. The success of the party's proselytizing efforts among the Old Whigs produced an oft-repeated prediction by the Republican National Committee that Tennessee would be "the first state to secede from the solid South." Expectations of the return of genuine two-party politics seemed to be borne out by election results. The steady decline of the Democratic margin of victory in gubernatorial contests from 43,600 in 1876 to 9,180 in 1884 cheered Republican efforts, and the 1888 victory of Republican industrialist H. Clay Evans for the third-district congressional seat added that district to the Republican column for the first time since Redemption.[21]

Increasingly, the Republican party valued African American members primarily for the votes they cast on election day. Disillusioned black Republicans held a convention meeting in Nashville in 1884 to call for the creation of local political clubs, since "our experience with political parties has taught us . . . that if we want anything done, we must do it ourselves." National black leaders urged African Americans to divide their votes between the two parties to achieve their demands, while local leaders called upon black voters to reject candidates imposed upon them by white political leaders.[22]

By threatening, or taking, independent action, blacks achieved a few modest successes. Eight African Americans served in the Tennessee General Assembly in the 1880s, whereas only one other black man had previously achieved that distinction. In 1886, John J. Irvine was elected as county court clerk in Chattanooga in a biracial effort that made use of his active membership in the Knights of Labor. Two years later, in Middle Tennessee's sixth congressional district, African Americans gained control of the Republican congressional nominating convention and nominated a black man, William H. Young, for congress. At the convention that nominated Young, blacks complained "in a bold and uncompromising" manner of the long years of service that had gone unrewarded by the Republican Party. The bitterness of African American voters toward the unresponsiveness of the Republican Party found its counterpart among the small-farmer element of the Democratic Party.[23]

Like African Americans in the Republican Party, white farmers in the Democratic Party loyally voted the party's slate of candidates year after year but realized few benefits. Throughout the 1870s the Democratic Party remained under the control of men with the industrialist leanings commonly associated with the advocacy of a progressive program of mining, manufacturing, railroad construction, and urbanization that would later receive the title of "New South." Antisecessionists initially, these men served in the Confederate army or Confederate government. In 1870, they figured prominently in the redemption of the state from Republican rule. The defeat of the party in the elections of 1880 ushered in a new era by raising the Bourbon wing to a position of prominence.[24]

Historians of the postbellum period have engaged in an endless debate over the relative influence of Bourbons and former Whigs in the rebuilding of the South. C. Vann Woodward set the tone of the debate with his assertion that "Old Whigs" captured the political advantage during Redemption and built a New South that broke with the old cotton South socially, economically, and politically. A bold and pathbreaking view, Woodward's thesis came under attack by a number of scholars who cited the support and investment of planters in the building of textile mills in the cotton fields. In addition to establishing the financial investment of Old South money, other students of the era carefully researched legislative genealogies to demonstrate the persistence of planter control in political affairs, leading to the suggestion that the South pursued a "Prussian Road" to modernity. More recently, James C. Cobb asserted that whatever economic differences separated New South industrialists from Bourbon planters, they remained united in their determination to maintain control over labor.[25]

Among Tennessee Democrats evidence exists to support all three views. An industrial-minded wing of the party included former Whigs as well as young, ambitious men dedicated to the new progressive ideas of professionalism. Coalescing around Nashville, Knoxville, and Chattanooga, they promoted mining, manufacturing, immigration, and scientific agriculture. Many, like Arthur S. Colyar of the *Nashville American*, used their newspapers to advocate a more industrial, modern South. At the same time, remnants of the planter aristocracy looked to Memphis as the economic center of the Mid-South cotton culture. With access to rail and river transportation, Memphis developed around a burgeoning cotton economy, claiming the title of the largest inland market in the world. As a result of their favorable geographic location, Memphis merchants enjoyed a virtual monopoly over the distribution of dry goods and groceries in the counties of West Tennessee, Eastern Arkansas, and North Mississippi. Nevertheless, Memphis economic and political elites, like Napoleon Hill, merchant and president of the Cotton Exchange, quietly invested in Birmingham's New South steel industry. And throughout the state, industrialists, planters, and merchants colluded to guarantee a ready supply of cheap, compliant labor.[26]

Thus, although factional fights erupted between the two views throughout the last decades of the nineteenth century, Democratic power brokers could be counted on to close ranks in the face of any threat against cheap labor, whether the danger emerged from conditions outside their control or came from within the party. For the farming element this closing of the ranks relegated participation in government to the exercise of the franchise. Democrats needed farmers' votes, but even as they made promises they would not keep and whipped up partisan support, they tempered radical change by constantly reminding farmers of the need to "make haste slowly." The familiar refrain characterized farmers as "the most law-abiding and conservative class in society" and encouraged them to "see to it that nothing is done that will result in the loss of this justly earned reputation." Perhaps hoping to frighten would-be radicals into compliance, political leaders frequently reminded farmers that they possessed "great power" but warned that that power "if not kept within its legitimate channel [may] prove to be a curse, instead of a blessing." The message was clear, whether it came from New South advocates or Old South Bourbons: the party needed the votes of the farmer, but he should not presume to operate outside his "legitimate channel" by asking more than the opportunity to cast his ballot, lest he disrupt the order of society.[27]

Tennessee politics flirted with the third-party movements of the day, and many Wheelers and Alliancemen gained their first reform experiences in the prohibition movement. In the 1880s, progressive farmers and the enterprising citizens of market towns and county seats united to demand the abolition of the manufacture and sale of alcohol in Tennessee. Focusing their efforts on a state constitutional amendment, prohibitionists watched in eager anticipation as Democrat John H. McDowell shepherded the enabling legislation through two sessions of the general assembly. McDowell's political acumen became apparent to both supporters and enemies as he negotiated the barriers post-Reconstruction "Redeemers" had erected to stymie efforts to amend the Constitution. After successfully gaining majority support for the amendment in two successive legislatures, prohibition supporters organized for the final step: ratification of the amendment through a statewide referendum. Both sides marshaled their forces to wage a battle for votes that was marked by vitriol and exaggeration. When voters failed to ratify the amendment, prohibitionists turned to local option and a weak Prohibition Party to field legislative and gubernatorial candidates supportive of their antialcohol position. The presence of the Prohibition Party often presented a wrenching decision for Wheelers and Alliancemen. As farmers moved into the political arena, many struggled to balance their devotion to rural economic uplift against their moral concerns.[28]

In fact, the Prohibition Party offered little competition to either the Democratic or the Republican Party, but it contributed to the reforming mentality that the Wheel and Alliance supported. Like Prohibitionists, Alliancemen organized to effect change, beginning with their own farms and rural communities and moving toward a transformation of state and national government. And, even as the Prohibition Party failed, it laid the groundwork for a more successful effort in the early twentieth century, one that transformed Tennessee politics and society. For men like John McDowell, prohibition provided a mechanism for continuing their push for rural progressive change. As such, its minor role in the agrarian reform of the late nineteenth century masks its greater influence in twentieth-century rural politics.

Controlling the Vote

Political control by ordinary citizens had been an elusive democratic dream long before industrial capitalism reinvigorated the fight. And, rural men and women were not alone in their complaints about cliques and courthouse rings. Editors of respectable daily newspapers routinely issued warnings against the abuses of power, and scholarly evidence supports the charge that party control rested on the perpetuation of static ideologies and electoral corruption corroborated by a complex network of friendship and kinship. Sons and nephews followed fathers and uncles into the courthouses and legislative halls of the state and nation in a manner guaranteed to stifle grassroots reform. What friendship and familial ties did not control, manipulation and electoral fraud could.[29]

Control over office holding began with the nomination of candidates. Although some counties were already selecting candidates through primary elections, Tennesseans generally nominated candidates for office through county, congressional district, and state conventions called by each party for that purpose. Manipulation of county nominating conventions had become a fine art by the time farmers entered the political fray.

Anxious to nominate their own candidates, farmers eagerly shared information about the various tactics county politicians had developed to control convention processes. In Perry County, local politicians attempted to "get up strife and confusion" within Wheel ranks by nominating two Wheelers to oppose each other for the same office. Warning others not to "allow our flock to scatter in county elections," the correspondent asked rhetorically, "how can we expect to unite and send men to represent us in the next General Assembly?"[30]

A Marshall County farmer offered another example of a common mechanism for controlling nominations. As he explained it, Lewisburg, the county seat, had its "town ring" of prominent Democrats who had "become firmly impressed with the idea that it was their especial prerogative to select a [convention] chairman and secretary of their own choosing." After selecting the presiding and recording officers, a motion would be made to allow the chairman to appoint a committee to determine a slate of candidates. In this way the will of the "ring" prevailed, and "it was generally understood that the ring and their little coterie of followers over the county would consider it exceedingly cheeky in any outsider to object to anything the bosses proposed." Agrarian leaders hoped their publication of these and other political tactics would educate farmers to the mechanisms that had long stymied rotation of officeholders. Politically educated farmers could marshal their anger and put it to good use to nominate and elect candidates to support their interests.[31]

On election day, fraudulent voting, ballot box stuffing, and false reporting were almost the norm. At every level of government the vote was manipulated to determine the outcome. Indeed, the redemption of Tennessee from Republican rule in 1869 rested on the willingness of Governor DeWitt C. Senter to replace election officials with men who would ignore registration laws and permit voting by disfranchised ex-Confederates. Seven years later Americans watched in fear and wonder as charges of voting fraud in the presidential election threatened to split the country into warring factions in the nation's centennial year. The "resolution" of the conflict, which stole the election from Samuel Tilden and gave it to Rutherford B. Hayes, while perhaps satisfying southern demands for an end to Reconstruction, did nothing to reassure voters that their ballots could be honestly cast and honestly counted. In 1890, the Obion County election to determine the location of the county seat was so corrupt that one of the competitors, Union City, reportedly adopted the slogan, "Win, fairly if you can, but win!" A tally of the votes determined that the victorious Union City won by "voting all the living and most of those who had been in their graves since the close of the civil war."[32]

Tennessee Democrats accused Republicans of the fraudulent use of African American votes, while Republicans threatened federal intervention against "the despicable methods" that characterized "Ku Klux democratic leadership." Regardless of party affiliation, few voters could have been unaware of the problems of obtaining an honest count. Widespread charges of fraud in the 1888 election led to the filing of suits in federal court and produced calls for voting restrictions that resulted in a series of disfranchising laws that operated against African American and poor white voters.[33]

Tennesseans discussed a variety of measures to cleanse the voting process. Many voters demanded primaries rather than mass meetings for choosing candidates, but others warned that the primary offered little relief since it was open to repeat votes and ballot box stuffing. Some critics argued for smaller ward or district meetings "where citizens meet together and are all known to each other" as the only legitimate solution. In such meetings, they believed, voters could choose delegates who would represent the will of the district. There, "the humblest voter has a fair chance of expressing himself in a district meeting, among his neighbors, without fear of brow-beating from ruffians and politicians."[34]

Champions of face-to-face political meetings had a point. Although defenders of primaries touted their expediency, fraud abounded because they were unregulated by general-election laws. In primary elections, as in general elections, some wards recorded more votes than potential voters. Tennessee politics remained a hodgepodge of nominating processes that varied from county to county with little oversight from the state. Every effort to end fraudulent election practices offered partisan advantage and produced little actual reform.[35]

The campaign abuses that characterized the election of 1888 brought to the forefront a reforming effort that threatened the political future of the farmers' movement. That election veered dramatically from the carnival atmosphere that had characterized the so-called War of the Roses in 1886. The earlier campaign pitted Democratic gubernatorial nominee Robert L. "Bob" Taylor against his Republican brother, Alf. In a canvass virtually devoid of issues, the two brothers joked and fiddled their way across the state, attracting huge crowds that came to enjoy the spectacle of siblings locked in harmless political combat. Although historian Daniel M. Robison viewed Robert Taylor as the harbinger of Populism and credited his crowd-pleasing antics with marshaling the rural vote for reform, Taylor's unwillingness to articulate a defined position on any issue suggests that his appeal to farmers was that of an entertainer.[36]

By 1888, however, Bob Taylor's close association with the New South wing of the Democratic Party produced the widespread perception that he took his instructions from Arthur S. Colyar, editor, industrialist, and would-be kingmaker. The *Toiler* proclaimed that Taylor "has no head of his own, but is supposed to wear the head of a certain fine old gentleman who heads the tariff monopoly and has a hand in all the combinations and speculative booms in Tennessee." Indeed, the relationship between Taylor and Colyar included the industrialist's support for "Our Bob's" lecture tours and the governor's investment in the Nashville Iron, Steel, and Charcoal Company.[37]

Taylor's friendship with Colyar produced a political backlash among many Democrats, including the wool-hat boys. Sensing Taylor's vulnerability, a number of would-be candidates mounted campaigns to deny the governor renomination for a second term. In a state gubernatorial convention one Taylor supporter called the most bitter in Tennessee history, delegates cast forty ballots before giving their nod to the governor. He won a second term, but in the view of Hart, Taylor "won votes not by rallying discontent, but by ignoring it."[38]

The contentious gubernatorial nomination provided a prelude to the hard-fought and violent general election. Farmers scrambled for every vote as they struggled to elect agrarian-minded candidates who would work to protect their cooperative state agency. The animosity engendered by the recent winner-take-all referendum vote on prohibition left voters searching for scapegoats. Indeed, many prohibitionists attributed the failure to ratify the antialcohol amendment to the African American vote, adding fuel to fiery calls for ballot box purification that focused on the perceived corruption of the franchise through African American voting. As candidates battled for seats in the general assembly, African American voters faced renewed intimidation at the polls. In Haywood County, armed men threatened voters and "guarded" ballot boxes. As a result, Haywood and Fayette counties, two black majority West Tennessee counties, where Republican candidates routinely enjoyed large majorities, reversed their voting patterns and overwhelmingly elected Democrats.

The combination of fraud, intimidation, and violence placed forty Alliancemen in the legislature and provided the Democratic Party with an unbreakable quorum in the 1889 general assembly for the first time since Reconstruction. The jubilant party of Jefferson and Jackson used the opportunity to undermine the

Republican electorate, secure in their knowledge that the minority party lacked the power to defeat disfranchising legislation through filibustering or quorum-breaking tactics. When Democrats first introduced disfranchising bills in 1885, Republicans had bolted the hall before the sergeant at arms could lock the door. Their quick action prevented a quorum and thus forestalled disfranchisement for that session. But in 1889, secure in their absolute majority, Democrats rallied to enact legislation to reduce or eliminate the votes of African Americans. Casting their work in the language of reform, the advocates for disfranchisement argued that eliminating black voters was the only way to guarantee the purity of the ballot box. The removal of African Americans from the political process promised to relegate the Republican Party to East Tennessee and ensure future Democratic control over state government. In addition to disfranchising a significant portion of the Republican Party, Democrats gerrymandered the Republican-dominated third congressional district to ensure its future control by the Democratic Party. At the same time, prompt action at the state level prevented federal intervention that might result from anticipated congressional debate over protection of black suffrage.[39]

Free to do whatever they wanted, Democrats ignored warnings that their actions would open a "Pandora's Box for the South" and enacted the Dortch Law, the Myers Law, and the Lea Law, reinstituted the poll tax, and redrew the congressional districts. Senator Joseph H. Dortch, a lawyer representing Fayette County, sponsored the bill that required each voter to mark and deposit his ballot in secret, with no assistance. Though not promoted as a literacy test, the law had the effect of disfranchising those who could not read or possessed minimal literacy skills. Initially the law applied only to Chattanooga, Knoxville, and Shelby and Davidson counties (the counties containing Memphis and Nashville). Although it was the most controversial of the bills under consideration, Democrats made no pretense of supporting it for any reason other than the disfranchisement of black Republicans. The *Memphis Avalanche* openly taunted Republicans by confirming that "the Dortch bill is for the benefit of the Democratic party," and commending it as the answer to the "Negro Question."[40]

The Myers Law, sponsored by East Tennessee Senator J. C. Myers, required voter registration in all towns and districts with populations of more than 2,500. The law covered some county seats in addition to the cities that fell under the jurisdiction of the Dortch Law. The Myers Law required voters to present a valid registration certificate at the time they cast their ballots.[41]

To complicate the voting process further, the general assembly also enacted the Lea Law, which provided for separate ballot boxes for state and federal elections. Introduced by Judge Benjamin Lea of Haywood County, a legislator McDowell identified as a Wheeler, the law represented a preventive measure if the much-discussed federal supervision of elections became a reality. The 1891 legislature repealed this act when congress failed to enact the so-called Force Bill.[42]

Democrats increased their political presence in the eastern portion of the state by gerrymandering the third congressional district. Legislators moved the two Republican counties of Cumberland and Marion to the heavily Democratic fourth district and added Democratic Cannon County to the third district. As hoped, the change returned the third district to the Democratic congressional column in the 1890 election. In 1870, provision for the implementation of a poll tax had been made in the 1870 Constitution, but despite the constitutional intent to limit suffrage through the use of the state's taxing power, the general assembly did not enact the enabling legislation until 1889.[43]

During the constitutional convention of 1870, the most vigorous protest against the inclusion of a poll tax provision in the state constitution had come from the friends and supporters of Andrew Johnson, the champion of the yeoman farmer. John Netherland, an associate of the former president, objected to the inclusion of a poll tax in the organic law of the state, fearing that it discriminated against the poor man. In his view, the poll tax requirement "renewed a system which the past has shown to be odious to the voters of the State." Netherland believed it gave the poor man cause to say "you require of me to pay the 'uttermost farthing' all that I own to the State, though I am poor, whilst my wealthy neighbor, who owns lands to the value of many thousand dollars, on which the taxes are due and unpaid . . . is allowed to vote."[44]

Andrew Johnson also complained in a letter to George W. Nixon that "the new constitution disfranchises every man in the state white or black, transfer[r]ing the elective franchise, which is the essence of sovereignty from the voter to the Legislature; converting it in fact into an article of merchandise to be bought and sold for a price fixed by law." Furthermore, he believed that "the effect of this policy will be to make a large number of voters dependent upon capital and a few opulent aristocrats." It would inevitably lead to "corruption, fraud and degeneration of the ballot box." Johnson refused to support the ratification of a constitution that "disfranchises every voter in the state." The constitution was easily ratified despite the concerns of Johnson and his supporters, although succeeding General Assemblies made no attempts to enact the tax for which the constitution made provision. Paradoxically, given his defense of the common man, if farmers knew of, or remembered, Johnson's concerns, they showed no evidence of it in the 1889 debates. Indeed, in the 1890 election campaign, the Alliance capitalized on the pro-disfranchising votes of farmer-legislators. Moreover, the timing of the legislation refutes the interpretation offered by historians of southern Populism that the poll tax represented an attempt to disfranchise Populists as well as African American voters.[45]

Historian Frank B. Williams argued against racism as the principal support for the poll tax, suggesting that educational improvement accounted for the measure's easy passage. And indeed, publicly, Alliance men tied their backing of the measure to demands for better education, pointing out that revenues generated by the tax were designated for schools. In an 1890 editorial, McDowell defended the poll tax, arguing that the law upheld the interests of the poor, who, unlike the wealthy, could not afford private school education for their children. Claiming that some 90,000 voters in Tennessee never paid "one cent to keep up the public schools," McDowell called the tax, which applied equally to all men, an example of Alliance ideals of "equality before the law." Although "opposed to the principle of restricting the right of suffrage" and "wish[ing] there was some other means of forcing every man to pay his poll tax," the editor offered the weakest of all defenses: the poll tax was the law of the land and must be paid until the constitution could be changed.[46]

Historians Joseph Sharp and Roger Hart refuted nonracial arguments and cited numerous editorial examples of support for the law as a mechanism for "making sure the rule of the white majority in this state." As they suggested, it would have been impossible for any legislator in the 1889 general assembly to remain unaware of the purpose of the acts. However, Alliance legislators never indicated that the poll tax bill presented the either/or dilemma posed by historians. Although the Alliance avoided race-baiting, it never repudiated the tenets of white supremacy. At the same time, overly taxed farmers supported almost any legislation that benefited schools *and* avoided additional land taxes. For Alliancemen, the poll tax offered two potential benefits: the assurance of white, Democratic rule and a mechanism for increasing the school fund that did not fall disproportionately on the farmers of the state. With relatively weak ties to East Tennessee, Wheelers and Alliancemen apparently worried little about the potential backlash from their Republican brothers. Perhaps they better understood that region's racial views than we do and expected few repercussions beyond standard hollow complaints from white Republican leaders.[47]

The assurance of white, Democratic control became apparent immediately. As Morgan Kousser noted, with the passage of the four electoral "reforms," the "most consistently competitive political system in the South" ended. In the thirteen West Tennessee counties of Crockett, Dyer, Fayette, Gibson, Haywood, Henry, Lake, Lauderdale, Madison, Obion, Shelby, Tipton, and Weakley, Republican votes dropped from 21,785 in 1886 to 7,068 in 1890. The educational benefits never materialized: poll tax revenues remained small, and the state's schools continued to be underfunded. Only after the Democratic Party successfully drove out white farmers would they recognize the peril to their own use of the franchise.[48]

Conflicts over Agricultural Institutions

In 1889, Middle and West Tennessee farmers had been instrumental in the successes of the Democratic Party, and some forty Wheelers and Alliancemen held seats in the general assembly. While not in a position to force their agenda through the legislature, they could no longer be ignored. Farmer-legislators sought reforms on both the state and national levels, and Tennessee's agrarian leaders differentiated between the two. For the most part, Alliance demands could only be achieved by federal and constitutional changes, but there were a number of issues at the state level that needed attention. The agrarians adopted positions on the state university, the state bureau of agriculture, private property laws, and taxation that were intertwined with historic divisions of geography, politics, and class.

Two agricultural institutions came under repeated attack during the final decades of the nineteenth century: the Tennessee Agricultural Bureau and the state land-grant college, the University of Tennessee. Both institutions had ignited controversy from their origins, and Wheelers and Alliancemen tapped into that longstanding discontent to marshal support for their demands. Many farmers believed the Agricultural Bureau and the University of Tennessee to be under the control of planters and industrialists and therefore of little benefit to them. Perceived as out of touch with ordinary farmers, agricultural leaders demanded explanations and a close accounting of the public funding for these institutions. Rural leaders repeatedly called for the abolition of the Agricultural Bureau and denied state funds for the University of Tennessee. Under Alliance pressure, farmers gained a bureau that focused on agricultural issues, but substantive changes in the operation of the University of Tennessee or its experiment station awaited a later day.

East Tennessee University (renamed University of Tennessee in 1879) traced its history with the federal government to 1807, when congress provided land for the establishment of two colleges, one in East Tennessee and one in West Tennessee. The former Blount College in Knoxville, founded in 1794, became East Tennessee College (University), and Cumberland College would serve students in what was then called West Tennessee.[49]

In 1862 congress passed the Morrill Act providing the sale of public land for the establishment of agricultural and mechanical colleges in each of the states. The legislation permitted states to endow an existing college, or create one or more new colleges to carry out the congressional intention. In 1869, the radical Republican-dominated Tennessee General Assembly acted to confer the funds on a single existing college, East Tennessee University. When Conservatives "redeemed" the state months later, the legislature revisited the decision to grant the funds to the Knoxville college.[50]

Before voting on the matter, legislators heard reports from two special committees, as well as a remonstrance from the East Tennessee University Board of Trustees. The report of the Special Committee on the East Tennessee Agricultural College, submitted by J. L. Russell, W. P. Caldwell, R. B. Cheatham, and B. A. Enloe, representatives from Middle and West Tennessee counties, faulted the decision to use the entire fund for the East Tennessee college and recommended that "either the whole of said fund be removed to some more central and accessible point in the State, or that it be divided, so as to make it available to every division of the State." The committee found the university to be in a state of disrepair and unable to meet the physical or fiscal requirements of the Morrill Act. They also worried that the funding of an agricultural college in East Tennessee thwarted the intention of the act "as the agricultural pursuits which would be taught in that extreme Eastern portion of the State, would not be adapted to the Middle and Western portion of the State." Finally, the report suggested that the agricultural college would serve few farm boys since they would be unable to afford the cost of transportation and the expense of living far from home. In short, funding an agricultural college in East Tennessee provided little benefit to the majority of the state's farmers.[51]

A second report by the Select Joint Committee on East Tennessee State University was submitted the following day. The four East Tennesseans and one West Tennessean came to similar conclusions. Although this committee found East Tennessee University physically and fiscally in compliance with the provisions of the Morrill Act, they also recommended that "full justice to *all the children of the State be had*" by dividing the fund and establishing a similar college in West Tennessee.[52]

The board of trustees offered a third view in a remonstrance submitted to the general assembly. The trustees based their support for the original appropriation on legal precedent, citing the Supreme Court rulings on the sanctity of contracts. Their tightly crafted legal defense persuaded few legislators, and the Russell bill to divide the funds between East, Middle, and West Tennessee passed the house, 44 to 26 and the senate, 13 to 10. Governor DeWitt Senter, a native of Grainger County in East Tennessee, vetoed the bill. An attempt to override the veto failed: voting along sectional lines, the representatives from Middle and West Tennessee fell five votes shy of achieving their goal.[53]

The controversy over the designation of the Morrill funds to East Tennessee University tapped into longstanding animosities between the grand divisions as well as realistic concerns about agricultural education in the state. East Tennessee claimed the funds as the beleaguered supporters of the Union cause, a position certain to generate animosity in the Redeemer legislature. Nevertheless, the objections voiced by Middle and West Tennesseans reflected legitimate concerns. From its founding, East Tennessee University promoted a liberal arts education. Its curriculum provided young men with a modest education in literature, history, and ancient languages: farmers might legitimately wonder how committed the university would be to an agricultural program. Moreover, the center of commercial agriculture then, as now, lay hundreds of miles to the west, calling into question the university's value to farmers. Although, as supporters of Senter's veto noted, division of the funds would have produced two or three weak colleges, farmers in Middle and West Tennessee continued to complain and to demand acknowledgment of their specific needs, especially when the federal Hatch Act (1887) funding experiment stations similarly resulted in no institutional presence outside the Great Valley of East Tennessee.

In 1890, John H. McDowell visited the university campus on one of his many trips to East Tennessee. He reported a courteous welcome and an informative visit, but declared that "the truth is, the way it [the university] is run is a farce, and benefits agriculture very little." He suggested that if "Professors [Charles W.] Dabney, [Charles F.] Plumb, and [L. P.] Brown would hold farmers' institutes over the State, and meet with practical farmers . . . this school might be made a power in building up and advancing the agricultural industry in Tennessee."[54]

McDowell was not alone in his criticism. A report in the *National Economist* complained that, across the country, land-grant colleges and agricultural experiment stations had failed to live up to their potential in part as the result of joint administration of the two. "Teachers are wanted in the faculty and scientific investigators on the staff of the station," advised the *Economist.* The situation at the University of Tennessee was a textbook example for the journal editorialists. The University of Tennessee Board of Trustees voted to establish an experiment station in 1882, one year before the passage of the federal legislation calling for such stations in all states. Funding from the Hatch Act ($15,000) began in 1887, the same year that Charles W. Dabney, a noted agriculturalist, arrived at UT to serve as president of the university and director of the experiment station. Having obtained complete administrative powers over all university operations and faculty appointments, Dabney set about the task of revamping the curriculum to reflect a new emphasis on agricultural and mechanical studies. As a result, he faced muted criticism from the board of trustees and near revolt from the faculty. Even Governor Robert L. Taylor complained when Dabney staffed the agricultural department with faculty drawn from colleges outside the state, noting that the university needed "a Tennessee farmer to teach Tennessee farmers."[55]

In 1889, farmer-legislators attempted to force the university to acknowledge their complaints when Dabney asked for an appropriation of $12,000 to fund new construction on the Knoxville campus. McDowell offered strenuous objections in the *Toiler*, charging that "every dollar given by the State is just that much thrown away." Calling the university's claim of "giving farmer boys of Tennessee a scientific education" a farce, he scoffed at the supposed benefits farmers received from the college or the experiment station. Charging that "in the past twenty years this school has drawn from the State and national government over a half million in cash," he asked, "what has gone of it?" Although McDowell inflated his figures, his criticism deserves attention because he tapped into a wellspring of discontent from farmers who found the university and its agricultural program a useless expense.[56]

McDowell demanded action. He proposed the immediate withdrawal of "every dollar of aid to this school unless a better showing can be made than we have yet seen." Second, he called for an amendment to the federal Hatch Act to place the annual $15,000 experiment station appropriation under state supervision. As McDowell's editorial was circulating, Gibson County farmers petitioned the state legislature to vote for a measure that would allow division of the experiment station funding to permit "the purchase of an experimental farm in West Tennessee where the soil is widely different from that of either of the other divisions of the State."[57]

The demands for greater accountability by the university and the simultaneous call for the development of an experimental farm in West Tennessee tapped into the wellspring of agrarian fear that the state's "best men" no longer understood or cared about the agricultural crisis facing virtually every farmer. McDowell assured readers of the *Toiler* that he would print "exchange[s] of sentiment and discussion vital to [agricultural] interest and prosperity." But he also promised to omit theories and experiments in farm operation "which are all well enough after the stumps are first cleaned out of the way."[58]

Others joined McDowell in criticizing agricultural experts. In his president's address to the National Farmers' Alliance, Leonidas L. Polk, a respected proponent of modern agricultural education, asserted that "the purely technical effort of improving our methods of farming . . . although it be praiseworthy, is not a force or remedy." Furthermore, he added, "the influences that tend to depress agriculture and render that occupation unprofitable have rapidly gained the ascendancy over and neutralized the beneficent effects that should have followed the introduction of wise methods and new machines."[59]

So-called "practical" farmers recognized both the potential benefits and the limitations of progressive agricultural reforms and demanded more thoroughgoing changes. Tennessee farmers called for the university's scientific expertise in addressing the peculiar conditions in every region of the state even as they realized the improved farming methods would not renegotiate the balance between capital and labor or secure the future of the landholding. Despite the efforts of the Alliance legislators, the Tennessee Experiment Station remained in East Tennessee until 1907, when a station was established in Middle Tennessee; the following year, West Tennessee gained an experiment station. The University of Tennessee continued to ignore calls for a West Tennessee campus. Finally, in 1927 the city of Martin and Weakley County voted to issue $100,000 in bonds to purchase the buildings of the failed Hall-Moody Institute, establishing the junior college that became the University of Tennessee, Martin.[60]

Farmers experienced similar problems with the Tennessee Bureau of Agriculture. Indeed, at times, the animosity between farmers and the bureau produced calls for the abolition of the state agency. Much of the conflict centered on the multiple tasks assigned to the agency and the relative importance of agriculture and mining in the state's economy. When the general assembly established the agency in 1871, it authorized a catchall commission to oversee development in agriculture, mining, and immigration. As secretary of the Bureau of Agriculture, Statistics, Mines, and Immigration, Joseph B. Killebrew was charged with the task of "publish[ing] annually such statistics and other information as the Bureau may deem necessary to advance the general interests of Agriculture and the material development of the State." In the same vein, Killebrew was ordered to include in his reports "such matters as will convey a proper idea of the resources of the State to practical men." Clearly, the responsibilities of the bureau exceeded development of agricultural interests, and Killebrew never doubted the importance of using the bureau to promote timber and mining interests and to entice immigrants into the state.[61]

Within months of his appointment, Killebrew met with Governor John C. Brown and other New South agriculturalists to establish the *Rural Sun*, a weekly agricultural journal, edited by the bureau secretary. Although Killebrew declared his desire to "make the *Rural Sun* represent the farmer's interests and reflect his sentiments," the paper did not neglect the advancement of the state's budding industries. Indeed, it advertised "the great agricultural advantages, mineral resources, unsurpassed water power, magnificent forests, and genial climate of the state and its surroundings" to promote immigration and investment. The journal brimmed with the latest scientific agricultural information on everything from kitchen gardens to tobacco, livestock, and poultry. Heralded as one of the best journals of its kind, the *Rural Sun* also included articles on mining, timber, and increasingly, manufacturing and industry.[62]

By the time Killebrew left the *Rural Sun* in 1873, his role in the Agricultural Bureau was raising questions. In 1877, legislators launched an unsuccessful attempt to remove Killebrew from his post as Commissioner of Agriculture. C. F. Vanderford, a professor of agriculture at the University of Tennessee, and B. M. Hord, the man who succeeded Killebrew as editor of the *Rural Sun*, and who eventually became commissioner of agriculture during the administration of Robert L. Taylor, led the fight. From Killebrew's perspective, the battle involved nothing more or less than jealousy stemming from Hord's imagined possession of "a world of knowledge about agriculture which no one else had." Killebrew's superior knowledge notwithstanding, a close look at the legislative journals reveals an alternative interpretation for the attempted removal of Killebrew.[63]

George W. Martin, state senator from Weakley County, introduced a bill to abolish the Agricultural Bureau (and thus deprive Killebrew of his position). By the time the bill arrived at its third reading (the last step toward passage), partisans for both sides were ready to present their views. Governor James D. Porter supported Killebrew's work and condemned attempts to abolish the bureau as "shortsighted." The state needed the bureau, according to Porter, because "capitalists look for information from official sources and not from parties who wish to sell." According to opponents of the bureau, Killebrew's "information" was the problem. Senator William A. Millikin of Fayette County "favored the repeal" on the grounds that the bureau acted more as an agent "for the sale of real estate of private individuals than one of agriculture." Indeed, he believed that "agricultural interests had been the last thing thought of." Furthermore, the publication of Killebrew's massive 1,200-page *Introduction to the Resources of Tennessee* was viewed by some senators as "such an exaggeration of the wealth and prosperity of the State as almost to defeat propositions for a compromise of the burdensome state debt." Senator Hugh M. McAdoo, representing Humphreys and Benton counties, favored "a fund for legitimate agricultural purposes." But, he noted "there was a prevalent impression that this Bureau had served rather the purposes of a real estate agency than for the benefit of agriculture." Andrew E. Garner, a senator from the Black Patch tobacco districts, cited the commissioner's report and claimed that it "virtually admits that it [the Bureau] cannot practically improve the agricultural condition." In hard times, Garner stressed, it seemed that this was "a good place to apply the pruning knife."[64]

Killebrew and the Bureau of Agriculture survived the threatened "pruning," but the views expressed by farmers who felt themselves ignored by their own agency were a source of ongoing discontent that resurfaced most importantly at the height of Wheel and Alliance political power. Farmers only became reconciled with their agency during the Taylor administration, when the governor appointed B. M. Hord to the position of commissioner. Ordinary farmers claimed Hord as a "brother" in first the Wheel and then the Alliance. He attended local and county meetings, listened to their concerns, offered advice, held farmers' institutes, offered premiums, and worked with their elected leaders. When the issue of funding arose in the 1889 legislative session, the *Toiler* reported that "the farmers in the house made a strong fight for the agricultural bureau." In 1891, the so-called farmer-dominated general assembly acted on the recommendation of Alliance Governor John P. Buchanan and enacted legislation that separated mining interests from agricultural concerns. The act created a Bureau of Agriculture and a separate Bureau of Labor, which was consolidated with the Bureau of Statistics and Mines.[65]

The ongoing legislative fight illustrates both the demands by farmers (particularly those from West Tennessee) for responsive agricultural institutions and the "class" differences that permeated the various approaches to agricultural issues. Analysis of the identifiable legislative opponents to Killebrew's management of the Bureau of Agriculture suggests that the frustration with a university situated far from the principal regions of commercial agriculture was repeated in the claims that the bureau existed to promote the sale of real estate and investment in mining. Both ventures promised greater benefits to East Tennessee than to the other grand divisions. Likewise, the emphasis on scientific agriculture, while not dismissed by members of the Wheel and Alliance, failed to address rising concerns about costs and commodity prices. The political elites who dominated government largely ignored the justified fears of small farmers that their land, their independence, and their way of life were in jeopardy. Men with large plantations or college degrees lectured small farmers on their perceived failings and assured themselves that the yearly losses in agriculture were not the result of economic forces beyond the control of individual cultivators. If farmers viewed the conflict in class terms, the Upper-Case (elite) Democrats did, too.

In the agricultural South, the upper class was defined by membership in an elite society that measured its status by standards that included, but were not limited to, wealth. Often linked by marriage and kinship, the "best men" admitted up-and-coming young men to their ranks, but only at their own discretion. Though not born into this upper class, Killebrew was a member as a result of education, marriage (and the acquired wealth that accompanied that marriage), and elite patronage. Hord and the leaders of the Wheel and Alliance, some of whom controlled modest-to-substantial wealth, did not belong to that class. Killebrew clearly felt betrayed by Hord's association with men of lesser rank, calling the commissioner's tenure in the bureau "a perfect farce." Unlike Killebrew, whose writings make no mention of the Wheel and Alliance, Hord played an active role. As Tennessee's commissioner of agriculture, Hord negotiated with the New Orleans Cotton Exchange on behalf of the agrarian organizations during the jute boycott. Killebrew earned the respect of planters and manufacturers, but ordinary farmers saved their praise for B. M. Hord, who they claimed as one of their own.

Defending Common Rights

While smaller, commercial-minded farmers struggled to focus the direction of the state's agricultural institutions, the poorest farmers battled with planters and policy makers over the regulation of common rights and the protection of property rights. Every meeting of the general assembly in the late nineteenth century dealt with the problems of fence laws, dog laws, and game and fish laws, but they failed to find acceptable and permanent solutions. At the heart of the fencing issue was the question of who would be held responsible and therefore bear the financial burden for the destruction of crops by wandering livestock. Open-range laws that characterized antebellum agriculture were no longer acceptable to commercial farmers, who complained of the high cost and waste associated with fencing agricultural land. On the other hand, stockmen and herders pointed to time-honored range laws and accused planters of elitism when they demanded that the owners of livestock fence in their animals.

In Tennessee the problem was compounded by the constitutional power given to the state legislature that inhibited county and city home rule. Whatever the original merits of this line of reasoning, the requirement precluded recognition of the differences in agricultural production across the state. As a result, the general assembly expended an extraordinary amount of time in every legislative session reformulating the fence laws and granting exemptions to particular counties. As one constitutional expert noted, "the Counties which are densely populated, need one kind of road and fence law, and the sparsely settled mountain Counties need another kind, but as it is, the populous suburbs of Nashville and of Memphis, must have the same road and fence laws as the trackless mountains of East Tennessee." Calls for constitutional reform greeted every legislative campaign, but Tennessee continued to struggle with an outdated system for the sake of continuity and political control.[66]

While the issue never attained the political importance that it achieved in Georgia, perhaps because it was not made a local option for the electorate, the fence law/stock law debate hovered around the periphery of agrarian concerns. Although the Grange strongly supported stock laws and dog laws, later agricultural organizations, while cognizant of the problem, refused to make these issues central to their political agenda. Instead, they focused on the ineptitude of lawyers in trying to determine agricultural policies. The impression their activities fostered is one of political expediency based on the recognition that demands for stock laws would reduce their support among tenant farmers and herders in East Tennessee. Therefore, at every legislative session, negotiations began anew to find a compromise that farmers could support, politically and financially.[67]

Fencing the land to keep livestock out of tilled acres, or constructing pens to keep livestock within farm boundaries, involved the agricultural community on political, social, and economic levels. Robert Cartmell tied the issue to the emancipation of slavery when he noted the proximity of freedom and the appearance of barbed wire. Without slave herders or human protectors of planted acreage, free-roaming cattle and hogs trampled corn and cotton, an economic loss market-oriented farmers could ill afford, especially as prices dropped. To farmers the answer was simple: require stock laws to keep livestock penned and make stockmen responsible for damages if gates were left open or animals broke free. Raisers of blooded cattle, horses, sheep, and hogs agreed, since contamination of the bloodlines by wandering animals threatened their livelihoods as well. But poorer stockmen pointed to the traditional rights of free range and argued in favor of fence laws that required farmers to fence in their crops to prevent the entry of roaming animals.[68]

For their part, small farmers criticized the waste inherent in fence laws that required the use of precious timber resources to fence crops against wandering livestock. They were joined by tenant farmers whose contracts often stipulated that they bear the expense of fencing in new fields. When Archelaus M. Hughes rented "about an acre of ground to plant in tobacco" to "Old Frank," he did so "on condition that he finishes cleaning it up and fences it." More than a decade later, as county Farmers' Alliances passed resolutions in favor of stock laws, a renter from Shelby County complained that such laws taxed the poor when landlords erected fences and charged tenants "from one to five dollars per month for pasturage." As much as market-oriented farmers wanted to reform the laws protecting agricultural property, the diversity of views and the need to create unity among farming interests prevented an easy settlement of this basic issue.[69]

Agitation for a dog law centered on a similar need to protect the sheep industry of the state. Agricultural leaders encouraged sheep raising as a profitable alternative to cotton or tobacco farming, noting the advantage offered by the protective tariff on wool. Proponents of the law attributed the decline of sheep to the threat posed by packs of unrestrained dogs that attacked unprotected animals. On the opposing side of the dog law stood the poor farmer and renter, who made up the majority of the electorate.[70]

When the general assembly met in 1875, the legislators garnered enough votes to pass an Act to Encourage Wool Growing, which made the keeping of dogs a taxable privilege. Tax assessors were given the responsibility of enumerating all dogs in the county, and tax collectors were charged with collecting fees of one dollar per dog and five dollars per unspayed bitch. As might be expected, the law raised strong opposition, and the next legislature returned to the question with the intention of repealing the law.

Defenders of the law charged that "a few demagogues, seeking the favor of the dear people, have in order to curry favor with the populace, persuaded themselves that the poor people were terribly oppressed by the dog law." The real problem, according to this view, was that "men of honesty and ability seeking election had not the courage to argue the question and convince the people, as could easily have been done, that a sheep was worth more than a dog." Having questioned the courage of the lawmakers, editorialists admonished the legislators to "lead public sentiment, direct it in the proper channels, and not be weakly led by the populace." It was clear to dog law advocates that they were fighting a losing battle, and when the legislature convened in January 1877, "four out of every five of the members had a bill stowed away in their coat-tail pockets to repeal the dog law." During a legislative break in February, the opponents of repeal encouraged lawmakers "to consult the [county] tax books." There, it would become plain that those who kept large numbers of dogs were not property owners. Supporters of the dog law pointed to a case where "a free and independent voter was taxed with five dogs and one poll—nothing else." Clearly, the line had been drawn between property owners on one side and advocates of diversification and the "free and independent" but unpropertied voters on the other. In this case, the poorer class won, and the law was repealed, undoing what at least one farmer saw as one of "the worst blunders of the last two legislatures." Pressure continued from commercial farmers and livestock men to enact a new dog law, but political expediency ruled against further legislation.[71]

The position of the Wheel and Alliance on the matter lends support to the perception that farmers with commercial views dominated the membership. Wheels and Alliances in counties as diverse as Stewart, Wilson, and Davidson passed resolutions in favor of dog laws. J. W. Taylor of White County, where sheep raising represented a significant portion of the economy, proposed "a premium on dog scalps" to encourage the killing of "worthless dogs." Supporters of the wool industry kept the public informed of the problem by periodically printing the tally of sheep lost to dogs.[72]

But, as with fence and stock laws, not every Wheeler or Allianceman joined the fight to tax dogs. A letter to the *Toiler* from Shelby County complained that dog laws operated in such a way as "to make the poor renter and share worker bear this burden." Shaping his argument to reflect a rural/urban split, the writer asserted that town folk could buy their meat even as they insisted on taxing the poor man who used his dogs to hunt "possums, coons, squirrels, and rabbits." Taxing dogs took meat from the tables of poor families and reduced their meager pleasures.[73]

The Wheel and Alliance printed county resolutions and opposing letters but took no definitive position, though the editorial comments make it clear that the leadership favored stock laws and dog laws. In the name of unity, they refocused the problem to a campaign against the predominance of lawyers in the general assembly and their historic inability to understand and deal with agricultural problems, rather than championing solutions that might disrupt the organization.

Farmers and Taxation

The fourth area of concern for the agricultural community was intimately connected to the other three and centered on the problem of tax assessment. Both the amount of the assessment and the mechanism for determining individual tax liability worried farmers. Since taxes on land and personal property remained the chief source of public revenue, periods of economic stagnation or decline fell with greater force on farmers than it did on manufacturing and mining interests. Even the most austere government expenditures came under intense public scrutiny. Small-town boosterism, anticipating a New South boom, argued in favor of new, centrally located schools, new courthouses, paved streets, and public utilities. But farmers, who realized few of the gains, not unexpectedly resented paying for such benefits.[74]

Of particular concern was the renewal of the right to issue public bonds to finance the construction of railroads, which had been denied following

Redemption due to the extraordinary growth in public debt during the Brownlow administration (1865–1869). In 1889, when the citizens of Nashville were given the opportunity to vote for railroad bonds to build a local line, the Agricultural Wheel urged defeat of the proposition. After giving assurances that the Wheel harbored no ill will toward railroads, and, indeed, "want more of them, and favor encouraging their building by every fair, just and proper means," the Wheel warned its constituents against a repetition of Tennessee's "bitter experience," and the "burdensome debt" that "hangs over us and will be handed down to our children."[75]

The argument against public funding of railroad construction posited the view that "if the merchants of Nashville will show to the capitalists, that the building of railroads will be an investment that will pay a reasonable profit, the roads will be built by those who are to own them and get the profits." Property owners who desired additional facilities should be encouraged to subscribe, but Wheelers asserted that the subscriber had "no right through the means of unjust taxation, to compel another property-holder to subscribe to an enterprise which he may honestly believe will not in any degree benefit him." The Sumner County Wheel deemed the whole bond effort "contrary to the principles of wheelerism." Farmers did not oppose progress, but they took exception to what they saw as "taxing individuals for the benefit of corporations."[76]

The concerns of the farmers were not misplaced. Despite the settlement of the state debt issue, Tennessee continued to struggle with chronic problems of public finance attributed to the deficiency of state revenue collections. In 1889 Governor Robert L. Taylor traveled to New York City to obtain loans to make the interest payments on the state bonds. As a result of the state needs and at the urging of the governor, legislators endeavored to close tax loopholes and streamline the process of tax collection.[77]

Under the postbellum system of tax assessment and collection, assessors selected from each voting district were charged with visually inspecting all property and obtaining from the property owner an estimate of the market value of the land and a declaration of personal property in excess of one thousand dollars. Constables from each voting district collected the taxes that had been assessed on the basis of the property owners' valuations. Such a system offered many opportunities for underreporting and lowered tax revenues. Assessors and constables were neighbors, and often kinfolk, to the property owners and were subject to personal and communal pressures to recognize the effects of economic hardship and agree to the underassessment of property.

Agitation to tighten tax collection and improve the state revenues had a lengthy history. In his message to the forty-third general assembly in 1883,

Republican Governor Alvin Hawkins alluded to the "serious complaints" that had come to his attention regarding the assertions "that much of the taxable property of the state ha[d] not been assessed for taxation." Furthermore, complainants claimed that "much of what ha[d] been assessed ha[d] been assessed at figures far below its true value." Hawkins declared it "manifestly unjust" that one property owner should pay taxes on an assessment at "true value" while "the property of another should be assessed at less than its true value." Governor William B. Bate, Hawkins's Democratic successor, more specifically condemned assessment laws. After citing the state constitutional requirement for equality, Bate noted that a significant portion of property owned by corporations had been "exempted for years by charter stipulations." Bate, a Bourbon Democrat, claimed to find no justification for this practice and recommended the search for a "practical and efficient mode of ascertaining the value" of business property in order to tax it "as the Constitution requires, according to its value."[78]

Farmers cheered this recognition of their plight as bearers of most of the responsibility for maintaining the fiscal health of the state. As historian Michael R. Hyman observed, "during the antebellum era, landed property had been lightly taxed and other property owned by small farmers had been generally exempt from taxation," with a large portion of public revenues being raised from a levy on slaves. After the war, the principal source of state and local revenue was the ad valorem tax on land. But, as Hyman noted, the increase in taxation felt by the agrarian class "did not translate into increased state services," but serviced the public debt. Throughout the period farmers pressed for equity in taxation and an end to the "special privileges" accorded to manufacturers and railroads.[79]

The demands of Hawkins and Bate produced no significant reforms, and when assessment reform passed during the administration of Governor Robert L. Taylor (1887–1891), it generated a new round of anguished cries for relief and elicited dire predictions of the domination of agricultural interests by those of the towns. Instead of reducing the tax burden of the farmers, reform, with its new mechanisms for assessment and collection, increased the annual levy just when farmers felt the pinch of declining prices. Under the new system of taxation, the state levy was set at thirty cents per one hundred dollars assessed value, with a school tax of fifteen cents per one hundred. Each county court was required to name a tax assessor to oversee assessment and collection for the entire county. As a result of the change, state revenue collection improved dramatically. In his 1889 message to the state assembly, Governor Taylor reported an increase in state revenues of $223,752, which he attributed to "closer collections and prompt settlements." He expected even greater returns in the following years due to the increased valuation of property under the new assessment law. Additional revenues amounting to $170,000 were anticipated for the coming year.[80]

Complaints against the assessment law filled the newspapers as the general assembly began its legislative session in 1889. When the Gibson County Wheel met on January 3, the assembled delegates passed a resolution demanding "a repeal of so much of the present assessment law as provides for the election of a county assessor." These farmers urged a return of the system whereby "the assessment of property and polls be made by the magistrates in their respective districts as heretofore, and that the collection of public taxes be made by the district constables." Gibson County farmers were joined in their protest by Wheels and Alliances in Chester, Stewart, and Wilson counties, among others, who made their demands known by passing and publishing similar resolutions. The Gibson farmers did not dismiss the new law entirely. They endorsed the creation of a county board of equalization and urged the creation of a state board, which would set assessments "based upon an average valuation derived from the actual sales of land in each county in order that taxation may be uniform and equal throughout the state dependent upon the ability of the citizens thereof."[81]

While farmers continued to press for equalization, the issue that now concerned them the most was the loss of local control to the courthouse-based tax assessor. Farmers already distrusted the "rings" and "cliques" that dominated county politics, and now they saw that the granting of the additional power of determining the valuation of property for taxation placed assessors beyond criticism or removal. This was one more concern to add to the list: the closing of convenient, local schools, the neglect of farm roads, and the rise of tax rates.

If magistrates and constables failed in their duties of assessing and collecting the full measure of taxes, they also reflected the views and concerns of their neighbors and constituents. Now assessments were to be made by town men, who were unknown to individual farmers and whose obligations and responsibilities lay elsewhere. The new tax assessment law confirmed the feeling expressed by A. E. Johnson of Wilson County, when he wrote, "truer and better men never lived than the farmer and to him this whole country belongs. By his ax, hoe, plow, maul, and wedge this country of ours has been changed into a land flowing with milk and honey. The time was once when the politicians and lawmakers had great respect for this class; they were sought after and looked upon as supporters of the world; but not so now."[82]

Like other Americans of the period, farmers welcomed the benefits of progress. But increasingly they recognized that the cost of progress included a devaluation of their contributions as producers and as political participants. The combination of political corruption and economic decline marginalized the agrarian community of smaller producers and left their political influence vulnerable to the demands of expediency as the two political parties vied for electoral prizes.

The events of the 1880s confirmed the common perception that money and power determined political campaigns centered on personalities rather than issues. Farmers did not need demagogues to convince them of legislative ineptitude; the inactivity of their own general assembly was ample proof. Nor did they need the fiery editorials of the new agrarian newspapers to understand the failure of the political parties to recognize and champion the interests of farmers; their inability to influence political activity at any level confirmed their impotence.

If they needed additional proof, three attempts at reform offered ample evidence of the need for greater agrarian input into the political process. Their apparent inability to make agricultural institutions responsive to their needs limited their options in meeting the demands of a national and global market economy. The lack of success in working through the problems of protecting the property rights and supporting the growth of the agricultural community substantiated the view that agricultural interests were not represented by the lawyers who filled the state legislature. And the creation of a new tax assessment law supported the suspicions of farmers who feared the consequences of the apparent shift in local political power from the yeoman to the townsman.

Agrarian organizations advised farmers to adopt a nonpartisan, antiparty position to effect necessary political changes. However, their modest successes in electing farmer-legislators would first lead Tennessee's farmers to attempt control of the Democratic Party, and then lead to the formation of the People's Party.

Shades of Jefferson, Madison, and Jackson, has it come to this that men call themselves Democrats, professing your faith and claiming to be your followers and disciples, enunciate such doctrines as these and inculcate them under their motto, "equal rights to all and special privileges to none?" Shame, shame, upon such wicked perversion! Truly have evil times come upon us, presaging wrath alike to citizens and government.

—WILLIAM B. BATE, 1891

CHAPTER FIVE
FROM NONPARTISAN POLITICS TO POPULISM

By 1890, the Alliance insurgency represented a political movement that acted within the parameters of a two-party system, an organized culture of reform, and a legal framework of disfranchisement. Within the context of these competing forces, Tennessee farmers moved beyond the rhetoric of antipartyism and attempted to control the Democratic Party in order to elect candidates supportive of agrarian objectives to state and national offices. The decision to engage in more politically aggressive action developed from the successes and failures of the state and national Farmers' Alliance. Year after year, from Cleburne to Shreveport, St. Louis, and Ocala, farmers met in national conventions and systematically refined their demands for action. With each meeting, as the limitations of local voluntary cooperativism became more apparent, calls intensified for government intervention on behalf of farmers and laborers. With a highly motivated grassroots organization in place, the momentum for political action gained traction.

The inability of cooperativism to alleviate economic constraints imposed by the "trusts and monopolies" nationally, and the local economic and social control imposed by the crop lien system led farmers to contemplate stronger measures. The lists of demands articulated in successive Wheel and Alliance conventions provided a framework for a radical reconceptualization of the political economy. Agrarian insurgents demanded government action within the existing political and economic paradigms to protect both the rights of the individual and the common good. In their minds, one could not advance without the other.

The successes Tennessee's Alliancemen experienced in marshaling support for the jute boycott and defending the cooperative system against political attacks demonstrated their "stickability" or perseverance and whetted their appetite for further action. Initially ignored or dismissed as ineffectual, the farmers utilized the political knowledge they had gained through Alliance education and hard-won battles against political bosses to mount state and congressional campaigns and elect men sympathetic to their demands. The rapid rise to state power in 1890 and the equally rapid demise of the agrarian insurgency led historians to a tacit agreement with Edward Carmack's characterization of the Populists as residents of the regions beneath the mudsill of hell. Little attention has been given to the congressional races, where fusion with Republican interests gave Populists a much stronger voice and influenced modern reform. A reconsideration of the political influence of the Farmers' Alliance and the People's Party in four elections suggests a more complex political threat than previously recognized.

Although highly visible political figures predicted dire consequences for the presence of agrarian candidates within the Democratic Party and the subsequent establishment of the People's Party, contemporaries quickly developed political amnesia as the threat of the rural insurgency receded. In 1911, following a new period of general unrest, Robert Gates delivered an address to Middle Tennessee farmers in which he called upon the "conservative forces in American politics—the farming class" to avoid the "entangling alliances" that might make the farmer a "cat's paw for selfish interests." Gates reminded his rural audience that their conservatism had served them well in 1890 when the "spirit of organization" had sowed "much bitter and dangerous economic and social seed." Characterizing farmers as "conservative by virtue of their calling, equally interested in the good of all classes of citizens, in capital and labor alike; removed by their environment and independence from the influences that inspire prejudice, with the most to lose and least to gain by disorder," he harkened back to time-honored agrarian rhetoric to proclaim them "the conservative force in the body politic of the Republic."[1]

In Gates's view, Tennessee's agrarian conservatism had saved the republic from intemperate radicalism in 1890; in a new era of excess, farmers could exert their influence again to man the bulwarks of democracy. Gates congratulated them on their common sense and promoted a revised history of late-nineteenth-century agrarianism that suggested Tennessee farmers had never supported radicalism. Although Tennessee agrarianism did not produce the inflammatory rhetoric or the violence evident in other states, Gates mistook civic order for complacency and ignored the farmers' demands for reform.

Historians Joseph A. Sharp and Roger L. Hart dealt more explicitly with the elections of 1890 and 1892, and the activities of the so-called Farmers' Legislature of 1891. Sharp viewed the movement as "a desperate attempt of agriculture to regain from industry its lost social, economic, financial, and political prestige." His work suggested that the agrarian involvement in politics divided the loyalties of the Alliance membership and doomed the movement to failure. Hart provided the most sophisticated examination of the politics of the agrarian movement. When his voter analysis failed to show evidence of a farmer bloc in the 1891 legislature, he concluded that although the Alliance attracted men with similar "value expression," it could not break the "iron grip on . . . voters' loyalties" claimed by the Democratic and Republican parties. Therefore, he suggested that the agrarian movement primarily attracted office seekers who had failed to obtain elective office within the more powerful Democratic Party. Suffering from the status anxiety displayed by men increasingly out of touch with modernization, these would-be leaders played upon the emotions of voters through a paranoid style of political persuasion, but made no radical demands beyond a transfer of state offices to men previously denied political power.[2]

Close attention to Tennessee's political culture suggests that the electoral strengths and weaknesses of the farmers were more complex and threatening than historians believed. At the state level, farmers demonstrated their newfound political acumen as they gained control over the governor's office and elected fifty-four men to the general assembly. Operating within the Democratic Party, they expected to add to their strength and consolidate their power by presenting themselves as the "true Democrats," the sons of Jefferson and Jackson who had not sold out to the "money power." As Democrats, the farmer-legislators expected to name the next U.S. senator when the office became vacant in 1893. But the 1890 success of the farmers united the Bourbons and New South Democrats, who pushed the insurgents out of the party and into the newly created People's Party. Operating from a weak position, the Populists never successfully fused with the Republican Party to press their cause at the state level. The failure stemmed as much from the factionalism of the Republicans as it did from the inexperience of the Populists.

A second, and largely ignored, political battleground emerged in the congressional races. Alliancemen clearly understood that the reforms they demanded required federal regulation and constitutional action. As early as 1888, farmers succeeded in holding congressional candidates to the Alliance yardstick and exercised considerable control over the 1890 congressional campaigns. When the insurgent wool-hat boys were pushed into the People's Party in 1892, an interesting realignment occurred. In congressional races, where candidates were freer to ignore the dictates of state party bosses, Populists and Republicans supported fusion candidates and sustained a strong opposition to Democratic control in Middle and West Tennessee. However, as the national debate moved toward the titan conflict between gold and silver, the Populists shifted between the two parties in order to advance their agenda—supporting silverites in both the Democratic and Republican parties. Ultimately, the silver question undermined Populist fusion with Republicans, and congressional races, like the insurgent efforts in state campaigns, sputtered and died.

The Election of 1890: An Alliance Triumph

The agrarian rise to political power took political insiders by surprise. In 1890, Democrats, confident of the power of disfranchisement, anticipated gains in the general assembly and the capture of the third congressional district seat from the GOP. When four contenders entered the contest for the gubernatorial nomination—Jere Baxter, Josiah Patterson, John M. Taylor, and John P. Buchanan—most assumed it would be a fight between Bourbon and New South forces.

Davidson County capitalist and raiser of thoroughbred race horses, Jere Baxter received the support of the New South wing of the Democratic Party. Former president of the Memphis and Charleston Railroad, Baxter had little in common with the one-gallused, wool-hat boys, but he attempted to win their votes by joining the Davidson County Alliance, which gained him little beyond joking references to "Farmer Jere." The best funded of the candidates, Baxter campaigned heavily in industrial East Tennessee and reportedly bought a considerable amount of advertising for his Maplewood estate in county newspapers, an expenditure Alliance leaders regarded as an attempt to "buy off" rural editors and undermine support for Buchanan. McDowell joked that Baxter would "spend a pile of money advertising his Shetland ponies and then get badly beaten.[3]

The Bourbon wing of the party supported Josiah Patterson, a Memphis attorney. An Alabama native, Patterson boasted an exemplary Confederate military career. He had served under General Nathan Bedford Forrest and, at age twenty-five, was promoted to full colonel and placed in command of the Fifth Alabama cavalry regiment. Patterson moved to Memphis in 1872 and entered Democratic politics ten years later. As he lined up delegates for the gubernatorial convention, Patterson campaigned extensively in the farm communities. Betraying disdain for both Bourbon Democracy and the Farmers'

Alliance, the editor of the *Knoxville Journal* chided Patterson for his pretense at agrarianism, "when he couldn't have told to save his soul whether turnips or potatoes grew on trees." Although the *Memphis Avalanche* rushed to defend the candidate with stories of his early life on the farm, he remained an Upper-Case Democrat. If Patterson had little in common with the farmers, he believed his political doctrines matched those of the agrarians. Patterson deplored the New South ideology and "preached . . . the good old doctrines of Democracy" and the rule of the white man's party. Supporters affectionately called Patterson "Uncle Josiah," and "the big-brained old champion of Bourbon Democracy." While many Democrats hailed Patterson's support of the tried-and-true party positions of low tariff and antipaternalism, others found his views dated.[4]

John M. Taylor, Henderson County attorney and former congressman from the eighth district, drew some support away from Buchanan's campaign among the farmers of West Tennessee. Taylor went on record as an opponent of the "visionary and impractical schemes" of the Alliance, but his campaign never materialized significantly beyond his old district. His principal contribution to the race was his success in tying up the nomination for twenty-four ballots.[5]

John P. Buchanan, president of the Tennessee Farmers' Alliance, owned a livestock farm in Rutherford County and had served two terms in the general assembly. The *Toiler* recommended him as a "self-made man," "a prosperous farmer and stock raiser," and a "splendid parliamentarian." Buchanan's brief political career had not elevated him to the councils of power but had gained him a reputation as a hard worker on behalf of agricultural interests. Farmers were especially grateful for his role in shepherding the state agency bill through the 1889 general assembly. Alliance men looked upon Buchanan as a candidate of the people, one who, in time-honored fashion, accepted the burdens of the gubernatorial race at the insistence of the state's farmers and laborers.[6]

Political observers discounted Buchanan's race, but as counties held their nominating conventions and primaries, it became apparent that the farmers had learned valuable lessons from their lectures on the political economy. The Lauderdale County Democratic convention instructed as a unit for Buchanan, after the farmers captured the proceedings in a lightning strike that placed "one of their own in the chair." Some speculated that the farmers had caucused in the morning and "secretly planned the schemes by which they would 'hoo-doo' the afternoon Democratic convention." Controlling the convention for Buchanan, the "Alliance men gaily cantered out to the rural precincts, laughing in their sleeves over the inglorious manner in which they had plotted in secret conclave against the untrammeled action of the Democratic Party." Likewise, in Madison County, the Buchanan forces "captured the caucus and effected a permanent organization," although observers expected a more heated battle, "after the arrival of the [John] Taylor reinforcements," who were expected to hurry from Jackson in response to the telegraphed pleas of their friends.[7]

As the number of Buchanan delegates mounted in West Tennessee, the *Avalanche* conceded that "with his strength [Buchanan] ought to be able to name the [gubernatorial] candidate," but discounted the possibility of anything more. "There are innumerable and indisputable evidences," the editor wrote, "that his forces are not well organized, that they lack generalship, and that after having cast a few ballots for him in an unavailing effort to secure his nomination it [the organized political effort] will go to pieces."[8]

The West Tennessee editor, like many others, misread the political signs. In every section of the state, the antiparty groundwork laid by the Wheel and Alliance in previous elections produced a groundswell of support for Buchanan. Urban and county newspapers reported local conventions that were "wild and woolly, and full of fleas" as farmers turned out in unprecedented numbers and refused to be bullied by "county rings" and "wire-pullers." Buchanan, McDowell, and State Lecturer R. W. Tucker crisscrossed the state on behalf of the Buchanan campaign. As the established political interests belatedly recognized the threat and focused their opposition on the agrarian campaign, McDowell advised farmers to be on their guard against wily politicians and their deceptive tactics. He suggested that farmer-controlled conventions instruct their delegates by adopting resolutions in support of Buchanan "as long as his name is before the convention."[9]

Seventeen Middle Tennessee counties instructed for Buchanan and provided 250 of his 759 votes on the first state convention ballot. A Smith County farmer optimistically predicted Buchanan's support, promising, "We will give Hon. J. P. Buchanan 90% of our vote." He identified himself as a Democrat, but noted "our Republican brethren say hurrah for Buchanan." A Robertson County farmer confessed, "I have been a Republican and a staunch Prohibitionist, but I have laid them all by until the great question which is of much more importance is settled." He urged all members of the Alliance to set aside their squabbles, warning "if we divide on a man when we come to vote, you may rest assured that we will never get men in office who will benefit the poor farmer." He encouraged his Alliance brothers to remember that they "have a power to fight with," but warned them "if we don't pull together we might as well quit."[10]

In East Tennessee, the Alliance Democrats rallied behind Buchanan to the surprise and chagrin of local political elites. When the Knox County Democratic Convention met, pandemonium reigned as Buchanan forces joined with the Baxter men to undo Patterson's expected control of the convention. In

Sullivan County, the Buchanan forces also controlled the county convention and instructed for the State Alliance president. Unable to overcome the insurgency, Patterson and Baxter supporters bolted the convention and instructed a separate list of delegates for Baxter. These "simon-pure" Democrats charged that the Buchanan followers masterminded the convention by allowing Alliance Republicans to vote for their candidate. Farmers presented a different view of the situation. A Sullivan County woman hailed the farmers for their united stand and called it "a sad day for 'the bosses,' their sheep had strayed off into the alliance, gone through a training school and come out with their eyes wide open as free men capable to think and act for themselves."[11]

By July 15, the opening day of the state convention, the agrarian interest in the gubernatorial race had attracted considerable attention and not a small amount of concern. The editor of the Nashville *American* evoked a variety of political and agrarian images that he believed would demonstrate his "folksy" good humor and marginalize Buchanan's support. Advising the delegates to act in the best interests of democracy and not sacrifice the party to the ambitions of one man, he reminded them that "the woods are full of small potatoes who regard themselves as some pumpkins." Moreover, he assured them that "there are too many good men to nominate a bad one, too many good Democrats to nominate a mugwump." In the end, editors and old-line politicians believed that the best hope for defeating Buchanan rested on prolonging the deliberations until the farmer-delegates became bored or felt compelled to return to their neglected crops. In 1890, however, the farmers came to stay.[12]

Buchanan led the balloting from the beginning, with more than twice the number of votes of his nearest competitor, Josiah Patterson. Nevertheless, he did not command the mandatory two-thirds to sew up the nomination. Alliance leaders roamed the floor, cajoling the laggards and stiffening the resolve of the weak; the farmers held out for four days and twenty-six ballots. As the balloting dragged on, pressure increased to discard the two-thirds rule, but resolution of the impasse was reached only after John M. Taylor withdrew on the twenty-fourth ballot, and Baxter and Patterson capitulated on the twenty-sixth. With the Alliance president now the party's choice, McDowell crowed that the political wire-pullers had been "outwitted and beaten by the 'hayseeds' in a fair, square and open fight." The *American* editor rushed to support the Buchanan nomination and unify the Democrats. Buchanan had been "carried to victory" by the organized support of the farmers of the state." Casting aside his earlier concerns about the effect of Alliance ideology on the candidate, the editor assured readers, "he is bound by no narrow ideas or prejudices, and has no policy in view in connection with his policy as Governor of Tennessee which will not be broad enough to encourage the whole people and every interest." For the first time in his career, Buchanan was embraced as an "Upper-Case Democrat."[13]

Democratic leaders agreed that "the result of the Convention was simply to make a sound Platform & to concede to the farmers vote the candidate." Party stalwarts anticipated limited damage because the farmers failed to dictate an Alliance platform or control the State Executive Committee, an outcome that left Tennessee Democrats in a better position than their counterparts in Georgia and South Carolina. Party elites quietly withdrew from the canvass, confident that the Alliance insurgency would be limited to a single term.[14]

The excitement of the gubernatorial convention barely died down before the more volatile congressional conventions got underway, as Democrats reconvened in August and September to nominate candidates for national office. Tennessee farmers clearly understood that the limited power of the state to control trusts and monopolies made congressional elections more important than gubernatorial campaigns. A. P. Smith of Lincoln County earlier wrote to the *Toiler* to offer his endorsement of Buchanan's candidacy, but noted, "a governor will not benefit us, as a Congressman would, and the honorable record which Buchanan has made as an indefatigable worker in the Legislature of Tennessee points him out fair and square as the man for the place in Washington." County Alliances passed resolutions in favor of term limits, popular election of U.S. senators, and the abolition of the electoral college—issues that expand popular control over national offices. Congressional candidates soon found themselves answering questions about their positions on Alliance demands, especially the subtreasury plan, as they struggled to measure up to the Alliance yardstick.[15]

The seventh and tenth districts offered the most spectacular battlegrounds for farmers to demonstrate their perseverance and force ballot after ballot to determine who would represent them in Congress. In the seventh district W. C. Whitthorne stood for reelection after serving seven congressional terms (1871–1883 and 1889–1891). His political service dated to the eve of the Civil War, when he stumped the state for Breckinridge's presidential ticket. Speaker of the Tennessee House in 1861 when the state struggled with secession, he served under the Confederate adjutant general of Tennessee. In 1890, he faced competition for the Democratic nomination as voters in peripheral counties of his district fought to control the nomination.[16]

The convention deadlocked immediately, with N. N. Cox, a Williamson County lawyer, leading but lacking the necessary two-thirds for nomination. For three days, delegates cast their ballots with no appreciable change in the outcome. Unable to obtain a nomination, the convention agreed to adjourn and reconvene in October.[17]

When the second convention also ended without a nomination, the delegates met for a third session on October 24. This time Cox supporters succeeded in suspending the two-thirds rule and obtained the nomination of their candidate, but Whitthorne's backers refused to follow tradition and make it unanimous. Although the Alliance took no obvious role in the convention, Whitthorne's men believed Cox won the nomination because of the "Damn[ed] Farmers' Alliance." The Alliance were not entirely responsible for events in the seventh district, but they had opened the door for other Democrats to move ahead. The old warhorses of the party had lost touch with the emerging issues, not only among Alliancemen, but within the party generally.[18]

The congressional nomination in the tenth district also demonstrated the Alliance strength and perseverance in a dramatic way. As in the seventh district, counties long relegated to the periphery of power demonstrated their displeasure with politics as usual and determined to support their candidate as long as necessary to win the nomination. In August, farmers in Tipton, Fayette, and Hardeman counties marshaled their forces in support of T. K. Riddick, a Fayette County lawyer. Riddick's profession precluded his membership in the Alliance, but he was friendly to Alliance demands. When rural Shelby County joined the insurgents, worried observers believed the agrarian coalition signaled an Alliance takeover of the Democratic Party. Such concerns notwithstanding, Memphis Bourbons dominated the voting and, in spite of an agrarian show of force, controlled the Shelby County convention. All 120 Shelby County delegates instructed for J. S. Galloway, a Memphis judge, giving him a majority of the tenth district convention votes.[19]

When the convention met on September 10, delegates soon locked in a stalemate. After numerous ballots (newspapers reported an astounding 5,000 ballots), neither candidate controlled the necessary two-thirds vote to gain the nomination. Ballot after ballot, Galloway received the 120 Shelby County votes and Riddick got the 82 votes of the other three counties. In desperation, the convention adjourned and reconvened on September 15 but remained at an impasse. In a third meeting, on September 29, after a period of similarly deadlocked balloting, both candidates agreed to withdraw. Now the Bourbon Democrats took control of the convention, and Memphis favorite son Josiah Patterson, the unsuccessful gubernatorial candidate, received the nomination. Tenth district farmers had persevered, but it remained to be seen whether they could force action on their demands.[20]

Of the eight remaining congressmen, the Alliance counted four as friendly and four as either unfriendly or lost to the Republicans. The four friendly congressmen were James D. Richardson, fifth district; Joseph E. Washington, sixth district; Benjamin A. Enloe, eighth district; and Rice A. Pierce, ninth district. Both Richardson and Washington claimed membership in the Alliance, but only Pierce supported the subtreasury scheme.

James Richardson had served in the U.S. House since 1885, after single terms in the Tennessee House and Senate in the 1870s. During the Civil War, Richardson rose from private to first lieutenant in the 45th Tennessee Infantry. In addition to his political interests, Richardson served as director of the Safe Deposit Trust and Savings Bank in Nashville, where the Wheel and Alliance State Agency kept its funds. Richardson chaired the state Democratic gubernatorial convention in July, and his brother, John Richardson, placed Buchanan's name in nomination for governor. In 1889, Richardson cosponsored the Alliance bill to require an enumeration of mortgage indebtedness. Clearly, Alliance voters could support Richardson with few qualms, despite his reluctance to embrace the subtreasury measure.[21]

Joseph E. Washington, of the sixth district, was born in 1851, too late to serve in the Confederate army. He graduated from Georgetown College in Washington, D.C., and Vanderbilt Law School. Washington was elected to the fiftieth U.S. Congress in 1887 and was running for his third term in 1890. Washington had never practiced law, instead returning to Wessyngton, the family's 13,000-acre tobacco plantation in Robertson County. He served as director for the Nashville, Chattanooga, and St. Louis Railway and the Nashville and Decatur Railroad. A member of the Farmers' Alliance, Washington ducked the question of support for the subtreasury, but it was generally believed that he would not back the scheme. Washington introduced a bill in the 51st Congress to abolish the internal revenue tax on the manufacture and sale of tobacco by farmers. He not only supported Richardson's census measure, but the *Toiler* credited him with the favorable committee report for the bill and with its subsequent enactment into law. Though clearly representing his own interests and those of other large planters, Washington's support of agrarian measures and his commitment to free silver enhanced his credibility among Alliance members.[22]

Benjamin A. Enloe, eighth-district congressman, was born in Carroll County in 1848 and attended Bethel College before earning a law degree from Cumberland University. Throughout his political career, Enloe held office during critical moments in Tennessee history: he was in the general assembly during the "redemption" of Tennessee state government in 1869 and 1870; in 1878, Governor Albert Marks appointed him to a commission to negotiate the settlement of the state debt; and he served in the U.S. Congress during the Alliance and Populist era. From 1874 until 1886, Enloe edited the influential Jackson *Tribune and Sun*. Enloe's politics reflected those of his constituents in the eighth district, shifting between the Bourbon stance against the so-called Force Bill and the McKinley Act, and issues more closely identified with the Alliance demands. In 1889, he and Richardson introduced the farm mortgage bill, and he joined with Washington in supporting the antitax bill for tobacco farmers. Enloe sponsored a joint resolution to add an antitrust, antiagricultural gambling amendment to the constitution, and also introduced a bill for the popular election of U.S. senators. Although he favored free coinage of silver, Enloe opposed the subtreasury, government ownership of railroads, and abolition of national banks. Once again, in Enloe, farmers knew they had support for some measures but faced opposition on more radical demands.[23]

Benton McMillin, fourth-district congressman, had been in the U.S. House for six consecutive terms, since 1878. As the recognized leader of the Tennessee congressional delegation, McMillin's position seemed unassailable. McMillin supported the farm census bill with the rest of his colleagues and advocated free silver, but beyond that offered no encouragement to the Alliance movement. Except for White, Smith, and parts of Putnam counties, the fourth district attracted few Alliance members, making a challenge to McMillin virtually impossible.[24]

Republicans dominated the three congressional districts of East Tennessee; only third-district congressman H. Clay Evans anticipated a serious fight. Evans faced Henry C. Snodgrass of Sparta, whose opposition to all Alliance programs was well known. However, Evans had little reason to engage in a battle for agrarian votes, since the economy of East Tennessee depended not on agriculture but on industrialization and resource extraction. While the Farmers' Alliance claimed union with the Knights of Labor, the tenuous ties between the two groups had not yet translated into a political commitment in Tennessee. Indeed, the East Tennessee Alliance movement had only begun to put down roots; it seemed unlikely that the members were ready to support a political battle with the Houk-Evans-Taylor Republicans. Farmers in the eastern section also found reasons to quarrel with their Middle and West Tennessee brothers. The subtreasury offered little relief to the hog producers of the mountainous regions. But, more important, their support of the Conger Lard Bill threatened the budding cottonseed oil lard compound industry important to cotton farmers in the west. In the end, one historian noted, "Republican Alliance men were better Republicans than Alliancemen, and hesitated to endorse . . . demands which were mostly repudiations of Republican policies."[25]

The most important congressman to the Alliance was Rice A. Pierce of the ninth district. Pierce, like Buchanan and McDowell, joined the Confederate army while still in his teens. He began law practice in Union City in 1869, and served his first congressional term in the 48th U.S Congress. In the summer of 1890, Pierce outlined his views to the *American*. "The passage of a law giving us free coinage of silver," he said, "would give the necessary relief now so urgently called for by the depressed agricultural interests of the country." If Congress failed to enact a free coinage measure, Pierce was "favorably inclined toward what is known as the sub-treasury bill," though he candidly voiced his doubts that either free coinage or the subtreasury could pass the House. But Pierce objected to any subtreasury scheme that did not place the appointment of warehouse managers in the hands of the voters. In addition to his cautious support for the subtreasury, Pierce voted for the farm mortgage bill and introduced measures providing for the equalization of taxes, a graduated income tax, and antitrust measures. Although his Democratic opponent, Frank P. Bond, accused Pierce of pledging his support of the subtreasury only to gain the farm vote, Pierce's identification with agrarian reform led him to become the only Tennessee congressman to run as an Independent (Populist) in the 1892 election.[26]

Gubernatorial campaigning got underway in September, when Buchanan made his opening speech at Franklin. Throughout the fall, most Democratic newspapers gave tepid support to the party's candidate or remained silent; a few sniped from the sidelines. But the editor of the Nashville's independent and iconoclastic *Banner* launched a full-scale crusade against the Alliance leader, foreshadowing the Bourbon attacks that followed the November election. From his opening speech, which the *Banner* labeled as "remarkable . . . for its brevity . . . [and] its lack of anything like original suggestions or vigorous discussion," the criticisms never stopped. In the editor's view, the speech appeared to be "a mere gathering up of trite and stereotyped platitudes . . . the work of one who has but a smattering of politics . . . [and] no proper conception of the fundamental principles involved in the questions discussed."[27]

Throughout the campaign, the editor referred to the Alliance as the fourth party, placing the farmers' political prospects behind those of the Prohibition Party's candidate, D. C. Kelley, and implying both the future political intentions and the insignificance of Alliance schemes. Buchanan made clear his determination to run as a Democrat on the Democratic platform, but the *Banner* constantly berated him for not discussing the subtreasury plan and clarifying his views on the proposal. A barrage of stories hinted at "dark lantern" activities in secret Alliance meetings and compared Alliance supporters to the Know-Nothings of the 1850s.

When county Alliances in Republican East Tennessee endorsed Buchanan and urged members to cross party lines and vote their interests, the *Banner* editor outlined what he perceived to be the significance of the 1890 election: an experiment for the National Alliance and an opportunity to evaluate the Kansas Plan and the Tennessee Plan. The Kansas plan called for the establishment of a third party, while the Tennessee plan aimed at control of the Democratic Party. "If the Tennessee plan works well," the editor predicted, "the Alliance all over the country will be rejoiced, and Brother McDowell will be praised as one the greatest political managers of the secret political order." Although the editor presented a more conspiratorial view of Alliance aims than the facts support, he correctly stated McDowell's intention to keep the farmers inside the Democratic Party, and his perception of the Alliance activities as part of a Tennessee Plan gained increasing credibility.[28]

The November election placed Buchanan in the governor's seat, making Tennessee and national history as he became one of six Alliance members elevated to gubernatorial positions in the 1890 election. In addition, fifty-four Alliancemen won seats in the general assembly, constituting a significant minority in both houses. In the congressional races, Alliance-supported candidates won with comfortable majorities. Just as importantly, the national Democratic Party rode a groundswell of public sentiment against the so-called billion-dollar Republican Congress of 1889 to take control of both the House and Senate. So complete was the revolution that some worried that the Republican Party faced collapse, much as the Whig Party did in the 1850s. Democratic stalwarts and Alliancemen interpreted the possibility in different ways.[29]

The 1890 election portended both triumph and failure for Tennessee Democrats. They had recently banished Republicans to the eastern reaches of the state through disfranchisement of African Americans, and now they anticipated national political plums through committee assignments, with the possibility of controlling federal patronage should they capture the White House in 1892.

The Alliance was the fly in the ointment. Alliance perseverance in the summer and fall conventions caught "Democrats" off guard. Not only did the "hayseeds" and "clodhoppers" capture the governor's seat, but they increased their numbers in the general assembly and forced congressional candidates to make some concessions to their demands in at least four of the ten districts. If they succeeded in portraying themselves as good Democrats, reelecting Buchanan in 1892, and adding to their numbers in the general assembly, it was possible for the Alliance to name the next U.S. senator, when William B. Bate's term ended in 1893. Bourbon Democrats could not let that happen.

A vocal minority within the Alliance viewed the Democratic triumph in the national elections as an opportunity to press for a third party, in effect re-creating the situation that led to the foundation of the Republican Party in the 1850s.

While few in number, these third-party advocates undermined McDowell's program of revolution from within. Third-party men anxiously waited in the wings, ready to fill any power vacuum that might develop.

The Alliance in Power, 1891–1893

Much to the annoyance of the Bourbons, who wanted a show of radicalism, and to the mystification of historians, the Alliancemen made virtually no efforts to enact "class legislation." As Hart's careful analysis shows, they generally voted with the Democratic majority. Aware that their behavior invited questions, Dr. J. E. Head attempted to explain the "embarrassing predicament of Alliance legislators." "Any departure from the dictates of Democratic bosses," he wrote to the *National Economist*, "is regarded as treason to the party." Powerful Democrats constantly reminded Alliancemen that "they owe their position to that party, and if they fail to carry out its behests, it is because they are under the control of an iron clad, oath bound, secret organization." If farmer-legislators voted with the party, the press labeled them as weak greenhorns, unknowledgeable in politics and government. If they voted against party-sponsored legislation, journalists denounced them as a dark lantern group. The conundrum Head outlined depicted one problem faced by the Alliancemen; legislative procedures and political trickery provided additional pitfalls.[30]

In many ways, the appearance of farmers in the legislature produced a standoff that limited their ability to advance the Alliance cause. Old-line Democrats and anti-Alliance legislators retained the advantage in three ways. Their constant refrain for conformity to the party platform cast every rural-supported reform effort in a negative light. At the same time, anti-Alliance lawmakers introduced some measures as "bait" to trap agrarian neophytes into support for "radical" and "class" legislation. Finally, and perhaps most importantly, the Bourbons retained control over the party machinery and chaired the most important committees, reducing the possibility that farmer-backed legislation would ever reach the floor. Supported by the urban press, party insiders were well poised to discredit the Alliance triumph and undermine reform. Three issues provided the Bourbons with the means to discredit the Alliance legislators and Buchanan: the distribution of patronage, railroad and tax legislation, and the miner's war of 1891–1892.

Buchanan's distribution of the few state patronage positions among his friends and supporters rankled anti-Alliance Democrats. Patronage positions were highly sought jobs everywhere in Gilded Age America, but in poor states, with few positions, the fight for appointments assumed greater importance. During her 1892 visit to the state, Populist orator Mary Lease recognized how much the region's poverty enhanced the value of patronage positions. Commenting on the "great horde of office seekers" she found, Lease attributed the "miserable scramble to obtain office" to the fact that there was "no other occupation under God's heaven that will pay the men or make them a respectable living." The loss of patronage to previously noncompetitive farmers, particularly for office seekers in the winning party, produced as much bitterness as the presence of the "hayseed" governor in the seat normally reserved for men of wealth and power.[31]

Appointment seekers besieged Buchanan with pleas for offices. Some even included newspaper clippings or letters of endorsement to "prove" their longstanding commitment to agricultural interests. It was clear, however, that the governor had already determined his appointment list. Buchanan named Eth B. Wade of Murfreesboro, an old friend, to the position of superintendent of prisons. Wade, who held the post of state secretary of the Farmers' Alliance, had long served as clerk of the Tennessee House of Representatives. D. G. Godwin, an Allianceman from Arlington, became commissioner of agriculture. W. R. Garrett of Nashville, president of the National Education Association, received the appointment of state school superintendent. Newspapers gave the new governor high marks for these appointments, but his selection of McDowell for the lucrative position of coal oil inspector supplied the Bourbons with ammunition for their charges of corruption and mismanagement of state offices. In Hart's view, the appointment brought McDowell "under the most severe public attack any Tennessee politician had suffered since the days of Governor W[illiam] G. Brownlow."[32]

The *Banner* used McDowell's appointment to turn the Alliance antiparty rhetoric on its head. The editor initiated a campaign to lower coal oil fees and invited other presses to join the clamor for reform. City and county newspapers across the state took up the cry for lower fees, arguing, as the *Banner* did, that although once necessary, the position of coal oil inspector had become a source of corruption and a method of political payoff. Though clearly directed at McDowell, the anticoal oil inspection campaign promoted an honorable issue: the elimination of an unnecessary and costly government position. The success of the campaign can be measured in the resolution passed by the Lauderdale County Alliance demanding that the legislature reduce the coal oil inspection fees "to an amount commensurate with the actual service performed." By focusing on the office awarded to McDowell, anti-Alliance editors sowed the first seeds of suspicion among rank-and-file Alliance leaders that, once in power, their leaders abandoned agrarian demands and pursued a policy of business as usual.[33]

With the strong Alliance presence in the general assembly, most Tennesseans expected a flurry of antirailroad legislation. Buchanan made no mention of the need for railroad regulation in his messages to the assembly (nor did the 1890 Democratic platform mention railroads), but legislators introduced forty-one railroad bills in the opening days of the 1891 general assembly. The most important regulatory bill was submitted by John A. Tipton, a Covington attorney. Alliance representatives split their votes 17 to 22 on the measure, with the majority voting against the bill. When combined with the house Republican vote, the antiregulation forces prevented the passage of the regulatory legislation.[34]

In the state senate, Vanleer Polk, a Democrat from Maury County, and Flourney Rivers, a Democrat from Giles County, introduced railroad commission bills. Neither man was a member of the Alliance; indeed, both actively opposed the farmers' organization. When the senate voted on Rivers' bill, nine Alliancemen voted against it, making them the largest group opposing the measure. The Alliance voting record raises important questions about the commitment of agrarian legislators to their organization's aims.[35]

Demands for state transportation regulation in 1891 arose from a rate war between the Louisville and Nashville Railroad and the Nashville Street Railway Company. The high cost of coal carried by the L&N from East Tennessee mines drove up costs for the municipal line, raising charges of unfair competition and fears of forced consolidations. The Alliance had nothing to gain in this battle between railroad interests; any regulatory agency emerging from these measures would not reflect the interests of farmers. Thus, for very different and equally pragmatic reasons, both Republicans and Alliancemen opposed the Rivers Bill.[36]

If the failure of railroad regulation caused Tennesseans to question the sincerity of Alliance aims, the 1891 Revenue Bill united the business interests in wails of dismay over what they perceived as agrarian tax gouging. The long-awaited bill, presented in the waning days of the legislative session, raised the privilege tax on corporations to exorbitant rates. In addition, the new rates discriminated against larger towns and counties by apportioning taxes on the basis of population. Critics felt they understood the motives behind the charges when they came to the provision that exempted agricultural associations from the privilege taxes. Clearly, the bill pitted agrarian interests against commercial and industrial concerns. Punitive and indefensible, both Democrats and Republicans demanded changes.[37]

No one wanted to take credit for the bill when it emerged from the Joint Finance Committee. The *Banner* blamed state comptroller J. W. Allen for the measure, but a house member refuted the accusation, asserting that the revenue bill was "entirely the result of the committee's action." Although no one named the Alliance legislators as contributors to the measure, the *Memphis Appeal-Avalanche* asserted that in "one of the most determined floor fights" ever made, the "70 farmers in the House . . . supported the committee in an attempt to get all possible out of the corporations." Fearing a backlash, five Alliance legislators took the unusual action of disassociating the state organization from the bill, noting that only one Alliance member sat on the committee and "he rarely attended its meetings."[38]

Defeated on the floor, the bill was sent back to committee and hastily rewritten. A second effort produced the necessary votes, and by the end of the month the legislature was ready to adjourn. A Memphis editor praised the second bill, noting that "in only one or two instances are the privileges higher, while in many they are reduced." One historian assessed the episode by blaming the messenger. "If not directly responsible for the 'guage [sic] revenue bill,'" he wrote, "the Alliance was at least largely responsible for the anti-corporation agitation in Tennessee which had given birth to the high privilege taxes in that bill." The timing of the bill (at the very end of the legislative session), the haste with which a new bill was prepared (after weeks of work on the original bill), the previous lack of radicalism on the part of Alliance legislators, and the complete capitulation of the measure in favor of reduced rates suggest that the first bill represented a political maneuver rather than a serious attempt to pass a revenue bill. Alliance foes successfully tied the organization to the measure while offering no evidence to support their accusations. In the minds of many Tennesseans, the Farmers' Alliance became the anti-industry, anticommerce, anti-urban naysayers that voters had always feared. When coupled with the events of the summer of 1891, the perception of corruption and radicalism became fact.[39]

On July 14, 1891, a coal miners' war broke out in East Tennessee, pitting armed and desperate miners against the governor and the most powerful corporate interests in the state. As the drama unfolded, the limitations of government to act decisively and the contradictions between Alliance ideology and agrarian pragmatism became apparent, undermining the tenuous farmer-laborer coalition and dooming agrarian political gains. In the end, neither the miners nor the Alliance benefited, and the Tennessee Coal and Iron Company continued to operate its mines with convict labor.[40]

The coal miners' anger erupted from long-simmering disputes dating back to Reconstruction and Redeemer politics. In the aftermath of the Civil War, successive Tennessee state legislatures passed a number of bills criminalizing activities, such as trespass and vagrancy, in an effort to bring black labor under control. In addition to producing the desired effect, enforcement of the laws also increased the prison population well beyond the capacity of state prisons. The most pragmatic solution for a revenue-poor state was the leasing of convicts to industrialists, who became responsible for prisoner maintenance and paid the state a fee for each prisoner leased under state-approved contracts. Arthur S. Colyar headed the industrial lobbyists who pressured the general assembly to adopt convict leasing. Tennessee negotiated its first contract in 1865; by 1891, the state was receiving $100,000 annually in convict leasing fees. Tennessee Coal, Iron and Railroad Company became the largest lessor of convict labor, placing prisoners in the mines at Briceville, Coal Creek, Inman, and Tracy City.[41]

Convict leasing generated criticism from the start. Reformers charged that the lease system operated as a mechanism for providing labor rather than rehabilitating criminal behavior—and indeed, men could be imprisoned for theft of eight cents worth of fence rails and leased to work under life-threatening conditions. Periodically, tales of poor living conditions and inhumane work practices produced public demands for an end to convict leasing. The most persistent calls for reform came from the labor press, which charged that convict leasing constituted unfair competition with free labor. Both the Wheel and Alliance adopted positions that demanded an end to the practice of convict leasing, although, as the jute boycott showed, farmers remained willing to utilize prisoners for their own benefit and purposes.[42]

The conflict in the coalfields pitted labor against industry in a dispute with little for compromise by either side. East Tennessee mining companies struggling to remain profitable in a period beset by labor unrest and price wars imposed illegal requirements for employment, ignored state law on wages and weighing, and sought additional ways to cut costs. Miners protested the payment of wages in scrip, the imposition of "iron-clad" or "yellow dog" contracts, the employment of company weighers, and the use of convict labor as impediments to their livelihood. The events that set the wars in motion began in April 1891, when the Tennessee Coal, Iron and Railroad Company closed its operation at Briceville for repairs. When the mine reopened two months later, the company demanded that miners sign iron-clad contracts; when miners refused, Tennessee Coal, Iron and Railroad brought in convict labor.[43]

Just after midnight on July 14, more than three hundred miners advanced on the recently constructed prison stockade, demanded the surrender of the convicts, and marched them to Coal Creek (now Lake City), where they were placed on a train and sent to Knoxville. The miners informed the governor that they had "come together to defend [their] families from starvation, [their] property from depreciation, and [their] people from contamination from the hands of convict labor." They called upon the governor to intervene to resolve the problem and prevent bloodshed.[44]

Responding to the labor issues as well as fears of civil disorder, Buchanan ordered three companies of militia to the scene and set out for Knoxville, where he met with the miners and urged them to settle their disputes through the courts rather than by armed confrontation. Miners countered by asking the governor why the state had not sent the militia to force the company compliance with state laws forbidding the use of scrip and iron-clad contracts. Buchanan's silence on the company's illegal actions undermined the miners' faith in the reliability of Alliance support.[45]

Many Tennesseans of all classes sympathized with the miners' problems and organized meetings to express their support, even as they praised the governor for his expeditious action to protect property and establish order. At the same time many of the same grateful property owners noted the shaky constitutional support for Buchanan's action in calling out the militia. Meanwhile, Commissioner of Labor G. W. Ford endeavored to negotiate a settlement, but the miners adamantly refused to compromise on the use of convict labor.[46]

In the early morning hours of July 20, the miners launched their second attack. Supported by comrades from across Anderson County and nearby Kentucky, they advanced on the stockade, where the militia commander wisely surrendered the convicts to the larger force. The insurrectionists marched the forty prisoners, along with the guards and soldiers, to Coal Creek, loaded them on the train, and sent them to Knoxville for the second time.[47]

In a second meeting with Buchanan, the miners presented a more formal petition, demanding a return to the status quo, the annulment of the lease contract, and the convening of a special session of the general assembly to repeal the convict lease law. Tennessee Coal, Iron and Railroad Company refused to negotiate, leaving Buchanan with few choices but enforcement of the law. He did promise to call a special session of the legislature, but he offered no guarantees on how the legislators would vote. Acting on his word, Buchanan called a special session of the legislature to consider several problems: legislation to empower the governor to call up the militia in periods of civil disturbance, repeal or modification of the convict lease law, prohibition of the use of scrip, and reform of the criminal procedures of the state to make counties responsible for criminals guilty of minor offenses and removal of the prison to a new location outside Nashville.[48]

The mandate of the extra session highlighted a number of divergent and competing interests that overlapped boundaries of race and class. Humanitarian concerns, property rights, labor rights, and political expediency influenced the deliberations. The complexity of the problem had even the state and national Alliance backing away from previous condemnations of convict leasing. The Tennessee Alliance met in August 1891 and adopted resolutions supporting the use of convict labor for road building and urging that leases not be renewed at their expiration in 1895—a tepid position that satisfied no one. The *National Economist* pointed to the labor unrest in the East Tennessee mines and also equivocated on the issue of ending convict leasing. "Every labor platform of modern times has announced against the competition of convict labor with free labor," wrote Macune. While "a candidate who desired popularity with the masses has simply announced against it to secure his election," the legislatures continued "to perpetuate the system" for fiscal reasons. The old Alliance reformer admitted that the question brought together three "fundamental principles": the duty of government to carry out the law, the rights of citizens, and the rights of criminals. Leasing appeared necessary to underwrite the cost of the prisons, but the editor deplored the profit earned at the expense of free labor. In other words, convict leasing needed reform, not necessarily abolition.[49]

When the general assembly met in September 1891, the Alliance members voted with the Democratic majority to retain the lease contract to its expiration and strengthen the powers of the governor to maintain law and order, but their votes split on the resolution introduced by J. L. Cochran that the general assembly would "never again lease its convicts at the expiration of the present lease contract." The failure to act decisively destroyed the compromise worked out by Buchanan in July and led to new unrest.[50]

In November, Anderson County miners again released six hundred prisoners. As state officials and local sheriffs offered rewards and hunted for the missing men, Robert Cartmell commented on the situation and, like others, found no easy solution to the dilemma. "Tenn. cannot afford to have the law trampled on in that way," he wrote. "If [it is] a bad law, until changed or repealed, [it] ought to be enforced." In January 1892, the Tennessee Coal, Iron and Railroad Company provided a temporary truce when the company decided to put the miners back to work at Briceville and place convicts in other mines.[51]

The Farmers' Alliance fared poorly in the contest over convict leasing. Buchanan's leadership ability, and even his manliness, came under public scrutiny. Bourbons, who harbored no great love for labor, reduced the complex issue to a question of the governor's willingness to act, characterizing him as a nice man, but one without the necessary qualities of leadership. The Alliance failure to support labor against capital undermined the stickability of East Tennesseans, who had always been stronger Republicans than Alliancemen.

Despite their rough introduction to Tennessee legislative politics, Alliancemen had made some gains with important future consequences. In his inaugural address, Buchanan had identified himself with the "Jeffersonian non-interference theory of government" but asserted that "progress [did] not mean a mushroom growth of 'booms' and wild speculations . . . but a substantial growth upon the solid basis of an increase in wealth, in taxable property, in resources and revenues." In order to facilitate the development of institutions and social infrastructure necessary for prosperity, he recommended an increase in the state appropriation for the agricultural department and expanded support for agricultural institutes. He asked the legislature to provide longer terms for schools and broaden the curriculum. He called on lawmakers to "stretch forth a helping hand" to the disabled Confederate veterans and to fund "the needed appropriation to the Confederate Veterans' Home." Finally, Buchanan's commitment to white supremacy, states' rights, and the Democratic Party moved him to call for the extension of the Dortch Law to "all places where registration applies under existing law." Purification of the ballot box, according to the governor, "leaves not a shadow of an excuse for Federal supervision."[52]

The legislature acted on his recommendations and passed an education bill establishing graded primary and secondary schools. Primary schools included the first through the fifth grades, and secondary schools extended from the sixth to the eighth grades. Completion of the secondary level entitled students to advance to high school. Secondary education included instruction in the "Elementary Principles of Agriculture." Although the restructuring of Tennessee education was not immediately undertaken, the passage of this act is generally viewed as the foundation for modern public education in the state. In an effort to improve the quality of education, the legislature increased its appropriation for Peabody Normal School and funded teachers' institutes. However, a uniform textbook act, a consistent Alliance demand, failed.[53]

The general assembly enacted a Confederate pension bill, providing twenty-five dollars per month to disabled former Confederates and smaller amounts to partially disabled veterans. They also appropriated $35,000 to construct and maintain a Confederate Soldiers' Home on the Hermitage property in Nashville.

The most important change in state government for Tennessee's farmers involved the separation of the Bureau of Agriculture, Statistics, Mines, and Immigration into two agencies. The new Bureau of Agriculture and Immigration provided the leadership in agriculture that farmers had long sought. Farmers could believe they had put their bureau on the right track and that they had the right man in charge. Agricultural Commissioner D. G. Godwin sponsored a Memphis convention in January 1892 to address falling cotton prices and develop a program to decrease cotton acreage. Two southern governors and four commissioners of agriculture attended the conference along with other state officials and private citizens. As the U.S. Department of Agriculture reported on the outcome of the program, four states—Virginia, North Carolina, Florida, and Louisiana—met or exceeded the proposed 20 percent decrease in cotton planting, and seven states reduced their cotton planting by at least 15 percent. As the leader in convening the southern agriculturalists and in securing the reduced acreage, Godwin and the Bureau of Agriculture projected an image of professionalism and efficiency.[54]

At the end of his first year in office, neither Buchanan nor his Alliance supporters had transformed state government, but they had made important reforms in education and agriculture, and they had provided additional support for the state's aging Confederate veterans. It could also be said that they had avoided the traps and snares set by the Bourbons and had handled a civil crisis without the bloodshed experienced in other Gilded Age labor disputes. Democratic Alliancemen expected to solidify their control of the party in the 1892 elections, by promoting themselves as the "purest" and "best" Democrats. With the goal of the U.S. Senate in mind, Alliance leaders worked to solidify their gains.

The Bourbons recognized the threat and took advantage of every opportunity to undermine the credibility of Alliance leaders, particularly McDowell. The campaign to reform the office of coal oil inspector chipped at the motives of all men in appointive office. As often as possible, anti-Alliance politicians placed the blame for the 1891 example of the general assembly's perennial ineptitude on the farmer-legislators. Finally, the Alliance contributed to its own problems by accommodating too easily to the Democratic platform. As a result, they had little to show for their efforts. Some legislators had come to resent the pressure exerted by Alliance leaders. They lost their most important opportunity, and future votes, in their failure to support the coal miners. Indeed, by the fall of 1891, many Alliancemen viewed the formation of a third party as the only possible mechanism for achieving agrarian goals.

The Election of 1892: The Rise of Populism

The election of 1892 demonstrated the strength of the reinvigorated and united Democratic Party and the weaknesses and divisions of the agrarian insurgency. The fear of another Alliance victory united Bourbon and New South leaders in an unusual show of force as Democrats mocked agrarian political claims.

The political failure of the Alliance after its recent successes suggests that the Tennessee Plan of operating through the established Democratic Party became a liability in the changing political climate of the 1892 election. Past assurances that the Tennessee Alliance would not support third-party formation cast doubt on the whole reform agenda when agrarian leaders transferred their allegiance to the People's Party. Unlike the Populist experience in other states, fusion with the Republican Party never fully materialized as GOP candidates focused on building and protecting personal patronage systems, to the exclusion of larger aims.

In May 1891, Martin VanBuren Ingram, the past editor of the *Toiler*, wrote a "strictly confidential" letter to Edward Carmack, outlining his insider's knowledge of the Alliance and offering his advice to the Democratic leadership on the best method for preventing a repeat of the Alliance 1890 victory. Focusing on McDowell as the Machiavelli behind the agrarian insurgency, Ingram assured Carmack that he "underst[ood] the man so well, and [knew] his weak points." He dramatically warned that "if let alone his methods are powerful for evil and his strength greater than you have ever supposed."[55]

Apparently Ingram's letter was prompted by McDowell's recent denials that he expected the position of U.S. senator as the price for leading the farmers to political victory. In April 1891, the Alliance president issued a statement declaring that he was "not a candidate for Senator" and had "not authorized any one to say so." Unwilling to close the door to his political future, however, McDowell equivocated. "I don't say my name has not been mentioned, for I have been approached by a number," he admitted, "but I have not authorized any one to say I am a candidate." McDowell's less-than-complete denial stirred Democratic Party fears that he was an office seeker busily organizing the farmers for his own benefit. Indeed, Ingram surmised that McDowell "was simply too previous in letting his purpose out and now wants it understood that he is not in it, so as to check the fire on him." Ingram predicted that McDowell intended "to hold the balance of power in the Democratic Party as was done in the gubernatorial campaign and strengthen the organization which will carry him in." The former Wheel editor further contended that McDowell "will either succeed or split the party and in either case he will have a soft place to fall on."[56]

Although the election was more than a year away, Ingram advised the prompt organization of Democratic clubs statewide. He further counseled care in the selection of party workers and the quick establishment of a platform favoring "tariff reform or revenue tariff and free coinage and admit no subtreasury heresy in it." Ingram believed that "the better class of farmers will gain this out [and] leave McDowell with so few that he will be powerless to make the fight."[57]

Ingram warned that McDowell's candidates were "already staked out," and he encouraged Carmack to make the fight early "as Georgians are making it," by putting senators and congressmen on the stump against the subtreasury scheme. He estimated the Alliance strength at 40,000, after subtracting Republicans, women, boys, and anti-subtreasury members. With planning and effort, the Democratic Party could overcome the Alliance insurgency, "but let alone they can use the whole force in county conventions and control the State Convention." Ingram ended his letter with the hope that Carmack would "call a private conference of prominent Democrats and take steps to organize the state at once, put backbone in men who are converging before the movement and the bulldozing tactics that are being use[d] to silence all oppositions to the grand Cyclops." Ingram's letter found a ready audience, and the party leaders launched an anti-Alliance crusade that followed his plan exactly.[58]

The political campaign of 1892 that united the New South and Bourbon wings of the Democratic Party and established the People's Party in Tennessee passed through several stages. For almost a year, from the early spring of 1891 until February 1892, the Democratic leadership limited its efforts to personal attacks against McDowell, refutations of the subtreasury scheme, and the organization of Democratic "clubs" that soon identified with the political ambitions of Chief Justice of the Tennessee Supreme Court Peter Turney. In February 1892, the opening of the canvass for the gubernatorial nomination signaled a new phase of the campaign. Turney and Buchanan supporters fought bitterly in county conventions and primaries across the state. Upper-Case Democrats wondered if the farmers would repeat their political victories of 1890, as Buchanan garnered considerable support.

In May the campaign entered its third phase, when radical agrarian reformers abandoned the Democratic Party, citing a history of fraud that prevented reform. With the establishment of the People's Party, Bourbons quickly assumed control and dominated the spring convention to select delegates for the Democratic National Convention. McDowell initially rejected the third-party move, but by the end of June he had joined the radicals. Left without his strongest Democratic supporters, Buchanan eventually withdrew his name from consideration for the Democratic gubernatorial nomination and ran as an Independent, all the while maintaining that he represented the true Democracy.

Populist leaders lacked direction from May until the August nominating convention of the People's Party, and their contradictory statements and actions confused some voters and alienated others. The August convention should have given the party the direction and momentum to reinvigorate their lagging campaign, but the East Tennessee miners' war flared up again just as the Populists met. This time the Bourbons and New South industrialists called for a violent suppression of the miners and mocked Buchanan's manhood when he failed to act swiftly and with as much force as they demanded.

Finally, in the months following the convention, as Buchanan's campaign stagnated, Democrats successfully rallied Civil War animosity against former Union General (and Populist presidential candidate) James B. Weaver to deny the People's Party credibility in their national canvass. Likewise, Democratic assertions of a Populist-Republican fusion bore fruit in the waning days of the campaign. In October, only days before the election, Democrats revealed a $15,000 campaign contribution by the Republicans to McDowell for the support of Populist legislative and congressional candidates. That final scandal ended Populist hopes for vindication at the polls.

Bourbon Democrats began their attack on John McDowell before the dust had settled on the 1891 legislative session. As Buchanan's friend and political manager, and the president of the State Alliance, McDowell's prominence offered anti-Alliance Democrats a perfect foil for undermining the governor's political cachet. Edward Carmack delighted in teasing the bewhiskered Alliance leader. When speculation about McDowell's future as a U.S. senator rocked the state, Carmack claimed that he had always known McDowell possessed senatorial qualities. After all, the editor wrote, "We know that God Almighty never gave him that mass of storm-stricken and insurrectionary whiskers and made him look like a weather-beaten tintype of [U.S.] Senator [William] Pfeffer for nothing." Marshaling his well-known sarcasm, Carmack hailed the agrarian leader, "United States Senator McDowell, noble knight of the horny hand and stone-bruised heel, we, the played-out Bourbon, moss-back, upper-case Democrats salute thee."[59]

Other criticisms of McDowell were less amusing. In the early summer of 1891, long before serious campaigning got underway, the Nashville *American* ran an exposé of McDowell's past. The Nashville newspaper sent a reporter to Desha County, Arkansas, where McDowell had lived for several years during the Reconstruction era, first renting and then buying land for cotton farming. During this period, the *American* claimed McDowell abandoned the Democratic Party and ran for justice of the peace with the support of the Radical Loyal League. Just as importantly in the view of Bourbon white supremacists, the newspaper charged that McDowell shared a meal with a black supporter.[60]

The Alliance leader denied all charges and accused the *American* of conspiring to divide his agrarian support with false claims. He obtained affidavits from fifty Arkansas citizens who vowed that McDowell had never been a Republican or a member of the Loyal League. Tennessee Congressman Rice Pierce confirmed his neighbor's status as a Democrat and a gentleman. Nevertheless, the damage had been done. The *American* named Jehazy Cole as the man with whom McDowell supposedly shared a meal. Thereafter, the anti-Alliance press frequently referred to McDowell as "Jehazy" McDowell, thus reminding readers of the questionable past of the Alliance president and, by implication, suggesting that a vote for Alliance candidates threatened white supremacy.[61]

Democratic Party leaders recognized that attacks on McDowell's character could not alone deny Buchanan a second term and destroy agrarian influence in the general assembly. Therefore a simultaneous campaign against the subtreasury, the centerpiece of the Alliance platform, raised questions about the organization's ability to speak for farmers' interests. Anti-Alliance politicians exploited agrarian divisions over the inclusion of land as a form of security, and used their concerns as a wedge to erode the "stickability" that brought the farmers to power in 1890.

The anti-subtreasury campaign provided a steady flow of editorials and speeches that picked at the ragged edges of the Alliance and questioned the unity the agrarians claimed. The Nashville *American* reported that thirty-six of the fifty-four Alliance members in the general assembly were opposed to the subtreasury scheme. Senators Isham G. Harris and William B. Bate warned farmers that the subtreasury plan would "allow farm owners to borrow at two percent and then loan to renters and sharecroppers at ten to twelve percent." Other Democrats charged that the subtreasury represented class legislation that benefited the few at the taxable expense of the many. The *Gibson County Herald* reprinted an article in which the writer speculated that the land-loan bill reflected the interests of railroad magnates, specifically Leland Stanford. Under the land provisions of the subtreasury bill, railroads and corporations owning land "could borrow money on it from the Government at two per cent as easy as could the farm owner, and in one-tenth the time." The writer warned that "the people would come to general chaos and bankruptcy quicker by this process than any yet proposed and the sooner be the serfs or slaves of corporations." All the antimonopoly rhetoric that inspired the farmers to organize now operated against their efforts to effect meaningful change. Farmers like the president of the Prospect (Giles County) Farmers' Alliance, who had heeded the organization's earlier subtreasury initiatives, now concluded, "we want too much if we want the Ocala demands in full."[62]

The Democrats followed their attacks on the subtreasury plan with the organization of Democratic Clubs that quickly became [Peter] Turney Clubs, once the Chief Justice agreed to carry the party banner against the agrarian foes. Sixty-five years old and of uncertain health, Peter Turney was an unlikely candidate for the rigorous campaign anticipated in 1892. After twenty years on the Supreme Court, he appeared better suited to retirement than a new career. An unrepentant rebel, his frequent declarations in favor of that discredited position met with cheers from some southerners and cries of dismay from others. In one speech, he said, "So strongly do I believe in the right of secession, that when I am dead, I want the word 'secession' carved as an epitaph on my monument."[63]

The son of a U.S. senator, Turney distinguished himself in 1861 by organizing a Confederate company and urging his home county of Franklin to secede from the state and join Alabama, when Tennessee first rejected secession. His appearance before the Cheatham Bivouac in 1888 where he reportedly defended the Confederate cause produced a barrage of criticism from the southern press. The *Chattanooga Times* reminded Turney that "the war is over and the questions which then convulsed our country have been forever settled." The editor characterized Turney as "prudent, conservative and thoughtful" but chastised him as well: "He belies this character by raising anew these questions and accomplishes nothing whatever except to provoke the condemnation of the thinking men of the South." The *Birmingham Herald* likewise condemned the speech and labeled Turney, "one of those southern Jackasses who are always certain to turn up in the nick of time to furnish the republicans with campaign material."[64]

Turney's plainspokenness, a political liability in most instances, became an asset to the Democratic Party in its efforts to draw Civil War veterans back to the party of the fathers. The old soldier's steadfast adherence to the past represented a sanctuary many rural voters found hard to resist. No one could argue that Peter Turney was not the same yesterday, today, and tomorrow. In his unrepentant defense of the Confederacy, "Ole Pete" represented undefiled Democracy at its finest.

Recognizing the potential pitfalls inherent in attacking Turney on his own turf, Alliance leaders shifted the campaign away from the Lost Cause to the less emotional ground of entrenched office holding and agrarian interests. The *Toiler* accused Turney of being "a continuous office-holder," who had "drawn a salary from the state treasury for twenty-one years." Now his selfish demand for office allowed him "to be used by the *Nashville American* crowd to split and destroy the Democratic party forever in Tennessee." Like the *Toiler*, the McMinnville *New Era* viewed Turney's candidacy as a class-based insult to the political rights of rural voters and argued that "the rule has always been to give governors a second term." Pointing to what farmers perceived as an elitist prejudice against agrarian interests, the editors contended that "now because for the first time in thirty years we have a farmer governor, the lawyers almost to a unit say he must be retired to give place to one of their own number."[65]

Contenders on both sides of the fight claimed that the lineup of pro- and anti-Buchanan factions clearly marked the class divisions within the Democratic Party. Commenting on the presence of Turney supporters in the city, the *Banner* declared the "presence of so many prominent Democrats . . . one of the gratifying indications in politics in Tennessee," and compared them with the "small men [who] have crowded to the front" in recent elections. While happy to see the return of men of prominence, the editor believed the party was reaping what it had sown. Having placed the Democratic stamp of approval on an unworthy man like Buchanan, they were now forced to deal with "all Alliancemen who were hugged and slobbered over in 1890." If the *Banner* took dark pleasure in watching party elites squirm as the result of their "hobnobbing and colleaguing with heresy," other Democrats adopted a "wild-eyed advocacy" that drove the Alliancemen from the party and concentrated political power beyond the threat of grassroots insurgency.[66]

In the opening days of the canvass for the Democratic nomination for governor, Upper-Case Democrats found little encouragement for their campaign. The Buchananites jumped to an early lead by calling a county convention in Shelby County even before the state executive committee had set a gubernatorial convention date. Angry over the "snap," Turney supporters predicted that the early actions of Buchanan's followers "stirred the Democracy . . . [and] roused them to a sense of danger which lurks in the secret methods of a secret society."[67]

Confident of success, Buchanan partisans pushed ahead in Middle and East Tennessee. Despite a pouring rain, farmers in Rutherford County turned out in force and ultimately controlled the convention there. When the supporters of "Ole Pete" failed to achieve their aims, they walked out. The *Banner* mocked the bolters and quipped, "We believe this is the first instance on record of a majority bolting in hot haste from the overpowering minority." In East Tennessee, Hamblen County Democrats selected a solid Buchanan delegation during a "turbulent" convention marked by "several fights." It appeared that Alliancemen were prepared to outlast the Bourbon–New South coalition again as Buchanan ended the first month of delegate selection well ahead of his 1890 tally.[68]

Predictably, Turney Democrats called for electoral changes, focusing on two alterations that cut at the heart of agrarian strength. Political insiders called for the elimination of stump speeches, and they demanded repudiation of the subtreasury as a requirement for voting in Democratic primaries and conventions.

Contemplating a long summer of canvassing, the editor of the *Gallatin Examiner* asserted that the speeches of gubernatorial candidates attracted "not ten per cent" of voters. In place of such unproductive effort, he suggested that "half a dozen speeches delivered in each of the three state divisions and well reported would fully inform the people of the views of the candidates." Denouncing "old-time stumping" as "a slow coach," the editor claimed "the press of Tennessee could speak daily and weekly to crowds ten times more numerous than those of the candidates." The editor buried his anti-Alliance agenda in the rhetoric of progress. Elimination of the stump restricted voting to the educated and denied country people the opportunity to size up candidates in person. McDowell and the Alliance leaders had built a powerful political force through rallies and public speeches, and in the process they mastered the art of public persuasion. Stump speeches clearly favored the Buchananites over the more polished and educated Upper-Case Democrats—a fact recognized by both sides. This and subsequent campaigns would experience fewer stump speeches of the "Lean" Jimmy Jones and Bob Taylor variety, and not coincidentally, public interest would vanish as well.[69]

The more overt anti-Alliance campaign began in the pro-Turney newspapers before advancing into party policy. Tactics ranged from outright denunciation to conspiratorial implications, to demands for repudiation of the subtreasury. The Nashville *Herald* denounced the Alliance as a "secret, oath-bound, one-class political party" and condemned the farmers' organization for its "insolence," insisting that Democrats had "too long tolerated" the Alliance as "a political factor." The *Fayetteville Sun* demanded that Democratic Alliancemen "must either abjure the eternal principles of Democracy . . . or absolve themselves of all connection with the organization of dangerous heresies and condemnable ideas." Occasionally, Democratic papers attempted to sow dissension among the Alliance adherents by printing the views of local, anonymous "farmers," who claimed that the agrarian organization leaders forced them to choose between the party of their fathers and loyalty to their secret society.[70]

Although some Turneyites cautioned that a too "flat-footed fight" could embitter farmers and improve their perseverance, the hysteria continued and escalated, as the party moved closer to expulsion of the insurgents. The *Herald* charged that the Alliance "has grown tyrannical, if not insolent and domineering, if not brazen." The Nashville paper warned that the agrarian organization had reached "formidable proportions" and intended to "make the Democratic party the tool with which to open up new possibilities to itself." The *American* declared its position, assuring fellow Democrats that "If a secret, oathbound, one-class political party captures the next Democratic convention they will realize that they have had a fight when they take the prize." The *Knoxville Tribune* asserted that "The Democracy should prefer to have its enemies openly arrayed against it rather than to have them inside its camp where under the guise of Democracy they may be enabled to work infinitely more harm." In the "flat-footed" manner decried by the *Banner*, the *Tribune* declared, "Those who are not for Democracy, pure and simple, are against it and must be fought with the same unrelenting and unwavering opposition that Democracy proposes to wage against Republicanism with all the sins of its infamies and its rottenness."[71]

Leading Democrats contributed to the abuse heaped upon the Alliance. Former governor Robert L. Taylor characterized the upcoming election as "the irrepressible conflict . . . for the life of the Democratic party in Tennessee." He urged "every Democratic orator in the state . . . to be upon the stump, warning the people against the wild heresies which have been hatched for the overthrow of Democratic government and the ruin of agriculture itself." Josiah Patterson denounced Alliance leaders as "the most arrogant set of knaves that ever masqueraded in any political guise or any so-called set of principles." He charged Alliance Democrats with being as "unfriendly to Democratic principles as the devil in Hell and all his angels are to purity and truth." Indeed, every able-bodied Democratic leader in the state took to the stump to denounce the Alliance "heresy" in similar terms. Sparing no epithets, they cursed the Alliance and McDowell "from hell to breakfast" in speeches across the state. Initially, the Alliancemen defended their rights as Democrats and determined to fight for the renomination of Buchanan as the Democratic candidate for governor. By mid-May many were having second thoughts.[72]

The Democratic State Executive Committee called for two separate state conventions, the first one meeting in May to select delegates to the presidential convention and the second meeting in August to nominate the party's candidate for governor. Democrats defended the calling of two conventions as necessary to determine the position of the national party on the issue of the gold standard before writing the state platform. Many simon-pure Democrats advocated free silver, but Grover Cleveland, a gold-standard advocate, seemed the likely Democratic presidential nominee. Tennessee party men, who had advocated free silver in their 1890 platform, postponed the state convention in order to prevent the potential embarrassment of proposing a state platform at odds with the national one.[73]

With Cleveland the probable presidential choice, Turney supporters anticipated a low turnout of Alliancemen at the May convention. They looked forward to the meeting to bolster enthusiasm for "Ole Pete" and to start a log-rolling movement for their candidate. The vote counts indicated that they needed a boom to invigorate their slow-moving campaign. On May 7, the *Banner* tallied Buchanan's undisputed delegates at 95 and Turney's at 65, with a total vote of 222 for the governor and 167 for the judge. Two days later, the paper divided the 533 delegates selected to date as 253 for Buchanan and 270 for Turney, but it noted that Buchanan was 18 votes ahead of his 1890 total for the same counties.[74]

The tense expectations that accompanied the almost daily updates on county votes and the anticipation of the upcoming convention were broken in early May by a startling announcement from the Alliance camp about the formation of a state People's Party. Faced with the choice of denying their economic interests or leaving the Democratic Party, some Alliance members had long advocated membership in the People's Party. Pushed by western members, third-party formation gained support with every meeting of the National Alliance. The strength of the movement became evident in 1891 when farmers convened in Cincinnati. The following February, at the St. Louis Convention, some Alliance members organized an extra session following the formal adjournment of the national meeting and formally endorsed a new People's Party. Charles Macune and most southern leaders expressed grave reservations about the move toward a political party, but the idea gained credibility among some Tennessee Alliancemen who had previously rejected the notion of a third party.

Early in May, the editor of the *Toiler*, L. K. Taylor, dropped all pretenses and came out squarely for the People's Party. Taylor resigned his membership in the Fifth Congressional Democratic Executive Committee and called for a state convention to select delegates to represent Tennessee at the July 4 National People's Party Convention in Omaha. Taylor apparently made the decision on his own, without consulting either Buchanan or McDowell. Nevertheless, as one historian noted, "the governor [suddenly] had one foot on the Democratic dock and the other in the Alliance rowboat." Democratic leaders and editors immediately recognized the difficulty of Buchanan's position and instigated an unrelenting campaign to force him to clarify his intentions.[75]

Buchanan vowed his "life-long allegiance to the Democratic Party." He reiterated his campaign position that he "was nominated and conducted the campaign of the Democratic Party in 1890" upon the party platform and maintained his support for the "principles of Jefferson, Jackson, Polk, and Calhoun." The governor made it clear that he had "no official connection with the *Toiler*" and exerted no influence in shaping its editorial policy.[76]

Editor Taylor also denied that he consulted Buchanan before his declaration for the third party and accepted full responsibility for his endorsement. He defended his move in very modern terms. He admitted distributing a questionnaire among the state's Alliance lodges to ascertain political support. He claimed that half the lodges responded to the question of whether individual members considered themselves Democrats, Republicans, or Independents ("those who hold principle dearer than party"), and among the respondents, "at least 90 per cent . . . reported as being independent." He concluded that "party collars are worn very loosely these days . . . [making] the very atmosphere . . . pregnant with the odors of independence." Since the Alliance claimed 80,000 members, Taylor believed "the third party had a very fair chance of success." The *Toiler* editor recognized the political reality that operated against the establishment of third parties, but he compared the "evident dissatisfaction" of his own time to the "fight which put the Republican party in power . . . in 1860." He upheld the Jeffersonian principles of the Democratic Party, but charged that "since the Whig party went to pieces, and a large element of it went to the Democratic party, we find that its bosses are almost exclusively aristocrats." As he surveyed the political field, Taylor concluded that the loss of Democratic principles and the statewide support for a third party portended success for the People's Party.[77]

The formation of the Populist Party initially brought mixed reviews from the press. The *Banner* editor suggested that the Populists did not want Buchanan's nomination but a "needlessly vituperative campaign" that would give the new party an "enthusiastic boost" and leave "as many people sore over [Buchanan's] defeat as possible" in order to draw them away from the Democrats. The *Memphis Ledger* worried that Turney's sudden "tidal wave of success" boded ill for the Democratic Party. The West Tennessee editor suggested that farmers were "discouraged and disgusted" with the Turney campaign and expressed their views by "refusing to vote in the primaries, preferring rather to take their chances with a third party." The *American* offered the amazing suggestion that the Democrats should nominate Buchanan as "a check to the wild rash of the impracticables into the third party." McDowell attributed the defections to the People's Party to "the bitter war which has been made on the Alliance and the vile slanders heaped upon our people." He believed some Alliance members would support Turney if he captured the gubernatorial nomination, but the "majority [would] not." Like the editor of the *American*, McDowell predicted the Democratic nomination of Buchanan would put an end to the third party."[78]

As anticipated, Turney men dominated the state presidential convention on May 26. The delegates quickly concluded their business and instructed for Cleveland. The considerable support Buchanan carried from the county delegations was not evident at this convention. Confident of their power, Bourbon

Democrats instituted the mechanisms to champion "simon-pure" Democracy by resolving to declare "unequivocally in [the Democratic] platform against the sub-treasury scheme, the land loan scheme, the purchase of the railroads and the telegraph lines by the general government, and against secret political organizations." As Roger Hart pointed out, the passage of this declaration, "in effect, changed the rules in the middle of the game." Democrats now required anti-subtreasury oaths from all would-be primary voters and convention delegates. The Alliancemen watched in frustration, unable to regain control within the Democratic Party or prevent the Populist shift.[79]

In late June, McDowell made his move to the People's Party. The *National Economist* printed his "Farewell Address," an essay that reflected his reluctance to abandon the party and his disgust with the fight just concluded. McDowell acknowledged that he had promised "in more than one hundred speeches . . . to make the fight for a redress of our grievances inside the two old parties." Nevertheless, "flagrant outrages," including the denial of "the right of suffrage at the primaries," forced him to change his mind. McDowell charged the Democratic executive committee with "misrepresentation, vilification and ridicule" directed against farmers "because [they] saw fit to join a farmers' organization and exercise the right to have an opinion on public questions." He reluctantly left the Democratic Party, but "in no wise depart[ed] from the pure principles of Jeffersonian Democracy." In a reference to the low-tax Democratic revolt of 1880, McDowell chastised the simon-pure Democrats who opposed him and reminded readers that "many of the very men who are now using the scare-crow of negro domination in the South" to warn against Populism, "had no fear of it a few years ago, when they bolted and left the Democratic party." He concluded that "the spirit of unrest and discontent . . . throughout the length and breadth of our land" demanded the union of reform forces and the formation of a new party.[80]

Buchanan never wavered from his adherence to the Democratic Party. On July 30, in response to the demands of "Buchanan Democrats," the governor withdrew his name from consideration at the upcoming party convention and ran as an Independent, Jeffersonian Democrat. When the Democratic Party convened August 9, Turney's nomination was a foregone conclusion, and he received the party's endorsement on the first ballot—though a few delegates registered protest votes for Buchanan. In his acceptance speech, Turney made a half-hearted attempt to heal party wounds, counseling Democrats to "talk to [Alliance members] not with epithets, but with argument," especially the argument that "the Democratic party has had no chance to bring them relief." Neither Turney's "argument" nor his belated olive branch persuaded the radicals who moved to the People's Party.[81]

The final phase of the 1892 campaign began with the People's Party convention in mid-August. A rousing convention followed by enthusiastic rallies, parades, and stump speeches still might have given the Populists a credible showing in the November elections. However, both internal conflicts and external events undermined the opportunity to reinvigorate the lagging campaign. Following the establishment of the People's Party, counties organized quickly but remained divided over how to proceed. The Giles County People's Party provided an example of the issues that divided Populists across the state. The Giles Pops resolved to support Buchanan's bid for reelection "in the event he sees proper to become our nominee for governor." The party was less willing to appoint an advisory committee "to meet the Republican executive committee to confer and decide on a joint slate for congressman, state senator, floater and representative." Those who opposed fusion declared it would kill the People's Party. They finally agreed to the substitution of the words "any other party" for "Republican," a ruse that fooled no one.[82]

The state meeting of the People's Party offered its support to Buchanan's candidacy without a single dissenting vote. Delegates selected presidential electors and endorsed the national candidates, General James B. Weaver for president and James B. Field for vice president. Weaver, the 1880 Greenback presidential candidate, a Union army general, commanded forces at Pulaski, Tennessee. James B. Field, a former Confederate major, although also called "General," had no previous ties with the agrarian movement.[83]

Though the convention reports offered no suggestion that Tennessee Populists objected to Weaver's nomination, his candidacy must have caused consternation among former Confederates. Democrats broadcast rumors tying him to mistreatment of prisoners and civilians under his jurisdiction and hinted at self aggrandizement during his command at Pulaski. His commander and even some under his command denied any high-handed and exploitive tactics, but the questions added another black mark to an already spotty Populist record.

The actions taken at the People's Party convention demonstrated both the limitations and the possibilities of the new political organization. Only weeks old, and faced with the prospect of creating legislative and congressional slates, the organization had little choice but to ratify existing candidacies for governor and president. Although party leaders could reasonably assess the desires of Alliance members who transferred their political loyalties to the new party, they had less experience with members from outside the agrarian organization. The influx of competing demands potentially threatened the party, although the convention harmony helped strengthen them for the inevitably bitter election campaign. The financial problems that plagued the Alliance transferred to the People's Party. The new political entity faced an immediate fight against a well-financed and well-organized party. Finally, the sectional and racial tensions represented in the selection of Weaver as the party's presidential candidate and the Alliance association with African American farmers confirmed the worst fears of the Bourbon Democrats and provided a wedge for separating Alliance members from the Populist Party.

While the Populists met in Nashville, the coal mines saw an escalation of labor violence. Miners at Tracy City in Grundy County grew angry after the Tennessee Coal and Iron Company laid off free miners and operated the mines with convict labor. On August 13, 1892, the miners confronted the superintendent and demanded equal working hours with the convicts. When the company rebuffed their demands, the miners left, but returned a short time later, burned the stockade, and sent the convicts to Nashville by train. Fifteen of the 362 convicts escaped during the ride to Nashville; one was killed and two others captured.[84]

No doubt remembering the attacks of his political enemies who condemned him for overstepping his authority in the first uprising, Buchanan refused to send the state militia until requested to do so by the appropriate county officials. This correct legal stance was not popular with voters. Moreover, it became a political liability when the miners repeated their actions at the Inman stockade, and tensions in Anderson County boiled over. On August 15, soldiers lynched a miner after a dispute at a local dance. The next day, an estimated one hundred miners attacked Fort Anderson, and a full-scale battle erupted. As news of the fighting spread, miners from across East Tennessee and Kentucky commandeered trains and rushed to the scene. The next attack came at Big Mount stockade, near Oliver Springs, where 1,500 miners overpowered the thirty soldiers and set fire to the structure.[85]

Pleading "nervous prostration," Buchanan remained in Nashville but urged the miners to exercise restraint. Nowhere did public opinion turn more decisively against the miners than in politically charged Nashville. Even as the People's Party convention met, two thousand "representative citizens" held a mass meeting at the Union Gospel Tabernacle (now the Ryman Auditorium) to address the trouble. Nashville Mayor George B. Guild chaired the meeting, but Arthur S. Colyar, former board member of the Tennessee Coal, Iron and Railroad Company, took the leading role. Guild accused the miners of "willingly and knowingly outrag[ing] the laws," and rejected the convict lease system as the immediate problem. "The question is," he demanded, "have we any law or a state?" Someone in the audience answered, "We have laws, but no Governor."[86]

Guild asked for no government assistance, assuring his audience that "we can and will put down that thing." His meaning immediately became clear when he essentially turned the meeting over to Colyar. In a speech with clear political implications, the iconoclastic Colyar painted a dark picture of "a calamity upon the state." He believed every man should feel disgraced by such a situation that had reached the point "where the people are willing to put a fool in the office of Governor." His remedy, which he believed "any man with sense enough to go in out of the rain would do," was to "get guns and shoot down the violators." Colyar informed the audience that he had called on Buchanan during the first day of trouble and advised him, "for God's sake go up there and kill a few men and come back and look up your statutes." He attributed "all this trouble . . . to a want of sense on the part of the Governor." The partisan crowd cheered as Colyar reached his finale: "I hope, for the memory of Tennesseans who sleep in their graves, that when this fellow gets out of the Governor's chair he will be rebuked, and every man who has been implicated in this outrage will be severely punished."[87]

The meeting ended with the ratification of a set of apparently predetermined resolutions that presented the state as "standing out before the entire American people in the craven attitude of surrender—a complete surrender—to an armed mob of a few hundred men." The "Indignity" such a situation produced resulted from the actions, or inaction, of the governor. The resolutions condemned Buchanan's refusal to strengthen the guard at Tracy City, his "temporizing policy," and the weakness of his call to "be patient a few days." The resolutions ended with no alternative provisions, but a repudiation of the governor and a "solemn protest against the people of Tennessee being held responsible in the eyes of the world for the calamity brought upon us and for the reproach which the Governor's course has brought upon us." Passage of the resolutions prompted calls for volunteers and promises of guns and cannon to "vindicate the majesty of the law" and throttle the "reckless insurrection . . . in its infancy."[88]

Although Nashvillians organized militias and commandeered arms, cooler heads prevailed, and the excited volunteers remained at home. Troops in East Tennessee arrested over five hundred miners and quickly restored order. The August 1892 uprising resulted in twenty-seven deaths and the incarceration of one scapegoat, labor leader D. B. Monroe. The miners failed to bring an immediate end to convict leasing, but the legislature refused to renew the leasing contract when it came up for renewal in 1895.[89]

This episode illustrates the political effect of the volatile labor problem on the 1892 campaign. Caught between the demands of the miners, whose votes he solicited as the friend of the working man, and the demands of the gubernatorial campaign under the disputed title of "Independent Democrat," Buchanan elected to restore order without acting decisively. It opened him to charges of treachery and indecision, and gave the Bourbon former Confederates an opportunity to question his manhood and his leadership. In the midst of excitement over the renewed miners' insurrection, the Populists ended their convention without the usual fanfare and public interest. The opportunity to regain the political momentum that fueled Buchanan's early spring campaign faded with the news from East Tennessee. The miners' war crippled Buchanan's campaign and affected Populist efforts in legislative and congressional canvasses.[90]

The final blow to the Populist hopes came only days before the November balloting, when the *Knoxville Journal* revealed the details of a Populist-Republican political fusion and the GOP contribution of $15,000 to John H. McDowell for support of Populist candidates. As the July Giles County People's Party resolution made clear, the fusion of Republican and Populist interests for political expediency had been discussed openly, and Tennessee Democrats worried that such an alliance would result in Democratic defeats in heretofore secure legislative and congressional districts. The union of interests between the two parties threatened to undo what the disfranchising laws had guaranteed. To prevent that outcome, Democrats fought for their political life.

The fusion politics of the Republicans and Populists was more limited than Democrats realized or perhaps cared to acknowledge. Democrats expected a last-minute Republican withdrawal of G. W. Winstead in favor of Buchanan, though the GOP held out for a minority victory like the one they achieved in 1880. Nor did the two parties agree on presidential candidates. The Republicans maintained their support for Benjamin Harrison, while the Populists rallied behind James G. Weaver. Neither party, however, denied fusion on congressional and legislative candidates. In districts where Republican strength offered a credible chance for election, Populists supported GOP candidates rather than field their own. Wherever Republicans stood little chance of success, they withdrew their candidates and supported the Populists. This expedient alliance offered the greatest opportunity to control the legislature and send like-minded representatives to Congress. Speculation ran rampant that Republican-Populist control of the general assembly would elevate John H. McDowell to the U.S. Senate.

On October 23, the Knoxville newspapers printed two letters that passed between Jonathan J. Ivins, editor of the *Knoxville Republican*, and D. H. Hill, Tennessee's Republican National Committeeman. The letters described the arrangement with the Populists in the fielding of candidates and the payment of $15,000 to McDowell. According to the letter writers, McDowell agreed to keep Buchanan in the governor's race to give Winstead a fighting chance in the November election. He also promised to support GOP congressional and legislative candidates and offer no opposition to the Harrison presidential ticket. If the Republicans and Populists gained control of the legislature, they would repeal the offensive election laws and elevate McDowell to the U.S. Senate. To make the cash-strapped Populist campaign viable, the National Republican Party provided McDowell with $15,000, some of which went to Buchanan's campaign.[91]

No one adequately explained how the newspaper obtained correspondence in which Ivins complained that "we find ourselves in with a man of insatiable greed, unblushing corruption, and the most determined political perfidy." He worried that "we, as Republicans, have agreed to put [McDowell] in the United States Senate, to the everlasting disgrace of Tennessee . . . and the utter ruin of Republican hopes in the state and the South." Hill agreed and speculated that the party had "now lost fifteen thousand dollars on McDowell and Buchanan and [had] gained nothing." Hill made clear the self-serving nature of the letters as he explained that the scandal could not injure John Houk or himself, since the two men were "let into it after the bargain was made."[92]

The best explanation for the release of the letters pointed to Republican factionalism between Houk and Evans supporters. John Houk took over the second congressional district party reins from his father, Leonidas Houk, at his death in 1891. The younger Houk operated the party machinery but commanded little of the power exercised by his father. In 1892, he faced a formidable challenge to his statewide control of federal patronage from H. Clay Evans, the third-district congressional candidate. More urbane, and with stronger national ties, Evans seemed the likely state leader if Harrison won reelection. Houk chafed at the knowledge that national party men provided financial support to the tight third-district race, while virtually ignoring his campaign (which Houk was expected to win handily). Houk's supporters frantically tried to defeat the Chattanooga "Carpetbagger." Houk's personal bitterness toward Evans and his frustration over the Populist congressional campaign of Frank P. Dickey (which Houk believed would split the Democrats and elect Evans) knew no bounds

The Houk camp remains the most probable source of the letters. Houk's close association with both Ivins and Hill and the clumsy attempt in the Hill letter to clear him of any complicity in the affair point to a political dirty trick. When Houk denied complicity, calling Ivins a "thief" and "traitor," the angry Hill issued a statement through the *Chattanooga Times*, in which he charged that "the express purpose of the publication [of the letters] was to injure H. Clay Evans." He maintained that "the letters were written with the full realization of the effect they would produce" after frequent consultations between Hill, Houk,

Table 2. Congressional Elections, 1892

District	*Democratic Vote*	*Republican Vote*	*Populist/Fusion Vote*
1	13,207	17,890	—
2	7,875	18,952	—
3	15,984	15,035	2,171
4	14,010	—	11,225
5	13,709	—	8,062
6	15,695	—	9,002
7	12,013	—	8,480
8	13,038	—	12,920
9	14,334	—	10,883
10	12,164	—	4,785

Source: "Congressional Directory," *Senate Journal*, 53rd Cong., 1893.

and Ivins. Whatever the source of the letters, Evans went down to defeat, but so too did Harrison, and the question of the distribution of patronage became a moot point for Republicans.

On election day, Peter Turney won with 126,348 votes to Winstead's 100,577 and Buchanan's 31,512. Nevertheless, the Democratic triumph carried a hint of defeat. The combined opposition of Winstead and Buchanan returned 5,741 more votes than Turney's total. Clearly, Bourbon Democracy faced sizable opposition from several sources, although historians point to the small Buchanan vote as indicative of the failure of Populism. The People's Party sent only six men to the general assembly, five representatives and one senator, a dramatic decline from the 54 Alliancemen of the previous legislature.[93]

In congressional races, Populists enjoyed a stronger showing. Not as tightly bound to the state political machines, congressional candidates demonstrated more independence in their actions. In the ninth district, congressman Rice Pierce bolted the Democratic Party and ran with the agrarians as an Independent. A strong showing by Populists in the eighth district almost defeated Benjamin Enloe, a powerful political figure in West Tennessee politics, who had previously claimed agrarian support. But voter turnout in the tenth district demonstrated the efficacy of the disfranchising laws and portended trouble for the Populists. Table 2 gives the voting tallies for the ten congressional districts in 1892.

Despite the state losses, the distribution of the vote for Buchanan suggested that Populism spoke for an important segment of the agricultural community. Buchanan garnered 11.9 percent of the vote statewide and did not gain a

Table 3. Counties Voting More than 11.9 Percent for Buchanan

Agricultural Category (crop value)	*No. Counties*	*No. Counties Voting >11.9%*	*No. Counties Voting >20%*
1. >$1 million	17	13	4
2. $500,000–$999,999	26	12	6
3. $250,000–$499,999	35	14	4
4. <$250,000	18	5	0

Source: U.S. Bureau of the Census, 1890 (11th) census.

majority in any county, but in forty-four of the ninety-six counties, Buchanan's percentage of the vote equaled or exceeded his statewide percentage. In fourteen of the forty-four counties, he received more than 20 percent of the vote. Table 3 gives the distribution of the Buchanan vote among the various agricultural categories.

The voting distribution suggests that the Alliance/Populist strength rested in the counties most affected by market forces. Having invested more heavily in tools, fertilizer, and improved seeds, these commercially oriented counties realized greater economic rewards. The insurgent agrarian message spoke to the farmers most concerned about the added risks of commercial farming as well as to their poorer neighbors whose access to more profitable agriculture was blocked. With a strong base among highly motivated farmers, the established record for agrarian perseverance, and the political savvy of John McDowell, Populists believed they could still prevail.

The Elections of 1894 and 1896: The Transformation of Tennessee Politics

Despite their confidence in future gains, Tennessee Populists had made their best showing. In the 1894 and 1896 campaigns, support for the Pops steadily declined. In a two-party state, farmers operated at a disadvantage; expelled from the Democratic Party, they could not effect statewide fusion with the factionalized Republican Party. Nevertheless, neither party could ignore the Populists entirely. The votes they pulled from the Democrats and Republicans were critical in the hotly contested races of 1894 and 1896. Both parties obliquely acknowledged the agrarian influence in their platforms by including currency, labor, and electoral planks straight out of the Wheel, Alliance, and Populist lists of demands. If the Populists could not elect men to office, they could, and did, transform the political dialogue.

In 1894, the Democrats ran Turney for a second term against the Republican nominee, H. Clay Evans, and the Populist candidate, A. L. Mims. Mims, who also served as state president of the Farmers' Alliance, headed a party deeply divided between the supporters of L. K. Taylor, John McDowell, and John Buchanan. Together, the three represented the full spectrum of Alliance and Populist ideology: Taylor, the early advocate for the People's Party, was an avowed enemy of McDowell, publicly characterizing him as "a moral coward, a spiteful villain, a treacherous wretch, a venomous reptile, and a voluminous liar." McDowell, the reluctant Populist, had embraced the third party emphatically by 1894 and worked behind the scenes as the would-be kingmaker. Buchanan still had not openly declared for Populism, an oversight that placed the former governor in opposition to McDowell, who dismissed the Buchanan faction as "the independent Democratic movement that was intended to overshadow, swallow up and obliterate the Populist party."[94]

Despite internal divisions, the Populists exuded confidence. Their sanguine expectations rested on potential Democratic weaknesses and a hoped-for fusion with Republicans rather than their own strengths. A number of New South candidates threatened to split the Democratic Party in order to deny Turney a second term. Moreover, the effects of the national economic depression (1893–1897) and the fallout from the congressional battle over the repeal of the Sherman Silver Purchase Act had forced many Democrats to reevaluate their position vis-à-vis the national party and President Grover Cleveland. Even the venerable Bourbon Isham Harris was leaning toward support of silver. Although Republicans tightly controlled the eastern counties and tolerated no Populist incursions, they had extended an opportunity for fusion in the statewide Supreme Court elections scheduled for the summer. Nominating only four candidates for the five positions, they effectively invited the Populists to join ranks with a fifth candidate of their own. It was a small concession, but one filled with possibility for future alignment if the Populists demonstrated sufficient electoral strength.

The platform adopted by the Populist delegates reflected the ideology of their national organization while also speaking to disaffected Tennessee Democrats and holding out the possibility for fusion with state Republicans. It supported their national party's currency and tax demands and characterized Cleveland as a pawn of Wall Street, a position that reflected both Populist sentiment and the views of silverites in the Democratic Party. The platform condemned the Tennessee Democrats for passage of $600,000 in penitentiary bonds, one of the primary sources of division between Turney and anti-Turney Democrats. In another nod to Jeffersonian and Jacksonian Democracy, the platform advocated strict economy in state government. Coming late to the table but now firmly committed to the position, Populists offered several planks that fit well with Republican views: abolition of the convict lease; repeal of the poll tax, the uniform ballot law, and the Dortch Law; and the elimination of double taxation through the so-called back tax law.[95]

By the time Democrats convened their gubernatorial nominating convention on August 15, the New South contenders for the nomination—James Head, Luke E. Wright, and Jere Baxter—had retired from the field. As Hart pointed out, the agrarian insurgency retrained the Democratic Party to settle its disputes outside public view. Democrats took less than six hours to renominate Peter Turney and agree on a platform, having worked out their differences ahead of time. Turney won nomination by unanimous acclamation on the first ballot. Praising "Ole Pete" for his possession of those "high qualities of American manhood that have made glorious the history of this republic," Edward Carmack conflated history to put Turney in the pivotal seat during the Convict Lease War and assure his audience that "[h]is very presence upon the scene of action stilled the storm and calmed the tumult of disorder." Carmack's speech and Turney's acceptance reinvigorated the assembled Democrats who responded with cheers and hat-waving enthusiasm.[96]

Because Democrats could count votes as well as the Populists and Republicans, they made concessions to the agrarian insurgents by incorporating farm and labor demands into their 1894 platform. After acknowledging allegiance to traditional Democratic principles and praising the Cleveland administration, they condemned trusts and monopolies and endorsed a federal income tax. The platform also supported the right of workers to combine for their protection and backed legislation to prohibit importation of foreign workers. The most contentious plank of the platform called for bimetallism and "the coinage by the United States, without reference to the policy of other nations, of both gold and silver in such manner as will maintain both metals in circulation at parity"—a troubling shift for Cleveland supporters and New South businessmen. On the one hand, the focus on national issues provided Tennessee Democrats with a low-risk platform to entice disaffected Pops back to the party of the fathers. Viewed from a different perspective, it foreshadowed the national conflict of 1896.[97]

Like the Democrats, the Republicans quickly set aside their factionalism to rally behind the candidacy of H. Clay Evans, the East Tennessee industrialist and former congressman. With promises that he could rally the party throughout the state and draw Populists and disaffected Democrats to the GOP banner, the Pennsylvania-born former Union soldier received a majority on the first ballot and became the Republican nominee.[98]

Evans and the Republicans mounted a vigorous campaign. Knowing that Populist insurgency and the failures of the Turney administration had weakened the Democratic Party, they anticipated a rare GOP victory. Evans presented himself as a hardworking, successful businessman with considerable political experience. He was backed by a Republican platform that generally upheld the party principles, but, like the Democrats, attempted to woo the disaffected with support for restricted immigration and bimetallism. In addition, the Republicans continued their strong condemnation of the 1889 election laws, especially the poll tax and the Dortch law. Finally, Evans promoted his own youth and vigor as a stark contrast to the enfeebled Turney. Indeed, "Ole Pete" made only one speech early in the campaign before abandoning the canvass to the more energetic Edward Carmack, who championed the Democratic cause.[99]

Each candidate canvassed the state separately. Evans conducted the most issue-oriented campaign, expressing sympathy for the suffering of wage earners displaced by the depression and promising an administration based on business principles. While acknowledging that the "condition of trade and commerce throughout the country, the dead furnaces, the idle factories and closed mines" warranted concern, he believed that thoughtful and careful investigation would provide "a remedy that may light the fires in these dead furnaces, set the wheels in the factories in motion, open the closed mines, restore trade and commerce to its accustomed channels, and thus find employment for the thousands of men who are being impoverished by idleness." He overturned the Alliance/Populist currency demand, asserting that banks had money to invest but refused to do so out of distrust for Democratic financial policies. And he claimed that the Democratic assumption that low prices resulted from overproduction would not withstand close scrutiny. The real culprit that destroyed prosperity was underconsumption by the "great army of wage earners [who] have been and are consuming less because their purchasing capacity (earnings) have been and are being reduced." Evans allied himself with Populist views when appropriate and used the insurgents' own words to support his position when he disagreed.[100]

Mims recognized the futility of the Populist canvass, although he conducted an active speaking tour. His speeches on free silver and corporate greed were generally treated with respect by the press, which praised his "clear-cut impressive style of speaking," although some made his education background a political liability by labeling him an impractical bookworm. The farmer-professor-politician gamely traversed the state canvassing for votes that were expected to hurt the Democratic chances for retaining the governorship more than the

Republican chances of gaining it. As the Gainesboro *Gazette* warned, "A vote for Mimms [sic] is half a vote for Evans."[101]

The election generated a new controversy that did nothing to enhance Populism's weak political position. By November 10, three days after the balloting, unofficial returns gave Evans the victory by a scant 750 votes and showed Republican gains in nearly every county. Democrats conceded nothing, however, as party elites across the state scrambled to control the situation. Jubilation in the Republican camp was short-lived when the secretary of state finally offered his official vote count: 104,356 votes for Turney, and 105,104 for Evans. The Republicans appeared to have won the election by a mere 748 votes.

Charges of voter fraud and court challenges by the Democrats resulted in a legislative investigation that decided questionable ballots along party lines. The majority report placed the final vote tally at 94,794 for Peter Turney, 92,440 for H. Clay Evans, and 23,088 for A. L. Mims. Peter Turney was the victor by 2,354 votes. As one historian noted, "When all [was] said and done, *power politics prevailed*." Evans's martyrdom got national attention and generated short-lived speculation about his elevation to the vice presidency in 1896. In their future campaign platforms, Populists and Republicans reminded voters of the miscarriage of justice and used Turney's manipulation of the vote and his second term as lessons in Democratic corruption.[102]

There was little cause for jubilation in the Populist camp. Mims's 23,092 votes amounted to less than three-fourths of Buchanan's tally in the 1892 campaign. In addition, only four Populists won seats in the general assembly. As in the previous election, Populist strength centered in those commercially minded agricultural counties in West Tennessee and in the tobacco counties of upper Middle Tennessee. Once again, however, the Populists acted as the spoiler, clearly taking votes from the major parties rather than winning positions from which to advance their demands.[103]

Tennessee voters again crossed party lines to vote for Populist demands on national issues. East Tennessee, a region where GOP leaders had successfully kept both the Alliance and the Populists at bay, remained staunchly Republican. Republican candidates won in the first three districts, and Democrat Benton McMillin retained his seat in the fourth, where the Alliance and Populism also made few inroads. Populists showed greater voting strength in three of the remaining six congressional districts. In the fifth district, James Richardson, a perennial congressman from Murfreesboro and an Alliance member, won his seat after a strong challenge from Populist W. E. Erwin. Joseph E. Washington, sixth-district congressman and a nominal Alliance member, also won his race, but the combined Republican and Populist votes suggested future problems

Table 4. Congressional Elections, 1894

District	*Democratic Vote*	*Republican Vote*	*Populist Vote*
1	8,542	18,017	2,622*
2	13,191	16,215	638
3	13,947	17,019	1,669**
4	11,965	10,582	—
5	11,440	320	9,543
6	11,234	4,798	4,783
7	9,098	6,366	1,844
8	12,161	13,071	—
9	10,634	—	7,983
10	6,654	1,955	1,454

**Populist/Prohibition fusion candidate* ***Prohibition candidate*

Source: "Congressional Directory," *Senate Journal,* 54th Cong., 1895, serial 3361.

for the Democrat. In the ninth congressional district, James C. McDearmon who had won his seat from Democrat-turned-Independent Rice Pierce, faced a strong challenge from Atwood Pierson, an Allianceman who had moved into the People's Party. Republicans won in the eighth district with John E. McCall defeating B. A. Enloe, the Populists having decided to abstain from nominating a candidate. The tenth district, where disfranchising laws had the greatest effect, demonstrated the relative weakness of both the GOP and the Populist Party. Table 4 provides a breakdown of the 1894 congressional vote.[104]

After early success as infiltrators in the Democratic Party, farmers were forced out into a weaker People's Party and suffered the fate of all third-party efforts. Many farmers, who had supported the nonpartisan, antipartyism of the Alliance, refused to countenance Populism. Although the Republicans tossed electoral sops to the Populists, their own factionalism and desire for unencumbered state control operated against a thoroughgoing fusion. In the congressional districts of Middle and West Tennessee, however, where Republicans held little expectation of winning office, fusion candidates were stronger. An analysis of the congressional races demonstrates both the continued influence of the agrarian insurgents and their pragmatism in establishing coalitions.

The bitterness of the 1894 election should not obscure important shifts in partisan positions. Thoughtful Democrats had long recognized the limitations of the Turney governorship. Early in the 1894 campaign, the Clarksville *Tobacco Leaf-Chronicle* characterized the Democratic platform as "little more than senitiment [sic]" and suggested that Turney "did not possess sufficient of the live business intentions that should have characterized governmental affairs of a progressive state like Tennessee." A weak governor elevated to office under suspicious circumstances, Turney exercised little control over the state or his party. Despite titular Bourbon leadership, the Tennessee Democratic Party moved rapidly toward free silver and other agrarian demands.[105]

The deepening economic crisis also generated a split in the national Democratic Party between supporters of Grover Cleveland's "sound-money" policies and a growing number of dissenters who championed bimetallism. The repeal of the 1890 Sherman Silver Purchase Act provided the first battleground. But the Memphis Silver Convention in June 1895 produced the infrastructure for capturing the 1896 Democratic Presidential Convention and the nomination of William Jennings Bryan.

Ironically, the Sound Money League of the tenth congressional district set the wheels in motion for the organization of a silver convention in Memphis. Meeting in May 1895, these West Tennessee gold standard Democrats invited several nationally acclaimed sound-money men to regale the crowd with speeches on the merits of maintaining a gold standard. Not to be outdone, supporters of silver organized the Central Bimetallic League of Shelby County and called for a June conference. Some League members favored a nonpartisan meeting that would include Populists and Silver Republicans. Those with visions of a larger conference gained control, but not without the adoption of a disclaimer that the convention represented the movement toward the creation or sanction of a third party.[106]

When the bimetallists convened on June 12, 1895, 2,207 delegates represented twenty-one states and all three parties as well as some nonpartisan silverites. The Tennessee delegation displayed an array of newly converted Bourbon, New South, and Independent Democrats, as well as Republicans and Populists. Both of Tennessee's Bourbon senators, William B. Bate and Isham Harris—who played a central role both in organizing the convention and creating the machinery for capturing control of the 1896 national convention—attended. E. W. Carmack, editor of the *Commercial Appeal*, Eth B. Wade, former Allianceman and member of the Buchanan administration, Rice Pierce, the former Democratic Congressman from the ninth district, John P. Buchanan, and John H. McDowell joined hands to support silver. Such nationally known figures as Ben Tillman of South Carolina, Marion Butler of North Carolina, Joseph C. Silbey of Pennsylvania, William M. Stewart of Nevada, J. K. Jones of Arkansas, J. Z. George of Mississippi, and William Jennings Bryan of Nebraska played active roles, chairing committees and whipping up enthusiasm with fiery, crowd-pleasing speeches.[107]

The convention adopted a platform calling for the coinage of silver at 16:1, a bimetallic standard that had behind it "the experience of the ages." The position was acceptable to silverites in all parties since each had been represented in the deliberations of the resolutions committee. Having been assured that Populists would have representatives on the resolutions committee but finding none named, John McDowell caused a brief stir by challenging the committee's composition. Eight Populists were quickly added to the committee, including two from Tennessee, McDowell and B. G. West, a Memphis cotton warehouseman who had been one of the founding members of the Farmers' Alliance Cotton Bureau. These first steps toward marshaling the silver sentiment produced a formidable organization within the national party and ultimately resulted in denying Cleveland another term.[108]

A year later, when the Democrats met in Chicago to nominate their presidential candidate, the worst fears of antifusion Populists were realized. Silverites carried the day and nominated William Jennings Bryan following his galvanizing "Cross of Gold" speech. But, selecting Arthur Sewall, a conservative banker from Maine, as his running mate, the Democrats signaled their commitment to hard currency. As Lawrence Goodwyn noted, "Greenback principles were anathema to the new crusade." The nomination of the popular, young, silverite shattered the political dreams of Populists, whose expectations of riding the issue of currency reform to victory faded.[109]

The Populist convention immediately followed that of the Democrats. With the Democrats having stolen the Populist thunder, many saw fusion as their only recourse. Southern Populists raised the strongest protests against fusion, citing their experiences with fraud, intimidation, and election-day violence—the standards of the Democratic Party in the South. Goodwyn suggests that the undemocratic actions of the South, having denied the election of credible, "big-name" Populists, ultimately left the convention at the mercy of the western fusionists. After days of spirited debate, chicanery, and deplorable fist fights, the fusionists won—William Jennings Bryan received the Populist nomination, with Tom Watson as the People's Party choice for vice president. As Watson feared, fusion would prove to be the undoing of the party, a case of the Populists' "Jonah" to the Democratic "whale."[110]

Among Tennessee's Populists, the decision of the national convention created a variety of conflicting options. Silver Democrats enthusiastically rallied behind Bryan and embraced their dissenting brothers, encouraging them to reunite with the party as the best opportunity to restore economic stability and achieve the goals of the Alliance and People's Party. In Shelby County, Populists joined their old nemesis, Edward W. Carmack, who led the silver battle against Josiah Patterson for the Democratic nomination for the congressional seat in the tenth district. Carmack openly solicited Populist votes, claiming that he was "ready and willing to swap any millionaire robber who would earn his living in the sweat of another man's face . . . for the wildest haired Populist that ever swept the plains of Kansas." Agrarian dissenters joyfully accepted the invitation to board the Carmack bandwagon. Even an appeal from McDowell failed to dissuade the West Tennessee Populists in their headlong rush back to the Democratic fold.[111]

Antifusion represented the minority view, and even McDowell claimed that the Bryan/Watson ticket was compatible with Populism's goals. Nevertheless, McDowell's position was not as clear as his assertion suggested: he simultaneously worked to retain his fusion ties with Tennessee Republicans; and he himself ran for the ninth-district congressional seat as a Populist (with Republican support) against Rice Pierce, the Silver Democrat. To complicate matters further, A. L. Mims, again the Populist candidate for governor, presided over a meeting with Bryan in early October, which resulted in the Davidson County Populist withdrawal of support for the third party's national ticket in favor of Bryan/Sewall. Across the state, Populists rushed to abandon Watson and join the Democratic boom. Support for the gold standard Republicans, who had proved reluctant fusionists in previous elections, became increasingly untenable. And, as McDowell feared, fusion with the silverites in the Democratic Party amounted to absorption. The debacle of 1896 proved the undoing of the Populist Party. Their strength in Tennessee had always been in rallying voters for support of national issues. Once the Democrats recaptured those issues, Populism was doomed.[112]

Democrats completed their sweep of Populism by returning to Robert L. Taylor as their gubernatorial candidate. Claiming that he had been out of the state during the 1894 Turney election debacle, Taylor avoided being painted with the broad brush of corruption that amounted to the Republicans' principal hope of victory. On election day, Taylor garnered 156,333 votes to Republican George N. Tillman's 149,374 and A. L. Mims's 11,966 votes. In the general assembly, Republicans were reduced to eight senators and twenty-five representatives, with the Populists claiming only four members of the house. As he had done in 1889, Governor Taylor shored up white, elite Democratic control by tightening and extending the reach of the state's election laws. By the early twentieth century, registration applied in all counties, and primary elections ensured the dominance of the entrenched leadership. Increasingly the poll tax, though enacted to remove African American suffrage, operated as Andrew Johnson had feared to limit the voting of poor white farmers. Robert L. Taylor, the so-called

Table 5. Congressional Elections, 1896

District	*Democratic Vote*	*Republican Vote*	*Populist Vote*
1	13,916	25,075	—
2	9,448	28,112	—
3	19,498*	17,716	133
4	18,070**	12,269	—
5	16,070	9,000	2,384
6	17,537*	12,102	661
7	15,434	10,744	1,795
8	16,568	13,619	1,130
9	19,138	—	10,714
10	11,024**	—	926
	10,556***		

**Silver Democrats* ***Democratic/Populist Fusion* ****Gold Democrat*

Source: "Congressional Directory," *Senate Journal*, 55th Cong., 1897, serial 3558.

friend of agricultural interests, had presided over most of the disfranchising efforts.[113]

In the congressional races, Populists clearly shifted their votes to the Silver Democrats, as table 5 demonstrates.

The effects of the Silver Democratic campaign were clearest in the fifth and sixth districts, where the Populist/Republican fusion broke down, drawing votes from the third-party candidates back to the Democrats. The third and the ninth districts were the most interesting for the Populists. In the third district, Populists convened after the Democratic nomination of Judge John Austin Moon and voted to support him in preference to naming a candidate of their own. Although the Democrats eschewed formal alliance with the Populists, Moon apparently retained third-party support, beating out his Republican opponent in the Evans/Sanders backyard. In the ninth district McDowell managed to keep the Populist/Republican fusion intact, though he lost to a much stronger Rice Pierce, who seemingly picked up a number of third-party ballots. The most controversial race took place in the tenth district, where a victorious Carmack squeezed by Josiah Patterson by a mere 468 votes.

The voting patterns in the three state and congressional elections conducted between 1892 and 1896 offer important clues to understanding who voted for the Populists and why. The pattern of support for Populist candidates for governor suggests that Populism resonated strongest in counties where commercial agriculture tied farmers to national and international markets.

Access to transportation, fertile soil, and progressive town development characterized these counties. For commercial-minded farmers, the Alliance and Populist organizations offered opportunities to expand their access to capital and control prices.

The presence of two active political parties limited the possibilities for the successful development of a third party. At the same time, the internal politics of both the Democratic and Republican parties denied the Populists the possibility of effective fusion. Democrats resisted Populism out-of-hand until the 1896 Presidential election. Republicans supported fusion only in areas of relative GOP weakness, but even there, Republican factionalism over patronage generated suspicion and limited the effect of fusion. In East Tennessee, where the Republican Party dominated elective offices, the party's opposition to disfranchisement and convict leasing as well as limited commercial agriculture undermined Wheel and Alliance organizing efforts, and hence support for the Populist Party.

Nevertheless, Tennessee's Populists did not pass from the stage unnoticed. Politicians of the era worried about Populist influence long after the party ceased to pose a threat. As late as 1904, the State Democratic Executive Committee polled county party leaders to assess the strength of Populist support. As Edward Carmack rose through the national political ranks, he attributed every setback to the work of the "Pops." The wool-hat boys had upset the political applecart in the 1890s, even if they could not claim electoral victories, and Democrats continued to keep a sharp eye out for agrarian insurgency.[114]

One by one, however, Populist leaders returned to the political fold. John McDowell became active in Confederate Veterans organizations, rising to the rank of major general. He returned to political life in 1905 when he was elected to the Tennessee General Assembly from Obion County. Remaining true to his earlier advocacy for Prohibition, McDowell campaigned with Carmack in the early twentieth-century efforts to end the manufacture and sale of alcohol in Tennessee. Ever the renegade reformer, however, the old political war horse also joined the Progressive Party for the 1912 presidential campaign. Seconding Teddy's Roosevelt's nomination, McDowell returned to an old theme: the need to set aside the wartime history that continued to divide former enemies, confidently proclaiming that "Col. Roosevelt is the only man in the country who can wipe out the Mason-Dixon line and make the North and South one people."[115]

Likewise, John Buchanan returned to the political scene, although his activities were limited to that of "elder statesman." His appearances with Benton McMillin and Malcolm Patterson (son of Josiah Patterson) suggest again that

Democrats needed the symbolic support of the old agrarian radicals. Moreover, Buchanan's backing of the son of his Bourbon enemy and McDowell's simultaneous efforts on behalf of Progressive candidates raise questions about the divisions within the old Populist camp. Although Buchanan frequently mouthed the Populist mantra, he perhaps remained closer to the Bourbon fold than appearances suggest.[116]

The decline of Populism after 1896 rested on a number of related political and economic factors—divisions within the People's Party that reinforced structural and fiscal weaknesses, the failure of both the Populists and the Republicans to effect a thoroughgoing fusion, and the unity of the Democratic Party in its determination to overcome the agrarian movement. The failure to capture political control did not end agrarian influence. Indeed, even as they fought their political battles, Tennessee's agrarian leaders laid the foundation for maintaining their clout in the emerging progressive bureaucracy.

If the future of our State is to reward the labors and sacrifices of the past, and the hopes and aspirations of the present, it must be placed and kept in touch with the conditions of modern progress. —EDWARD W. CARMACK, 1897

CHAPTER SIX
UP FROM THE MUDSILLS OF HELL
FARMERS IN THE PROGRESSIVE ERA

The political upheavals of the 1890s and the economic depression that overlapped them fractured whatever unity farmers had claimed through the Alliance and Populism. Convinced that mass movements offered no gains, farmers never again attempted to organize a party based on agrarianism, and many remained skeptical of all political action. The success of the Democratic Party in ousting the radical insurgents did not address the fundamental problems that produced the Wheel, Alliance, and People's Party. Hard times worsened with the 1893–1897 depression, and even substantial planters were at the mercy of the new corporate monopolies that dominated agriculture in the early twentieth century. Once again, farmers organized, but the rise in tenancy, the more precarious status of landholders, and the unwillingness to countenance any suggestion of mass agrarian solidarity left such organizations weaker.

There was a more fundamental difference between the new agrarian associations and the old ones that had important social consequences and ultimately affected agriculture economically as well. The world envisioned by nineteenth-century agrarians was more democratic, less hierarchical, and less competitive than what emerged with the triumph of industrial capitalism. Farmers had accepted the fact of industrialization but had organized to ensure that men could choose to engage in agriculture and live a productive life; agriculture would operate on an equal footing with industry in the world they desired. Moreover, farming would not be profitable only for the well-capitalized few but would constitute a livelihood and a way of life for any man with land, tools, seed, and skill.

That goal required the broadest organization of farmers into cooperative associations for buying and marketing. It also required that government mediate between industrial and agricultural interests. Mass organization to achieve those goals demanded that farmers set aside many of the issues that divided them—race, a bitter Civil War history, religious differences, gender, and class. The political defeat of the agrarian insurgency destroyed the tentative bridges that made stickability possible.

Destroying the mass movement did not resolve the agrarian crisis and did not remove the need to organize. As a result, in the early twentieth century, many farmers organized more narrowly as cotton farmers, tobacco planters, and peanut growers. They lost their earlier visionary agenda and concentrated on immediate need. Bound together by an economic or productive commonality, they had no need to negotiate social or historic differences. Rather, they demanded absolute conformity to community standards in order to shore up their fears and create a unified front. Their associations were more rigid and their actions more violent, as in the cases of Black Patch and Reelfoot Lake actions.

The Farmers' Union was the exception to the limited organizations favored by many farmers. Organized with practical and visionary goals, it sought both professionalism of agriculture and rural uplift. Dissuaded from mass political movement, Union members allied with Tennessee's new Agricultural Bureau and the emerging extension programs. Southern bureaucracies and institutions were not yet as professionally staffed as they would later become. In the first decade of the twentieth century, the door was still open for men with practical experience and determination. The Union men worked in and with the bureau, the experiment station, and the University of Tennessee to make farming successful for the small producer. They staffed the agency offices, maintained demonstration plots, served as county agents, and advocated good roads and improved education. Not willing to be motivated by profit alone, they endeavored to apply the Golden Rule to their personal and business practices. The moment for the Union men to act was very brief. By the 1920s, credentialed professionals filled the positions they claimed. Nevertheless, the farmers who had struggled for more than forty years prevailed, as they helped to shape the legislation that defined twentieth-century agriculture, monetary, and regulatory policy.

The foundations for the transformation of Tennessee agriculture and the shift from farmer interest in the canvass to a more bureaucratic and institutional approach began during the administration of Governor John P. Buchanan. The separation of the Bureau of Agriculture from the mining and labor interests freed the agency to pursue issues important to farmers. It was a transformation farmers had long demanded, and the vitality of the bureau was immediately evident. Biennial reports to the general assembly suggested that successive commissioners engaged in increasingly complex regulatory and educational programs that targeted small commercial farms rather than the planters of the past. Forging liaisons with the USDA, the University of Tennessee, and regional industries, these commissioners promoted cooperative marketing and improvements in farming methods. From the outset, agrarian leaders with Wheel, Alliance, and Populist ties contributed to this transformation by offering their services as organizers and their experience as progressive farmers.

In 1892, Commissioner D. G. Godwin set the tone for the new agency by launching a statewide effort to establish Farmers' Institutes modeled on the twenty-year success of the East Tennessee Farmers' Convention. The Allianceman and commissioner captured the convention from the "better men" and brought the institutes home to small farmers in Middle and West Tennessee. Claiming a desire to bring the Bureau of Agriculture in closer contact with the state's farmers, Godwin and his assistants attended fairs and stock shows to promote the organization of county and regional Farmers' Institutes. Organizing without a legislative appropriation to sustain the work, Godwin expressed the hope that "the seed thus sown would bear fruit in the future."[1]

Two years later, under Commissioner T. F. Allison, the Bureau of Agriculture established agricultural institutes in all three grand divisions. East and West Tennessee held annual conventions, while Middle Tennessee counties divided into two groups. In 1893, Knoxville hosted the eastern institute, Jackson held the western meeting, and McMinnville and Pulaski were the sites of the middle institutes. Like the earlier East Tennessee Farmers' Convention, the institutes built on a county structure, encouraged the adoption of modern agricultural practices, and promoted progressive rural social development. In many counties Wheelers and Alliancemen shifted their allegiance to local institutes as the more radical organizations declined.[2]

The institutes held three-day sessions during which prominent local farmers, experiment station staff, and university faculty presented papers outlining their experiences and research. Cotton and tobacco continued to appear on the list of topics, but the institutes focused on a more diversified agriculture that included horticulture, livestock, poultry, and grasses and encouraged the establishment of associations of like-minded producers in each of the specialty areas. Farmers also discussed the need for improving farm-to-market roads and debated the fence laws. In 1899, the general assembly appropriated a biennial allotment of $5,000 to support the work of the Farmers' Institutes. By that time the Farmers' Institutes were riddled with former Alliancemen and Populists, including John McDowell, J. M. Coulter, T. B. Harwell, and John P. Buchanan.[3]

By 1900, Tennessee's agrarian radicals and Populists no longer posed a threat to either political party, although unsuccessful Democratic candidates frequently attributed their defeats to the "Pops." Overcoming political radicalism, however, provided no solutions to the market problems that had fueled earlier agrarian insurgency. Indeed, the agricultural crisis worsened for Tennessee's commercial-minded farmers, ensnaring big planters as well as smaller producers as advantaged corporations engaged in a second and more extensive wave of mergers that further limited market options.

As in previous decades, farmers met the challenges through organization. Utilizing the familiar rhetoric of community, producerism, and Jeffersonian agrarianism, the new organizations nevertheless departed from the Grange, the Wheel, and the Alliance in important ways. More narrowly focused than earlier farmer organizations, these unions sought limited objectives and eschewed political activism. Defined by geography and commodity, the newly organized efforts to seek economic redress operated within the context of a national rhetoric of reform historians have identified as Progressivism.

Adopting the language and ideology of science and efficiency, Progressive agrarians displayed a comfort with the emerging bureaucratic institutions of government that separated many rural leaders from their followers. The new reformers relocated the arena of conflict from the electoral canvass and the stump speech to the court and legislative investigative committees, where men based their decisions on the arguments made by those who possessed expertise in scientific agriculture, finance, and the law.

Not all agrarian leaders made the transition, of course. Some would-be reformers found themselves on the margins of change, followers rather than leaders. Others outright rejected Progressive bureaucracy and fought harder to protect rural community life against perceived threats posed by industrial and urban growth. Perhaps, as a result, the agrarian movement of the first decade of the twentieth century was more violent than the earlier Grange, the Wheel, and the Alliance movement.

Three examples of agrarian activism marked this period. First, in the tobacco region, elite planters organized resistance to the monopolistic buying practices of the American Tobacco Company (ATC) and the Italian Regie. Launched as a cooperative effort to control the tobacco crop, the Planters' Protective Association straddled the line between modern agricultural practices and traditional community rights and was marked by intimidation and violence. In the second example, poorer farmers and landless whites in Obion County also turned to night riding after failing in their legal efforts to stave off attempts to end the common rights of settlers around Reelfoot Lake. In the last example of rural activism, middling-level cotton producers in West Tennessee organized rural cooperatives that adopted business methods and integrated into the emerging bureaucratic institutions of the state and nation.

Taken together, these independent examples of renewed agrarian action demonstrate the continuity of agricultural concerns. Like Sisyphus with his stone, farmers confronted the recurring problems of dropping prices and the need for more capital—a year or two of improved prices was inevitably followed by crop failure, market collapse, or some other natural or manmade disaster. But farmers continued to press for reforms to improve their opportunities to engage in commercial agriculture and provide social justice to men and women living in the country.

Two important changes separated these twentieth-century farmers from their predecessors. First, confrontations were more violent than had been true during the Alliance and Populist insurgency. The gulf between modern business practices and traditional exchange economies widened visibly after the depression of 1893. With fewer options, many farmers reacted with arson, night riding, and murder in an effort to maintain, or reinstitute, traditional social and economic practices. Tennessee had a well-earned reputation for violence, but the pressures of the early twentieth century produced a level of violence unusual even for the Volunteer State.

Second, a well-known cadre of former Grangers, Wheelers, Alliancemen, and Populists charted a different, less confrontational route to rural reform. These familiar agricultural leaders quietly returned to political office, organized the Farmers' Union, testified before congressional committees, and inserted themselves in the emerging state agricultural bureaucracy. They supported Progressive rural reform and laid the foundations for twentieth-century agricultural policy.

Farmers at the Century's End: The Promise and the Reality

Tennesseans ended the nineteenth century on an upbeat note despite the economic and agricultural depression that gripped the state and the nation. The year 1896 marked the centennial of the state's entry into the Union, and New

South boosters battled the perennially recalcitrant state legislators to celebrate the event with an exposition designed to attract potential investors and create short-term profits for hard-pressed railroads. The habitual hostility of the general assembly toward any progressive initiatives, the competition among the three grand divisions for the honor of hosting the event, and the pecuniary embarrassment of the state's coffers prevented an 1896 celebration funded by the state. Unwilling to be deterred, Nashville businessmen, with the help of the Louisville and Nashville Railroad, organized a centennial committee and raised the necessary funds to stage the event. When the Tennessee Centennial Exposition opened in May 1897, the stylized exhibits reflected a sanitized, harmonious vision of the past and an optimistic, progressive view of the future.[4]

Although the wealthy industrialists and railroad men of the New South controlled every aspect of the celebration, at least one well-known Populist made it into the ranks of the Centennial Executive Committee: John H. McDowell. As the former state president of the Tennessee Farmers' Alliance and the editor of the *Toiler*, McDowell's presence among the "best men" of the state reflected a recognition both of his ongoing power in agricultural and political circles and his longstanding advocacy for economic growth in business, industry, and agriculture. Although McDowell viewed currency and credit reforms as more important for agricultural advancement, he had always extolled the merits of scientific agriculture and had never repudiated business development. As he campaigned against monopolistic practices, McDowell recognized the complexity of sustaining family farming in an age of corporate expansion. He actively supported the opportunity to present the state at its best and thereby induce capital investment. Economic growth for McDowell represented new markets for agricultural products.[5]

In 1896, the Union City *Independent* published a letter from McDowell soliciting a $1,500 contribution for an exhibit extolling the county's merits; McDowell had turned to the newspaper after the Obion County Court refused his appeal on the grounds the money might wind up in a general fund rather than be used to promote local interests. The editor's letter reassured skeptics that the funding for the construction of exhibit buildings had already been secured, and the county's money would be used as the citizens directed. Like many Progressive businessmen, McDowell recognized the importance of attracting capital investment to the cash-poor South. Fairs and expositions figured prominently in most campaigns for economic growth by offering a cost-efficient means of advertising local advantages to a wide audience.[6]

McDowell argued that support for the exposition made good business sense, and Obion County farmers and businessmen need look no farther than

Georgia to recognize the benefits. Invoking the "Spirit of Atlanta," he claimed that the 1881 and 1892 cotton expositions in that city had attracted "thousands of enterprising capitalists whose investments amount to over fifty millions of dollars." Pointing to the recent construction of a cotton mill in Dalton, Georgia, McDowell asserted that "the great West has filled up, and the tide of emigration and capital . . . is now turning South." Mixing his metaphors, he claimed that those progressive states ready to act would "catch the tide and reap the benefits." While McDowell believed townsmen would readily understand the potential, he specifically directed his argument to farmers, writing that the construction of factories in small towns like Union City would increase farm income through consumer demand for dairy, poultry, and garden products.[7]

When the fair finally opened almost a year after the state anniversary, agriculture played a prominent role. The exposition's historian described the Agricultural Building as "the most ornate" of the exhibit halls. Over five hundred feet long, the building was also among the largest. Built in Renaissance style, it featured four small domes surrounding a central dome of gold-colored glasslike substance that filtered light into the display halls. From a commanding position at the north end of the grounds, the Agricultural Building could be seen from anywhere on the site and attracted many visitors.[8]

In keeping with the objectives of the promoters and the realities of Tennessee agriculture, the exhibits featured the state's principal commodities—corn, cotton, tobacco, grains, and grasses—in a well-designed display of crop diversity. "Practical object lessons" in the form of demonstration fields of cotton, tobacco, and hay supplemented the interior exhibits. Visitors saw the possibilities for increasing their production through the use of improved agricultural methods as well as the expanding uses of farm products. In addition to the exhibits available in the Agricultural Building, railroads displayed farm products in the Transportation Building. In both cases the exhibitors emphasized technology and diversity as the hallmarks of progressive farming. Congressman Edward W. Carmack's Inaugural Day speech neatly summarized the message of the agricultural exhibits: "We can not sow the old and reap the harvest of the future."[9]

Carmack's admonition provided a foretaste of the century at hand. Tennessee farmers faced a new day that demanded new technology, new marketing techniques, and new ways of organizing the old communal networks that had characterized the post–Civil War rural South. As disfranchising laws, the institution of primary elections, and the concerted efforts of the executive machinery of the two major parties rendered political insurgency and third-party formation untenable, farmers sought new ways to press their demands. The strength of modern consumerism and the integration of local, regional, national, and international agricultural markets created a cash nexus that undermined home production. As the twentieth century opened and a new merger movement consolidated industries further, both large planters and smaller commercial farmers must have felt like Sisyphus struggling with his stone. The choices they made in addressing their problems included "sowing the old" and "reaping the new."

The economic depression of 1893–1897 undermined the modest gains small commercial farmers had made in the decades following the Civil War and Reconstruction. The average value in land and buildings per farm fell from $1,392 in 1890 to $1,180 by 1900. Statewide, tenancy rose from 30.8 percent in 1890 to 40.6 percent as marginally capitalized farmers lost control of their land. Likewise, average farm size decreased from 115 acres in 1890 to 90.6 acres by the end of the century. Although most farmers still worked their own land, the prospects for improvements in farm equipment and production, and even the expectation for future control over individual farms, appeared grim in Tennessee's centennial year. The number of farms carrying mortgage debt rose 8.3 percent during the 1890s. Moreover, as farm value declined, the ratio of debt-to-value climbed ominously. As had been the case throughout the postwar era, the most productive counties, those most closely tied to the agricultural markets experienced the harshest effects of the depression.[10]

The first decade of the twentieth century brought modest improvements and crystallized ongoing agrarian problems. On the positive side, Tennessee farm values rose significantly, as they did throughout the United States. The U.S. Census Bureau calculated the average value per Tennessee farm in 1910 at $2,490, a welcome improvement over the 1890 valuation of $1,794 per farm. Much of the increased value derived from rising land prices and farm improvements: $1,953 came from land and buildings, $450 represented livestock, and $87 came from implements and machinery. For risk-taking commercial farmers, the improvement in farm values meant that debt-to-value ratios fell from the 1890 high of 40.1 percent to more acceptable 26.7 percent. Although land values rose, southern farmers, unlike their counterparts in other areas of the country, found little relief in commodity prices.[11]

Across the South, despite years of falling prices and repeated crop reduction campaigns, the area devoted to the cultivation of cotton increased 20.3 percent from 1889 to 1899. The 1900 U.S. Census report declared the 1899 cotton crop the largest of the nineteenth century at 9,534,707 bales. The role of cotton in Tennessee's agricultural economy followed a more erratic path as farmers assessed the risks of cotton versus diversification. During the last decade of the nineteenth century, as cotton prices fell to a low of 4.59¢ per pound, Tennessee farmers shifted production away from cotton. In 1899 only 23.8 percent of the state's farms reported cotton harvests. By 1909, an improvement in prices had lured many farmers back into the cotton market. In that year, over 27 percent of Tennessee farms produced cotton, and acreage rose 26 percent to 787,516 acres.

The four counties of the tenth congressional district remained the most firmly wedded to cotton and tenancy and accounted for the largest increases in production. Owner-operators in other cotton-producing counties appeared to allocate acreage to production of the fiber based on anticipated market prices, combining cotton with other crops or livestock production.

Seasonal cotton prices over the period 1880–1910 averaged 8.66¢ per pound, but the range of prices could frighten even the most resolute cotton planter; prices ranged from a low of 4.59¢ per pound in 1893–94 to a high of 13.96¢ per pound in 1909–10. In the first decade of the twentieth century, as cotton prices rose and remained relatively high despite record-breaking crops, Tennessee farmers shifted back into cotton production. As the census statisticians noted, although corn continued to dominate the state's yields, cotton, with one-fourth the acreage of corn, constituted one-third of the overall state crop value. In 1910, the average crop value per farm in the eight largest cotton-producing counties in Tennessee was $711 ($568 if Lake County's unusually high return is discounted). By comparison, the state average value per farm was $490. For many farmers, it still paid to plant cotton.[12]

Tobacco also presented a complex picture of gains and losses for farmers who raised the commodity. Dark-fired tobacco dominated the Tennessee production. With dark-fired tobacco grown in seventeen counties near the Kentucky border from the Cumberland Plateau to the Mississippi River, production more clearly reflected the changing market than cotton. Industrialization of cigarette manufacturing and monopolization of the market through mergers transformed the world of tobacco production. Technological changes shifted the tobacco market from its traditional reliance on snuff and chewing tobacco toward cigarettes, with a demand for lighter tobaccos. And finally, buyers for the heavy, dark-fired tobacco Tennessee farmers produced complained that poor farming methods, including failure to fertilize overworked soil adequately, produced an increasingly marginal leaf.

Like cotton, the acreage devoted to tobacco production expanded over the course of thirty years, from 1880 to 1910. By the turn of the century, Tennessee ranked fourth in the nation in the number of acres devoted to tobacco, and eighth in production of the crop. Because of tobacco's labor-intensive nature, the actual number of acres in tobacco accounted for only about 10 percent of Tennessee farmland. Like cotton, however, the return on tobacco remained high relative to other crops. In 1910, the state's tobacco farmers produced yields averaging 760 pounds per acre, with some counties producing more than 800 pounds per acre. Despite low prices that reigned throughout the period, the value of production per farm for the thirteen largest tobacco-producing counties in 1910 averaged $566. The paradoxical circumstances of soil depletion, poor farming methods, and relative high return meant that the number of acres in tobacco cultivation rose substantially. At the turn of the century, 26,960 Tennessee farms produced tobacco on 71,849 acres, up from 51,471 acres in 1889 and 41,532 acres in 1879. In 1910 the state reported 90,468 acres in tobacco. In thirty years, tobacco acreage had more than doubled, with the largest increase recorded in the last decade of the nineteenth century.[13]

Tobacco producers faced two related problems: persistently low prices and large labor needs. In 1900, dark-fired tobacco was grown using the same labor-intensive methods that had characterized tobacco cultivation since the colonial era. Largely dependent on black sharecroppers and tenant farmers, wealthier planters feared the downward spiral of tobacco prices would drive labor away. For smaller farmers, who relied on family labor to produce the crop, low prices reduced their ability to repay their debts and thus endangered their future as landholders. The persistence of their problems and the dynamics of modern industrialization bridged the class differences that separated planter from yeoman and tenant to create an agrarian organization that roiled the Black Patch area for years.

The Black Patch War

Black Patch planters, large and small, had a long history of organization. Granges, Wheels, and Alliances had promoted cooperativism, organized crop reduction programs, and constructed warehouses in an attempt to control the crop and dictate prices. Each, in turn, had collapsed in the face of internal disputes and external pressure from better capitalized and more powerful tobacco buyers. Grangers and Wheelers had organized short-lived tobacco warehouses. Tobacco farmers had successfully opposed a planned consolidation of the Louisville and Cincinnati Tobacco Boards of Trade in 1890, unsuccessfully petitioned the U.S. Congress to investigate the influence of European Regis on tobacco prices in 1897, and maintained steady pressure to repeal the federal 6¢ tobacco tax. So when farmers in the dark-fired district confronted falling prices and reduced market options in the early twentieth century, they once again organized to defend their economic position in what they perceived as a war between planters and monopoly.

Organizers of the Planters' Protective Association (PPA) drew on a history of regional agrarian unrest and consciously incorporated the localism, cooperativism, and a reliance on communal networks that had supported earlier farm movements. They were aided in their organizing efforts by a general familiarity with rural reform among the region's farmers. As the PPA took shape, the men previously associated with Grange and Alliance cooperativism assumed highly visible roles. Charles P. Warfield, one of the most important leaders in the 1888 failed Wheel effort to organize a cooperative tobacco warehouse, joined with other well-known reformers, such as newspaperman Martin VanBuren Ingram, to press for a merger of town and rural interests against the tobacco buyers.[14]

Nevertheless, the organization of the Black Patch differed from earlier efforts in several important ways. First, a handful of the wealthiest and most powerful planters in the region quickly dominated the PPA. Although the association claimed to speak for dark-fired tobacco farmers from the lowliest sharecropper to the largest planter, it didn't pretend to be an egalitarian organization: the organizers determined the principles under which the association would operate, and decisions about the marketing of the tobacco crop were made at the top without the input of smaller farmers. The organization also utilized pressure tactics characteristic of progressive reformers in other areas of American life. They focused on crop reduction and withholding of tobacco to force higher prices from the principal buyers of dark-fired tobacco, the ATC and the Italian Regie. The association pressed for tax relief but made no efforts to influence banking and currency laws as previous agrarian movements had done. Likewise, PPA organizers studiously avoided political claims to, or criticism of, the American electoral process. Nonetheless, they demonstrated a familiarity with the bureaucratic institutions of the Progressive era and pressed for government investigations of buying practices and enforcement of antimonopoly and antitrust legislation. Finally, the association exhibited the paradoxical behavior common to the Progressive era, employing traditional community pressures to conform. As association leaders met resistance to their organizational initiatives, events in the Black Patch veered onto a more violent course than had earlier Tennessee agrarian reform efforts. Indeed, most students of history focus on the Black Patch's episodes of night riding that resulted in assaults, arson, and murder.[15]

After several abortive attempts at organization, the Planters' Protective Association emerged from a local union established by tobacco growers in Robertson County in 1902. Farmers there established marketing cooperatives as a mechanism for raising tobacco prices. Until the turn of the century, landholding farmers countered falling prices by shopping for the highest bids from a variety of buyers, including warehousemen and local merchants. By 1900, changes in the uses of tobacco and in the marketing of the leaf reduced individual grower's options.

The transformation of the tobacco market centered on the business decisions of James B. Duke of Durham, North Carolina. In the early 1880s, Duke contracted with James Bonsack for use of his new cigarette rolling machine and revolutionized the tobacco industry. By investing heavily in advertising and engaging in cutthroat competition, Duke dominated the cigarette industry by 1889 when he merged the five largest firms into the ATC. In quick succession, Duke absorbed or drove out of business more than 250 smaller tobacco manufacturers. Accordingly, his control over tobacco output in snuff (the principal use for dark-fired tobacco) rose from 2.9 percent in 1891 to 67.5 by 1900. The ATC's control over tobacco output increased over the first decade of the twentieth century to 79.7 percent for plug and twist, 96.1 percent for snuff, and 71.2 percent for cigarettes.[16]

The elimination of competition left Black Patch growers at the mercy of the ATC in two important ways. Most dark-fired tobacco was sold on the international market, and in 1901 the ATC eliminated competition with the British through the organization of the British-American Tobacco Company. This merger reduced international sales in the Black Patch to two principals: the ATC and the Italian Regie. By 1902 the ATC and the Regie had worked out an agreement to avoid competition altogether by dividing the Black Patch into noncompeting districts in which only a single buyer would offer prices for all grades of tobacco. The two companies agreed that farmers would be prevented from selling outside the territory in which they grew tobacco, leaving individual growers with only one source for marketing their crops.[17]

The Tennessee organizers of the Planters' Protective Association—Joseph E. Washington, Felix Ewing, Joel B. Fort, and Charles Fort—represented the oldest, wealthiest, and most politically powerful families in the dark-fired district. For generations these families produced the pungent tobacco used for snuff, first with slave labor and then with tenants and sharecroppers. Intermarriage consolidated their economic and political power rather than dissipating it through inheritance. Wealth and paternalism pushed these families to the political forefront as sons followed fathers and grandfathers in elected and appointed offices at the local, state, and national levels of government. When Black Patch farmers organized their Wheels and Alliances in the 1880s, the elected officials among these planter families added their names to the rolls; they supported the least threatening of the agrarian demands but never became an integral part of the movement—neither did they presume to speak for the Wheel or the Alliance. As the Wheel and Alliance movements became more political, the planter-politicians gave no quarter in the battle against the farmers for control of the Democratic Party. Having defeated the agrarian insurgency once, the leaders of the PPA carefully avoided reactivating political expectations.[18]

After months of local and county organizing, PPA leaders presented their plan of action to area farmers at a rally and mass meeting at Guthrie, Kentucky, on September 24, 1905. The organizers unveiled their agenda before an audience of five thousand men, women, and children, representing tobacco growers of every economic class from sixteen counties in Tennessee and Kentucky. John Wesley Gaines, a Tennessee congressman, and Benton McMillin, former Tennessee congressman and governor, spoke to the crowd. Newspapers hailed the unity of such a vast throng of "solid yeomanry" and proclaimed the occasion "the beginning of the struggle for their homes, for life itself." The characterization of the struggle as a personal battle for their future as yeomen resonated with Black Patch farmers, and the Guthrie rally became a touchstone in their war against the monopolistic practices of the ATC and the Regie. The anniversary date was celebrated annually with ever-larger crowds of enthusiastic farmers and townsmen.[19]

As the organizers made clear, the purpose of the PPA was to raise tobacco prices. They advocated a plan of action that called upon members to pledge to market their crop through the association, signing an agreement designating the PPA as the "sole agent" to negotiate and sell their crops. Once 70 percent of the region's tobacco had been pledged, the association leaders believed they would be in a position to dictate prices.

Recognizing that success depended on the poorest farmers and bending to the counsel of earlier reformers, PPA leaders approached area bankers and obtained commitments to provide association members with low-interest loans using warehouse receipts as collateral. PPA leaders used their economic and political clout to marshal local support to fund the scheme, successfully arguing that low tobacco prices stymied industry and commerce as well as agriculture. Adoption of the plan unified town and country in a concerted action against the market hegemony of the ATC and pragmatically provided cash-strapped small cultivators with the means to sustain the boycott. For the good of the town businesses and the benefit of area farmers, warehouses and banks in Springfield and Clarksville strongly supported the PPA.

From an initial operation of twenty-two warehouses, the PPA expanded to thirty-seven in the second year, and fifty-two by 1906. Some thirty banks accepted association warehouse receipts. Likewise, the PPA retained the full support of area newspapers, which provided a steady drumbeat of support for the goals of the association. With the enthusiastic backing of so many business and professional men, tobacco planters expected a quick resolution to their troubles. The growth in farm members to an estimated 40,000 by 1906, and the enthusiastic and elaborate annual rallies suggested greater success than the records indicate. But the promised quick gains failed to materialize.[20]

Despite the careful planning and the support of business interests, the PPA faced many of the same barriers to success that farmer unions encountered earlier. Perhaps most important, the association assumed that rising tobacco prices would solve the region's economic problems, when, in fact, earlier agrarian organizations more realistically recognized the systemic difficulties that undermined rural reform: a lack of capital and a social structure that institutionalized poverty. Moreover, at the bargaining table, buyers had the advantage over sellers. The ATC and the Regie easily created conflict within the farming communities by offering higher prices to planters who refused to sell through the association. Buyers sowed discord and distrust where farmers needed the assurance of mutual support. Since dark-fired tobacco represented a decreasing percentage of the ATC's production, buyers could afford to wait.

Second, although PPA leaders avoided the appearance of seeking political resolution to the problems of the Black Patch, they expected government to hear their concerns and invoke antitrust laws against the ATC. Acting through Tennessee Congressman John Wesley Gaines, the PPA appealed to the federal Bureau of Corporations to investigate the ATC and prevailed upon U.S. Attorney General Philander Knox to open an investigation into the conditions in the Black Patch. Kentucky Congressmen Augustus O. Stanley and Ollie M. James joined Gaines in simultaneously pressing for repeal of the tobacco tax. At the state level, farmers in Tennessee and Kentucky called upon their state legislatures to enact laws designed to protect the cooperative tobacco markets. Their lobbying efforts provided a mixture of successes and failures.[21]

Both the U.S. Justice Department and the Bureau of Corporations conducted investigations into the business practices of the ATC in the Black Patch. And both investigations substantiated the claims of the farmers that the ATC had operated to eliminate competition among tobacco buyers. Nevertheless, association farmers found little relief in what they perceived as the limited actions of government. The Bureau of Corporations uncovered evidence of illegal buying practices but lacked the authority to act. The actions of the Justice Department suggested that the events in the Black Patch were not of particular national concern.

Although Attorney General Philander Knox doubted there was sufficient evidence to warrant a grand jury investigation, he permitted the collection of information by the U.S. attorney for Middle Tennessee, Abraham Tillman.

Congressman Gaines took offense at the limited federal action and mounted a letter-writing campaign to hound Knox and his successor, William Henry Moody, into complying with his demands for a full-fledged investigation by a federal grand jury. His verbal duel with the Justice Department eventually made its way into the Tennessee press. The intercession of Felix Ewing provided a compromise in which Judge Horace H. Lurton convened a grand jury in July 1906. After two weeks of testimony, the grand jury adjourned without issuing an indictment. Contrary to what Black Patch planters expected, the investigators never intended the probe to result in indictments. The Black Patch evidence was transferred to New York, where a larger investigation into the business practices of the ATC resulted in federal action and the breakup of the company in 1911. Ultimately, the demise of the ATC had little effect on area farmers and the prices they received.[22]

The effort to repeal the federal 6¢ tax on tobacco also produced mixed success. Tobacco growers had long believed that the tax unfairly penalized them and contributed to the demise of competition by effectively limiting home tobacco manufacturing, but their protests failed to move the U.S. Congress. In 1904, the House of Representatives passed a repeal bill, but Nelson Aldrich, the powerful chair of the Senate Finance Committee, saw to it that the bill never got out of committee. Spurred on by the encouragement of their constituents and area newspapers, Congressmen Stanley and Gaines introduced bills for tax repeal during the first session of the 59th Congress. A compromise bill again passed the House but again died in the Senate Finance Committee. A similar fate met repeal efforts in the 60th Congress. Success came with a change in strategy. In 1909, President William Howard Taft called a special session of Congress to deal with tariff revisions. Tennessee Congressmen Joseph Byrns and Cordell Hull, along with Kentucky Congressmen Augustus Stanley and Ollie James, introduced the repeal legislation as an amendment to the Payne tariff bill. Tennessee and Kentucky senators rallied to protect the amendment, and when the final measure passed, it included a repeal of the tobacco tax. However, repeal brought little relief to Tennessee farmers: home manufacturers of tobacco products were still required to register as dealers and to pay a manufacturer's tax, a prohibitive levy that prevented the development of local small-scale tobacco markets. Ironically, southerners had attached the repeal to a tariff bill they could not support themselves.[23]

Negotiations with the ATC and Regie buyers, political pressure to investigate the monopolistic practices of tobacco buyers, and lobbying to repeal the tobacco tax raised expectations for greater changes. But association leaders could not address the more systemic problems of smaller, landholding farmers or the tenants and sharecroppers who produced for the great planters. As one historian noted, the failures of the tobacco war were rooted in "an age-old problem that has dogged agriculturalists in virtually every society—credit." Exploited by creditors, small farmers and tenants faced far greater risks in withholding their crops than did larger planters. With PPA leadership concentrated within a tight community of wealthy growers, no one spoke for the smaller producers. The association raised expectations, but even with the support of local banks, most farmers could not hold out for long.[24]

By 1905, vigilantes were operating in the dark-fired district in an effort to coerce independent farmers into the association and speed the promised results. PPA leaders had set the tone for intimidation in their calls for community ostracism against those who withheld their tobacco and offered it for sale to ATC and Regie buyers. Events quickly moved to violence following an October 1905 meeting of association members at the Stainback School in Robertson County. Thirty-two members of the PPA formed the Possum Hunters Committee, whose purpose was to "visit" the "hillbillies" (a name given to those who sold tobacco independently) and force them to join the association. One of the first acts of violence occurred two months later in Trenton, Kentucky, when night riders burned the tobacco of independent buyers James and B. D. Chesnut and Joseph Russell.[25]

Violence has a long history in the Black Patch and in the South. Historians and social scientists struggle to explain the prevailing culture of violence that has characterized and continues to characterize the South. Slavery, the defense of honor, and Celtic heritage have all been offered as explanations. Whatever the root cause, the Civil War did nothing to ameliorate the culture of violence, and the tobacco counties were not spared, despite the fact that most of the area came under federal occupation early in the war. Guerrilla bands during the war and white cappers during Reconstruction were part of a legacy of violence against those who defied community standards and values.[26]

Although Redemption supposedly ended Klan activities, violence bubbled along under the surface as men continued to defend their property, their honor, and their political rights. Tennesseans of all classes accepted an individual and community right to enforce social standards when the law failed to act. As historian Christopher Waldrep noted, vigilantism condemned in one case might be considered acceptable in another. As the night riders in the Black Patch moved from verbal coercion to property damage and physical violence, local and state governments proved incapable of restoring order. Property damage ranged from the scraping of tobacco beds to the destruction of barns and fences to the burning of tobacco warehouses holding tens of thousands of dollars' worth of tobacco. Tracy Campbell documented 337 instances of night riding in the Black Patch and Burley tobacco regions of Kentucky and Tennessee. In what was perhaps the most spectacular episode, five hundred vigilantes advanced on Hopkinsville, Kentucky, a town of ten thousand inhabitants, and set fire to warehouses full of tobacco. The raid on Hopkinsville drew national attention as newspapers reported the $200,000 in damages and published pictures of the burned-out buildings.[27]

The demise of the PPA began in 1907 and was clearly in evidence by 1909. Bankers became more reluctant to support farmers after the brief depression of 1907. Farmers increasingly criticized the business practices of the association and withdrew their tobacco from PPA control. As public support for the vigilantes dwindled, state and federal authorities finally stepped in to restore order.

The efforts to organize the Black Patch tobacco farmers against the ATC and the Italian Regie buyers suggest both traditional and modern economic and social methodologies at work. The organization of a farmers' trust to resist a buyers' trust rested on a defense of traditional face-to-face exchanges and Jeffersonian republican values. Even as PPA leaders adopted interest group tactics that included testimony before congressional investigative committees, pressure for action by federal bureaucracies and cabinet-level agencies, and legislative lobbying for favorable laws for tax relief, older forms of community coercion continued unabated. With little public condemnation, tobacco farmers organized mass meetings and engaged in intimidation and vigilantism to gain their objectives. In the Black Patch, the modern and the traditional vied for dominance in resolving agrarian problems.

The Reelfoot Lake Uprising: Defending Common Rights

A conflict with comparable violent components arose on the border of Obion and Lake counties in northwest Tennessee. Known as the Reelfoot Lake Uprising, this agrarian battle bore similarities to the struggle in the Black Patch. Like the tobacco farmers, the Obion County residents first attempted to resolve their dispute through organization and legal action. Also like their counterparts in the tobacco district, the farmers and fishermen in the Reelfoot Lake region turned to night riding to gain their objectives.

Reelfoot Lake is a shallow estuary formed by the New Madrid earthquake of 1811–1812. With its cypress trees and stumps growing out of the lake bed, the lake presented a ghostly appearance and quickly became the stuff of folklore. On the western shore of the 25,000-acre lake, a rich, flat plain stretched to the Mississippi River. Periodically fertilized by the river's overflow, tiny Lake County contained some of the richest cotton land in the state. On the eastern shore, a steep bluff rose from the water's edge before leveling out to form rolling hills. There farmers raised their families on small, subsistence farms, marketing surplus corn and cutting timber in the off-season. The lake itself provided locals with an abundance of game and fish to supplement diets and earn cash through the sale of furs and fish. For settlers on the eastern shore, the lake represented an important economic asset they used in common.[28]

To James C. Harris, Reelfoot Lake held financial opportunity. Arriving in Lake County after the Civil War, Harris established himself as a planter and timber operator of considerable importance. Within two decades he acquired the reputation of a prosperous and hardworking entrepreneur. Harris's business practices elicited public condemnation from a disgruntled Allianceman who found his enthusiasm for work morally questionable because it included labor on Sunday. Calling Harris a "big merchant and rich landowner" who "has been enriched from [the] hard earning" of his many tenants who "have dragged along in poverty," the letter writer counseled against supporting the planter's 1888 bid for a seat in the general assembly. However, to progressive merchants and professionals in the nearby towns of Union City and Troy, Harris epitomized the forward-looking, community-minded man of the day. Thus, the "one-eyed Cotton King" of Lake County enjoyed a mixed reputation, that of an astute businessman and that of a grasping wealth seeker.[29]

In the 1890s Harris quietly bought up claims to Reelfoot Lake, anticipating the day when he could drain the lake to expose its rich soil bed for cotton production. Already the proprietor of some 15,000 acres, Harris was easily the wealthiest man in Lake County. His interest in the lake and the salvaging of farmland also included a scheme to construct a system of levees to prevent the annual inundation of low-lying regions by the flooding Mississippi River. Harris served in the Tennessee House of Representatives (1895–1897) and supported the creation of a local taxing district to fund the construction of a levee system to protect Lake County from flooding. His first effort met defeat when the Tennessee Supreme Court ruled the enabling legislation unconstitutional. A more progressive court supported a second attempt in the early twentieth century.[30]

Undeterred, Harris returned from the legislature and began acquiring property rights to the lakebed. But Harris was not the only one interested in acquiring rights to the lake. In 1897, the first of what would prove to be a series of escalating confrontations between claimants and settlers took place. When a sporting club, Reelfoot Outing Company, bought waterside property to be developed for urban vacationers, the lake people made known their feelings about the intrusive "sports." Disruption of hunting and fishing expeditions and harassment of surveyors convinced investors that Reelfoot Lake would never provide the vacation pleasure they anticipated, and they sold the property to Harris in 1898.[31]

By 1899 Harris had acquired ownership of most of the lake and announced his intention to construct a canal to drain the water. Assembling teams of mules, dredging equipment, and dynamite, Harris prepared for the arduous task. Such preparations convinced lake people that Harris was serious. At first they attempted to meet the threat of loss of fishing and hunting rights through legal action.[32]

In 1902, J. C. Burdick, a lakeside dock owner and operator, led his neighbors in litigation to stop Harris from draining the lake and undermining their common rights. In a suit that incorporated both traditional and modern interpretations of the public good, the litigants claimed riparian rights to the use of the lake waters by arguing that Reelfoot Lake was a navigable waterway and therefore not subject to drainage. In a more modern view of the public good, they argued that the draining of the lake would endanger public health.[33]

Harris countered that the shallow depth of the lake prevented navigation and presented his titles as legal proof of ownership of the land beneath the water. In Union City, Judge W. H. Swiggart ruled for the plaintiffs and enjoined Harris from proceeding with his drainage project. Harris appealed, and the Tennessee Supreme Court issued a complex ruling: The court upheld the injunction of the lower court and questioned the validity of Harris's land claims. But the court also ruled the lake unnavigable and therefore subject to purchase (and presumably future drainage). As one historian noted, "the decision amounted to an open proposal to J. C. Harris to perfect his claim.[34]

Harris died in 1903, but two years later, the Harris heir purchased the remaining lake bottom and most of the shoreline. When the new owner informed the residents they could no longer use the lake without permission, a second lawsuit was filed on behalf of dock owners John Shaw and Walter Pleasant. They questioned Harris's right to dictate terms of usage without having full ownership of the shore. The courts found in their favor. The interests of Shaw and Pleasant were defended by James Deason, Quentin Rankin, and Robert Z. Taylor, who had served as the elder Harris's attorneys in the 1902 lawsuit.[35]

In 1907 the three attorneys who had represented Shaw and Pleasant in their most recent court case recognized economic potential in the property, bought the remaining holdings, and forced Harris into a corporation. Capitalized at $125,000, the West Tennessee Land Company anticipated control over fishing rights, a potential income of $400 per month with all commercial fishing filtered through Burdick's dock. The one-time champion of native claims to the rich fishing grounds, Burdick paid for the right to sell fish from the lake and met his obligation to the land company by lowering the prices he paid to local fishermen. When lake people balked at Burdick's prices, he pressed legal action.[36]

The land company also adversely affected area farmers, who had used the surrounding woods and marshes as common pasturage for cattle and horses; now they were required to sign leases in order to continue the practice. In the minds of the lake people—farmers and fishermen—the progressive and educated businessmen and lawyers had betrayed their trust and the rights inherent in the use of common lands. With legal titles, contracts, and corporate charters in hand, Harris demanded and obtained the backing of the courts for the rents and prices offered by the land company. Lake people had neither the time nor the money for a protracted fight in the local and state courts. Moreover, they had no confidence that the men they hired to defend their rights would deal with them honestly. With the nearby example of vigilantism in the Black Patch, they took similar action.

The Reelfoot Lake night-riding episode was intense, violent, and brief. Like the eruption in the Black Patch, some aspects of the vigilantism associated with the night riders suggested the reestablishment of communal standards. From April to October 1908, the vigilantes adopted the paraphernalia of terrorism: the wearing of masks, the use of secret passwords, and engagement in intimidation and violence. The "homemade" night riders threatened and whipped those who defied racial or gender-based community expectations. Emma Johnson's efforts to obtain a divorce from her husband, Joe, earned her a beating by the night riders. Other women were whipped because they were perceived as immoral or because they quarreled with neighbors. In the worst example of racial vigilantism, the night riders crossed the border into Kentucky and murdered Dave Walker, his wife, and two of their five children; the African American farmer had reportedly acted in an arrogant and defiant manner. Although the night riders had many defenders, the editor of the *Nashville Banner* argued that their actions could not be justified by romantic notions that they acted "in the spirit of knighthood" arrayed against oppression. They were not "as St. George [slaying] the dragon or as Anglo-Saxon freemen battling against a foreign tyranny," he wrote. Their actions were those of "organized lawlessness."[37]

The editor's interpretation denounced any tendency to romanticize violence, but he had misjudged the degree of oppression. The potential loss of fishing and common rights to the land surrounding the lake affected poor families by removing an essential source of protein from their diets and closing off opportunities to supplement their meager incomes through harvesting of fish, small animal pelts, and timber. The violence was a manifestation of the oppression they experienced, and the most consistent focus of the vigilantes remained the battle between the lakers and the developers. Night riders issued repeated threats to merchants and business leaders in Tiptonville and Lake County. Lawyers and officers of the courts in Union City likewise received numerous warnings that the vigilantes would exact justice. After Burdick ignored repeated warnings, arsonists attacked his dock and destroyed the structures. The fight for common rights focused on two issues—fishing rights for lakeside residents with higher market prices for their catches, and a permanent stop to the scheme to drain the lake for cotton production.[38]

The event that gained the night riders statewide and national attention occurred on October 17, 1908. Masked men kidnapped two attorneys associated with the West Tennessee Land Company, Robert Z. Taylor and Quentin Rankin, from Ward's Hotel at Walnut Log on the shore of the lake. Taking the two into the nearby swamps, the night riders threatened the men and attempted to extract a pledge to reopen the lake to fishing. When Rankin resisted their demands, the vigilantes fitted a noose around the lawyer's neck, hoisting and lowering him several times before a shotgun blast ripped through his suspended body. After a volley of shots, Rankin's lifeless body was lowered to the ground, and the masked men debated the fate of Taylor. Sensing his only chance for survival, the sixty-two-year-old Taylor bolted for the murky water of a nearby slough and swam for his life. The pursuing night riders fired into the water. A partially submerged log shielded the attorney, and he survived. Confident of Taylor's death, the vigilantes dispersed, and Taylor, cold and disoriented, wandered through the woods and swamps for three days before finding help.[39]

Like events in the Black Patch tobacco district, this Reelfoot Lake conflict between traditional rights of community and modern commercial and capitalist endeavors provided the basis for violence. Farmers and landless hunters, trappers, and fishermen depended on access to the lake's abundant wildlife to maintain an independent livelihood. The monopoly over fish prices or the draining of the lake for commercial agriculture threatened the future of lake families just as the buying practices of the ATC compromised the livelihoods of tobacco growers.

Lakers initially attempted to use modern methods to gain their objectives and successfully defended their traditional rights through the state's courts, as had Black Patch growers. Two issues—control over the shore line and lake bottom and determination of the navigability of the lake—governed the validity of claims for both sides. Harris and his partners bought up all competing warrants to render ownership a mute point. More important, Harris co-opted the most articulate and strongest contenders—Burdick, Rankin, and Taylor—who had previously defended the lakers and guided them through the modern bureaucratic processes of justice. Conflicting rulings on the navigability of Reelfoot Lake left lakers only with an expensive and protracted court battle. Intimidation through night riding or justice through the exertion of the will of the community appeared to be their only option.

Governor Malcolm Patterson acted forcefully within his full authority as governor to end the troubles at Reelfoot Lake. He posted a $10,000 reward for the apprehension and conviction of those responsible for Rankin's murder. A grand jury convened at Union City to hear evidence regarding the murder and night-riding activities. Finally, Patterson utilized the expanded power granted to governors after the Coal Creek violence of the 1890s and called out the state militia to arrest the night riders and restore order.[40]

Patterson's strong action in Obion County contrasted starkly with his lukewarm response to night riding in the Black Patch—a contrast political opponents quickly exploited. Initially, newspapers applauded his quick action. New York and Atlanta editors noted his "prompt and firm stand for law and order" and suggested that Patterson "started out right in taking the night riders by the throat." Some smugly noted that 1908 was an election year, and Patterson's forceful response garnered headlines other candidates could not match. Patterson was an astute politician, and the impulse to grandstand cannot be dismissed. Perhaps more important to the governor's decision was the perception that the Black Patch vigilantes enjoyed the sanction of men of wealth and political power. The Planters' Protective Association marshaled local backing for its fight against a powerful national and international monopoly, a worthy cause that enjoyed broad public support. In contrast, the Reelfoot Lake night riders represented poorer farmers and landless lakers who battled local, powerful elites. For a governor in a tight reelection fight, the choice was clear.[41]

Given the relative power of the two sides, the outcome also suggests both a step toward modernity and a reestablishment of local justice; a progressive use of state power to establish order and find a resolution to the conflict and an unwillingness to punish men who enforced communal rights and values. The militia quickly took control of the lake and began rounding up suspected night riders. Officials soon identified the ring leaders, arrested them, and brought them to trial at Union City. Rice Pierce, local attorney, congressman, and would-be champion of the common man, defended the lakers. Pierce's flamboyant courtroom demeanor and his previous flirtation with Populism perhaps marked his suitability as the defense attorney.

After several tension-filled days of testimony, the Union City jury convicted six men identified as vigilante leaders of first-degree murder with "mitigating circumstances," and recommended sentences of twenty years. Two men, Bud Morris and Bob Huffman, were convicted of second-degree murder. Final sentencing was left to Judge Joseph E. Jones, who had no alternative but to accept the sentences for Morris and Huffman. In the case of those convicted of first-degree murder, he imposed a sentence of death by hanging. On appeal, the Tennessee Supreme Court ruled in a 3 to 2 decision that significant errors had occurred in the original trial. The high court ordered a new trial and permitted the defendants to post bail. The night riders returned to their homes, with no apparent reprisals, and no future legal action.[42]

Calling the decision a "triumph for murder and lawlessness," the *Memphis Commercial-Appeal* characterized the action of the court as "hair-splitting exaggeration of the importance of trivial things." "By the stroke of a pen," the *Atlanta Constitution* told its readers, "the Supreme Court nullified good work done in bringing the murderers to justice." Local Tennessee papers declared the people "outraged" and called the decision a "most severe slap in the face," and "little short of a crime," predicting that voters would remember the ruling at the next election. But the anticipated retribution failed to appear at the polls or among the citizens of Union and Lake counties.[43]

In the long run, the Tennessee courts and the general assembly provided the most comprehensive solution to the problem. In its 1909 biennial meeting, the Tennessee General Assembly conducted hearings in Union City and at various Reelfoot Lake sites pursuant to purchase of the lake from the West Tennessee Land Company. Upon receipt of the committee's report, the state entered into protracted negotiations with the land company to purchase their claims. Unable to reach a settlement agreement, the state sought resolution in the courts. In 1913, in a split decision, the Tennessee Supreme Court declared Reelfoot Lake a navigable waterway in the public domain and thus not subject to drainage. The state took possession of the lake and initiated condemnation suits against all remaining claims. In 1925, the lake and surrounding area were set aside as a state park, establishing the common right of all Tennesseans to recreational use of the lake.[44]

The Reelfoot Lake uprising offers another window into the reordering of rural society in the Progressive Era. Commercialization of agriculture promised new opportunities to men with capital. Exploitation of the timber and wildlife associated with the lake also provided security for men with limited access to cash and fewer opportunities to climb the economic ladder of the new century. The clash between men-on-the-make and established, communal rights occurred within a new political context that consolidated the advantages of capital by limiting the traditional power of the voter, even as it empowered the state to resolve social issues through bureaucratic and judicial action.

Lakers faced a political world in transformation as they sought justice. In the 1880s, Obion County farmers defeated James Harris in roiling political campaigns that questioned his economic morality using a traditional yardstick of community expectations within a network of exchange relations. By 1908, changes in campaigning and the effects of disfranchisement rendered the stump speech and the public canvass ineffective in reigning in the power of elites and resolving differences between competing economic and social views. The evolving Progressive-era bureaucratic government worked to the advantage of those with knowledge and expertise—men and women who understood the boundaries of the overlapping agencies and commanded the wealth to pursue justice through legislative committees and courts. Lakers recognized the changes and initially sought redress through these avenues, but they quickly exhausted their options. Like many Americans caught in this moment of transformation, they struck out against visible evidence of change—African Americans, and women whose actions questioned the traditional power of white men. They used the same vigilante methods to defend their economic rights. Donning masks and mounting their horses, the night riders exacted "justice" through intimidation and violence. Ironically, justice finally came through the state, which acted to secure access to the lake and prohibit the exploitation of the region for the benefit of private investors.

Farmers' Union: Business Agriculture and the Golden Rule

As Obion County's economically marginal farmers acted to prevent the spread of the cotton economy, commercial-minded farmers and large planters also organized to protect their interests as cotton production far exceeded the capacity of textile manufacturing, once again driving down the prices. The organization of cotton farmers reemerged at the turn of the century as producers recognized the detrimental effects the glut of cotton had on the world market. By 1900 the United States produced approximately 80 percent of the global supply of cotton. The estimated American crop for 1898–99 was projected to be the largest in history, an astounding 11 million bales. In such a productive year, southern planters netted a total of only $283 million. For many farmers, the alternatives to bankruptcy appeared to be withholding of the cotton crop or voluntary crop reduction—policies tried by the Alliancemen cotton associations earlier in the decade with limited results. To implement these options, cotton producers across the South organized first as the short-lived Cotton Growers' Protective Association and the Southern Cotton Growers' Association, and then as the more democratic and progressive Farmers' Union.[45]

Although Tennessee's contribution to cotton production lagged far behind that of Georgia, Alabama, and Mississippi, the state played a key role in several aspects of cotton marketing. The Memphis Cotton Exchange dominated the inland spot market for cotton. Moreover, the city's investment in wholesale grocery and distribution firms affected farmers and planters in three states. Finally, the development of cotton-related manufacturing, particularly the rise of the cottonseed oil business, expanded the market for cotton and cotton by-products and provided area farmers with much-needed cash.[46]

In addition, in Tennessee, as elsewhere, the agrarian organizations of the first decade of the twentieth century provided an important transition from the cooperative insurgency of the Alliance and Populism to the government-directed and state-regulated farm programs of the post–World War I era. The Cotton Growers' Association and the Farmers' Union incorporated the basic tenets of cooperativism and the subtreasury scheme in an apolitical, voluntary effort that appeared less threatening than earlier attempts to effect change. Farmers no longer challenged the political hegemony of the large planters and the New South industrialists, but engaged in indirect pressure politics that nonetheless reminded officeholders of their concerns and their potential voting strength. More important, the incipient agricultural bureaucracies relied on men of proven talents, both in farming and organizational skills, to staff positions that would later go to professional agriculturalists with degrees from land-grant universities. Likewise, state legislators and congressmen called on practical farmers to provide the information needed to write the enabling acts that shaped modern American agricultural law. Thus, the eventual development of regulatory agencies and policies occurred within a context of agrarian input, but not an overt agrarian threat to established political hierarchies.[47]

Memphis first emerged as a center of post-Populist organizational activity when the American Cotton Growers' Protective Association, led by Harvie P. Jordon of Georgia, convened in the city in January 1896 as the familiar refrain of diversification and cotton acreage reduction filled the air. Both the attendance lists and the roster of speakers suggested that control of the convention, like the organization, lay with the larger planters and the Memphis Cotton Exchange. The two most visible Tennessee delegates were John R. Godwin, former legislator, banker, and onetime president of the Memphis Cotton Exchange, and Thomas F. B. Allison, Commissioner of Agriculture under Governor Peter Turney. Convention speakers opined that there were too many farmers in cotton, namely small producers, whom they advised to practice self sufficiency and look for alternative commercial crops. As planters and cottonmen recognized, the success of efforts to withhold cotton, either through reductions in planting or stockpiling at the end of the season, rested on the shaky commitment of marginally financed small producers, who jumped back into the market as soon as prices showed modest gains. Largely a "letter-head" organization, the American Cotton Growers' Protective Association offered more in rhetoric than it did in concrete results and soon faded from view.[48]

By 1905, fears of a return to 5¢ cotton once again spurred growers, bankers, ginners, and railroad men to renewed action. Harvie Jordon again took the initiative. On January 24, 1905, a reported three thousand representatives from the cotton-producing states met in New Orleans in a much-heralded convention, at which they proclaimed the need for diversification and crop reduction. Convention leaders organized around a plan to reduce cotton acreage by 10 percent and withhold cotton from the market in a scheme heralded as the cotton states' version of Standard Oil's tactics to maintain prices.

Like the earlier Alliancemen and the Cotton Growers' Protective Association, the Southern Cotton Growers' Association recognized that no scheme to control the crop could survive without adequate financial backing. Jordon called upon bankers to support the actions of cotton planters to break the monopolistic power of the national and global markets arrayed against the South. Convention leaders exhorted small farmers to diversify their crops and meet their need for capital with self-sufficiency, noting that "the strongest financial institution for every farmer is a well-fixed corncrib and smokehouse." As Theodore Saloutos stated in his analysis of the convention, "Diversification promised to strengthen the position of the producer and make him independent in the sale of this product; but it was still necessary for the bankers of the South to stand behind the producers." To the most observers, it was clear that the Cotton Growers' Association's alliance with nonagricultural interests—bankers, merchants, and politicians—appealed primarily to large planters.[49]

Nevertheless, every effort was made to subsume class or productive differences in a bravado of southern patriotism. The *Daily Picayune* described the convention as a gathering of "all Dixie in a real rebel convention." As the bands played southern favorites, and flags waved, this "second Confederacy" vowed "to protect the whole people of the South from ruin by safeguarding the great crop on which the South lives." Shouting "Down with Speculators!" the delegates called for acreage reduction, withholding of cotton to achieve a price of 10¢ per pound, and diversification. Central to the plan was a "uniform system of bonded warehouses" that would be financed by southern banks.

After three days of speeches and planning, the New Orleans convention adjourned to implement a banker-supported acreage-reduction program and 2 million bale cotton holdout.[50]

At about the same time, former Grangers, Alliancemen, and Populists also organized anew to address the needs of smaller commercial farmers. As with earlier efforts, the parent organization originated in the hardscrabble hills of East Texas. There, in 1902, Newton Gresham, a former Alliance organizer from Alabama, brought together nine disgruntled agrarians in Rains County to discuss the problems of cotton. First organized at Smyrna schoolhouse, the Farmers Educational and Cooperative Union, better known as the Farmers' Union, spread rapidly among men with prior reform proclivities. Indeed, the speed of its growth quickly created organizational problems that almost led to the Union's demise even as it raised concerns among southern politicians that a resurgent neo-Populism threatened the recently achieved political peace.

More successfully than the Southern Cotton Growers' Association, the Farmers' Union operated to create a price floor for cotton by controlling the crop through a combination of acreage reduction and withholding of the crop. The Farmers' Union's achievement lay in the establishment of a system of county, state, and regional warehouses where growers could store their cotton for future price increases and use warehouse receipts as collateral to obtain short-term loans. Reviving the cooperative system, the Farmers' Union developed a scheme of vertical integration that included implement manufacturing, fertilizer production, warehousing, and marketing. At the same time the organization undertook lobbying efforts to pressure state legislatures and congress to enact measures to end futures trading and outlaw so-called "bucket shops."

Taking a cue from European agricultural practices, the Farmers' Union promoted the creation of voluntary associations to meet the needs of farm families, including credit unions, rural banks, telephone companies, and mutual insurance companies. Union leaders advocated improved rural education through the creation of agricultural high schools, rural libraries, Boys' Corn Clubs, and Girls' Canning Clubs. According to Saloutos, the Farmers' Union promoted "more efficient methods of marketing and production . . . lent support to farm demonstration work in its infancy . . . perpetuated the idea of acreage reduction as a relief measure . . . urged cheaper and more abundant credit facilities . . . [and] encouraged the construction of local warehouses."[51]

Among Tennessee's small commercial farmers, a convergence of voluntary, educational, and government programs provided a transition from older mechanisms of agricultural organization to modern rural progressivism. The arrival of the Farmers' Union in Tennessee coincided with the professional restructuring of the Tennessee Bureau of Agriculture and the University of Tennessee's College of Agriculture. The Farmers' Union approached agriculture as a business and organized to make farming profitable. Access to markets and farm credit, good roads, and modern rural schools dominated the Union agenda. Improved finances and a renewed commitment to agriculture enabled the University of Tennessee through the USDA Experiment Station to engage in research to improve crop yields and address the problems of hog cholera and tick fever that plagued livestock raisers. Finally, with biennial appropriations from the general assembly, the Tennessee Bureau of Agriculture and Immigration coordinated and publicized efforts to educate farmers through Farmers' Institutes, regular newspaper articles, and the publication of a farm journal. In addition, the bureau enforced the regulatory laws that called for inspection and certification of fertilizers, inoculation of hogs, and enforcement of livestock quarantines.

In 1904, J. W. Morris organized Tennessee's first Farmers' Union at Big Springs in Madison County. Within days, several additional unions appeared, but state organization did not occur until 1906, when Newt Gresham, one of the founders of the National Farmers' Union, arrived to supervise the establishment of the Tennessee Farmers' Union. Rallying behind the motto "To Secure Equity, Establish Justice and Apply the Golden Rule," cotton farmers in West Tennessee joined the Union. Pledging to support programs to place a price floor under cotton, they vowed to withhold 10 percent of the crop and reduce their cotton acreage. Tennessee Union men canvassed the cotton-producing counties in a house-to-house effort to obtain pledges from members and nonmembers to plow under and market at least a portion of the crop through the Union agency. By 1914, the Farmers' Union had spread to sixty-five counties and claimed 25,000 members. Like earlier agrarian efforts, the Farmers' Union also looked beyond prices to advocate rural uplift and achieve social justice. Thus, while their marketing efforts mimicked and overlapped those of the Southern Cotton Growers' Association, they also supported practical social reforms such as consolidated schools, tax relief, and social uplift more characteristic of the Progressive Country Life Movement.[52]

The leadership of the Farmers' Union included a number of familiar faces. Seven of the twenty-two men identified in the official history of the Farmers' Union had played important roles in the Grange, the Wheel, and the Alliance before turning to the Union. Two Union men had run for congressional seats on the Populist ticket in 1896. Christopher Columbus (C. C.) Adams, Wheeler and Allianceman and former state legislator (1889–1891), signed up with the Farmers' Union. The fifty-year-old Weakley County farmer who had served as the Wheel's state lecturer put his talents to work as a member of the Union

Lecture Bureau for Middle and West Tennessee. Another Weakley County native, J. E. Montgomery, became the first state president of the Tennessee Farmers' Union. Described as a "successful and progressive farmer," he had taught in the public schools and served multiple terms as justice of the peace. Like many other Union members, Montgomery had been both a Wheeler and an Allianceman.[53]

Foremost among the Union men was John H. McDowell, the inveterate organizer and agricultural progressive. Not surprisingly, Charles Barrett, national president and historian of the organization, claimed that McDowell joined as soon as the Farmers' Union appeared in Tennessee. President of the Obion County Union, he also served as a delegate to the 1909 National Convention. Less visible than previously, McDowell volunteered his considerable political acumen to work for Union goals. He successfully ran for a seat in the Tennessee General Assembly in 1904, ending a legislative career that included two terms in the house (1883–1885, 1905–1907) and two terms in the senate (1885–1889). In 1909 he put his political skills to work on direct behalf of the Farmers' Union when he served as a member of a three-man committee to lobby for a law to abolish bucket shops in Tennessee. At the same time he accepted a position as the state livestock inspector, an overwhelming job with little legislative support. Tennessee, under pressure from the USDA, engaged in a tick-fever eradication program, as well as fights against cattle tuberculosis, sheep scabies, and hog cholera. In his report to the Commissioner of Agriculture, McDowell complained about the lack of funding for the task, comparing Tennessee's paltry appropriation of $5,000 to Mississippi's $40,000.[54]

The most influential leader in the Tennessee Farmers' Union was Thomas J. Brooks. Born in Carroll County in 1870, Brooks became an active member of the Populist Party and ran an unsuccessful campaign for the eighth congressional seat in 1896. His early political failure did not reduce his commitment to rural uplift or to politics. In 1911–1912, he served one term as an Independent in the Tennessee Senate. A tireless Union man, Brooks held several state and national posts simultaneously. As national lecturer and organizer, he visited forty states extolling the merits of the Farmers' Union. As secretary of the Tennessee Farmers' Union and a member of the Board of Trustees for the Farmers' Union Cotton Warehouse in Memphis, he laid what he hoped would be a permanent foundation for the cooperative marketing of southern cotton.[55]

In 1912, Brooks accepted a position at Mississippi Agricultural and Mechanical College, where his friend and fellow Union advocate, George Robert Hightower, was the newly installed president. There, Brooks established what may have been the nation's first department of agricultural marketing. A popular professor, Brooks realized there were no textbooks to guide his students, and in 1915 he published *Markets and Rural Economics* to advocate reforms in stock and commodity exchanges, cooperative marketing, and rural credit systems. He testified before congressional investigative committees on a variety of agricultural subjects including the proposed Federal Reserve bill, the Rural Credits bill, and the futures market. His growing national reputation earned him presidential appointments to at least three international agricultural conferences.[56]

In 1918, Brooks moved to Florida, where two years later he became assistant commissioner of agriculture. The Brooks's family influence over Tennessee politics and agriculture extended into the second half of the twentieth century. Brooks's sister, Melissa, married Jim Browning. One of their sons was Gordon Browning, three-term governor and six-term congressman.[57]

Although former agrarian insurgents played prominent roles in the Farmers' Union, the organization was never simply the Alliance redux. Everything from the official portraits of Union leaders to the policies the organization pursued suggested modern business practices. The men who stare out from the pages of Barrett's history are barbered and shaved. They wear suits and ties of the latest fashion. Only occasionally does a bushy beard remind us of the Union's radical heritage. Indeed, as one Mississippi farmer proclaimed, there was a time when "a Farmer was a man who wore billy-goat whiskers" and his long hair "contained much hay seed." Now, as a member of the Union, "he wears 'Prince of Wales' style whiskers, and parts his hair in the middle." Moreover, this business-minded farmer "has money loaned to the bank and he drives in from the farm with a pair of high steppers."[58]

In 1908, at its third annual convention, the Tennessee Farmers' Union voted to federate all Union warehouses under a single head with a three-man board of control. B. G. West of Memphis, a long-time friend of cooperative movements and "one of the best-informed men in the United States on the subject of handling cotton," managed the Farmers' Union Cotton Company at 68–70 Randolph Building and oversaw the movement of cotton from local storage facilities to the Memphis warehouse. Established in 1907, the company accepted cotton from West Tennessee, North Mississippi, and East Arkansas. The warehouse operated "as a medium through which [members could] sell their cotton, but by members and unions taking all its stock as an investment." The Union warehouse sold cotton by sample in lots of fifty or more bales. The Memphis warehouse made "liberal advances on cotton shipped to . . . Memphis, Manchester, England, and Liverpool, England." If farmers chose to leave their cotton in a local warehouse, the Memphis cottonmen would assist in acquiring advances from local banks. Perhaps more important, the Union Cotton Company established a cotton grading school to standardize and modernize sales. Initially resisted by the New York Exchange, this standardized method of grading was finally promoted by the USDA and accepted by all exchanges in 1911.[59]

Charles Barrett, the long-serving national president of the Farmers' Union, thought Tennessee critical to the success of the organization. He anticipated that the Memphis cotton company would become "no small factor in controlling the price of the staple product throughout the South." He also believed Nashville would rise as a central shipping point for grain into the Deep South. Indeed, he anticipated Nashville would one day be "the base of supply for the states that lie East of Tennessee." In the minds of rural reformers, Tennessee's river and rail system "furnish[ed] ample facilities for cheap transportation," offering the Union an opportunity to profit from "the advantageous situation" of its geography. In assessing the state's potential, Barrett echoed the claims of earlier Alliancemen and New South proponents like Killebrew and DeBow as they too had looked forward to the day when Nashville would become the South's Minneapolis, and Memphis would handle most of the region's cotton. However, recognizing the potential and marshaling the capital and the legislative support to fulfill the expectation were altogether different problems.[60]

Although the Farmers' Union failed to meet its highest expectations in Tennessee, it did make some impressive gains and fostered a new relationship with the state's Department of Agriculture. In his 1914 biennial report to Governor Ben Hooper, Tennessee Commissioner of Agriculture Thomas F. Peck testified to the effectiveness of the Farmers' Union. Acknowledging his own membership in the organization, Peck claimed that "the small farmers, who collectively cultivate the largest area of our farm land" were the natural clients of the agricultural department. In a rare moment of candor, the commissioner confessed that past efforts to provide farmers with up-to-date information had "been presented . . . in such way that only the best equipped farmers could avail themselves of the information." Indeed, most of the suggestions "have been beyond the small farmers." This confession suggested a transformation in Tennessee agriculture: the farmer of modest means rather than the largest planter would control the agenda. Under his administration, and with the assistance of the Farmers' Union, Peck claimed the state agency directed its advice toward smaller producers and promoted improvements that realistically met the needs of commercial-minded farmers.[61]

Citing the warehouses, banks, fertilizer plants, cooperative creameries, and lime crushers owned by the Farmers' Union as examples of rural efficiency and modern organization, Peck proclaimed that rural unionism promised "better roads, better schools, better farms, happier homes, and brighter firesides." To ensure the continued growth of rural cooperativism, Peck's agency held Farmers' Institutes in each of the three grand divisions, organized community cooperatives, supplied weekly Talks to Farmers columns to rural newspapers, and published a monthly farm journal, *Tennessee Agriculture*. In each of these initiatives Peck advocated more diversified agriculture of fruit and vegetable crops, beef cattle production, dairy farming, and poultry and eggs. Essays on cotton and tobacco occupied increasingly smaller space in the biennial reports over the course of the first decade of the twentieth century, and in Peck's report, they appear only in the month crop information.[62]

Peck's enthusiasm captured the resurgent reform spirit of the new century, but he overestimated the climate for change. Progressives in the open country and in Tennessee's cities joined hands to improve home and community life. State and federal agencies established common ground for farm programs with voluntary associations in a blending of interests unknown in past agrarian efforts. Nevertheless, Tennessee's actual accomplishments fell short of those of other states as localism, rural skepticism, political intransigence, and divisive history limited agrarian achievements to more modest gains.

A number of historians of agriculture and progressivism have expressed reservations about the turn-of-the-century efforts to improve farming and the rural home. Indeed David B. Danbom condemned the Country Life Movement as the "industrialization of agriculture," an urban reform that promoted efficiency as a means "to make the farmer a productive supplement to the increasingly industrial sector of the nation." He and others, including Tennessee historians Mary Hoffschwelle and Jeanette Keith, suggested that farmers resisted the reformers' schemes or, at best, accepted only those ideas that offered immediate practical results or could be readily adapted to traditional folkways. And indeed, most Tennessee farmers continued to live their lives as their fathers had. But not every farmer resisted, and more recent historiography has reevaluated the effect of Progressive and Extension efforts.[63]

In 1908, the progressively activist president, Theodore Roosevelt, penned a letter to Liberty Hyde Bailey, Cornell University professor and foremost agriculturalist of the day, appointing him to chair a fact-finding commission on rural America. In his letter, the president expressed the traditional view that national greatness rested upon the "well-being of the farmer class." Although Roosevelt acknowledged data suggesting that "farmers in general are better off today than they ever were before," he worried that "the social and economic institutions of the open country [particularly in the South] are not keeping pace with the development of the nation as a whole." In commissioning the investigative body, Roosevelt hoped to "secure . . . such information and advice as will enable me to make recommendations to Congress" on the "question of securing better business and better living on the farm." Through the establishment of a comprehensive plan of cooperative marketing and social uplift the president anticipated country life would become "more gainful, more attractive, and fuller of opportunities, pleasures, and rewards for men, women, and children of the farms." Roosevelt believed, as did many Progressive reformers, that a revitalized country life would reverse the out-migration to the cities and secure order and stability in a nation in the throes of economic and social upheaval.[64]

Finally submitting to pressure, Bailey accepted Roosevelt's appointment to chair the blue-ribbon panel made up of the most important national figures in agriculture and rural life, including Henry Wallace, Kenyon Butterfield, Gifford Pinchot, and Walter Hines Page. When Charles Barrett, National Farmers' Union president, discovered he was not among the "elect," he successfully lobbied Roosevelt for a spot on the commission. Barrett later countered criticism that the commission "sought to teach the farmer his business or make over creation while-you-wait," by reiterating Roosevelt's view that it "was founded on the supposition that the farmer is at a fatal disadvantage with the rest of society," a position claimed by earlier agrarian movements. Barrett reassured his readers that the commission only "aimed at investigating and discovering these inequalities and striving to remove or mitigate them." The Georgia farmer was proud of his service to the Country Life Commission and suggested that Union members shared his enthusiasm for this rural example of the "square deal."[65]

The Country Life Commission obtained its data through a variety of means. Government agencies and rural organizations supplied names to whom the commission mailed 500,000 questionnaires; some 115,000 were answered and returned. In addition to these documents, "great numbers of letters and carefully written statements" were addressed to the commission from farmers everywhere. State superintendents of school were encouraged to hold public meetings to invite discussion of the questions; reports from the meetings were transmitted to the commission. Finally, the investigators held formal hearings at thirty cities, including Knoxville. In the language of cooperativism and community-building, the final report attacked traditional localism, outdated farming methods, and three-R schooling. In place of these it offered a modern blueprint for the creation of social institutions to anchor the rural family, improve agricultural efficiency, and promote intellectual and social growth. Whereas earlier rural reformers centered their efforts on the country church, the one-room school, and the ballot, the Country Life Movement that emerged from the commission report posited a new trinity: the consolidated school, the agricultural expert, and the modern home. In Tennessee, the Farmers' Union, the Department of Agriculture, and the experiment station absorbed the Country Life recommendations and worked to see them implemented in law and in daily experience.[66]

When the commissioners met in Knoxville, the audience of "prominent [Tennessee] farmers and educators" gave special attention to the need for good roads and improvements in education. John M. Thompson, commissioner of agriculture, testified that despite an 1891 act requiring the teaching of agriculture in the public schools, only a few counties complied. In response to queries about the availability of instructors with credentials to teach agriculture, Thompson suggested that county courts might subsidize a course of study at the experiment station for county superintendents, who could then train local teachers. Subsequent speakers supported agricultural education, particularly as a mechanism for stemming the flow of farm boys to the city, and they generally saw country schools as deficient but offered few concrete suggestions for improvement.[67]

Though no one mentioned it at the hearing, an example of agricultural education in action could be found within driving distance of the Market Hall, where the commission met. In 1902, Knox County farmers met with Charles Dabney, president of the University of Tennessee, Philander Claxton, professor of education, and Wallace Buttrick, executive secretary of the General Education Board, to devise a plan for the construction of an agricultural school. A subscription drive raised $5,000 for the endeavor, which the GEB matched. The school won the support of the University of Tennessee as a model for other rural communities and a laboratory school for their own future teachers. The farmers purchased twelve acres near Concord on which they constructed a six-room school, an assembly hall, a barn, and a house for the principal and his family. The combination primary and high school opened in February 1904.

A fire in 1906 destroyed much of the school and tested the mettle of the community. Housing the classes in temporary quarters, farmers used the insurance money and borrowed additional funds to construct a two-story brick building with indoor plumbing. In 1914, *Tennessee Agriculture* reported that the school had "an up-to-date poultry yard" and kept "a fine thoroughbred mare for breeding purposes." The school principal, Adams Phillips, who held a degree in agricultural science, directed the students (both boys and girls) in demonstration plots following the experiments outlined by the Tennessee Department of Agriculture and the USDA. The school boasted 150 primary students and 90 high-school students. Phillips reported that, although agriculture was not compulsory, 90 percent of the students studied it, with the girls substituting "domestic science for part of the agricultural work."[68]

Few communities enjoyed the commitment farmers made to agricultural education that inspired Concord to undertake a project of such magnitude. Nevertheless, in varying degrees, many counties offered some agricultural education. *Tennessee Agriculture* published pictures of agricultural classes at Hixon High School in East Tennessee and Paris High School in West Tennessee. Local newspapers encouraged the formation of Boys' Corn Clubs; by 1913, almost four thousand boys had joined such clubs. On the Cumberland Plateau, the success of the Boys' Potato Club enticed area farmers to plant the tuber where they had previously raised ever-smaller crops of corn. As an incentive to support agricultural education, the Tennessee State Fair hosted an annual Farm Boys' Encampment. County winners of a "Why I Want to Go to the Fair" essay contest spent a week in Nashville, viewing the fair exhibits, attending lectures presented by noted political and agricultural speakers, and sightseeing in Nashville. Sponsors of the event believed "the boys went home with a better knowledge of the resources of Tennessee and a better acquaintance with the people of the State." It was expected that they would "talk about what they saw and make others anxious to see the fair next year, and will doubtless put into practice when they go back to work on the farm much of the excellent advice given them in the lectures heard." Tennessee was one of the first states to develop girls' canning clubs. Under the direction of Virginia Moore, and with the sponsorship of the Department of Education and the General Education Board, girls grew and canned tomatoes, competing for prizes and learning valuable skills. Myrlie Hardin of Benton County gained statewide fame for her tomato and canning efforts. In 1913, she earned a $100 profit on the tomatoes she grew on one-tenth of an acre. The women's division of the Farmers' Institutes devoted considerable attention to girls like Hardin and publicized their productive, domestic, and earning skills.[69]

The Country Life Commission report found that the need for good roads vied with education as "the two needs most frequently mentioned in the hearings." In that regard, Tennesseans who appeared before the investigative committee behaved no differently from their counterparts elsewhere. Like other states, Tennesseans depended on the corvee system to maintain roads that often were little more than paths stretching from isolated farms to crossroads villages and county seats. No integrated highway system existed despite repeated calls for improved farm-to-market roads. A succession of agricultural commissioners, the leadership of the Farmers' Union, and individual farmers wrote letters to state and local newspapers, petitioned legislators, and addressed agricultural conventions on the issue of good roads, to no avail. Farm journals routinely printed drawings and instructions for making log drags to use in road maintenance, but as one historian noted, the use of the drag created a paradox by offering farmers "the possibility of continued self government by preempting the need for complicated organization, bonded debt, burdensome taxes, and high salaried civil engineers"; it failed to "foment a rural-based revolution in the administration of roads or forestall the urban-based revolution already underway." No one found a way to resolve the dilemma of rural eagerness for improved roads and rural reluctance to vote taxes to construct such highways.[70]

During the heat of election campaigns, congressional candidates held out the possibility of federal funding for highway construction, but most progressives looked closer to home for relief. In 1910, James Key of Jackson provided the annual Farmers' Institute with a detailed explanation of the financing and construction of almost sixty miles of macadamized roads in Madison County. The Good Roads Movement of Madison County obtained enabling legislation from the 1903 general assembly to authorize a bond issue for the construction. The resulting work provided approximately "five miles of graded and graveled roads on each of the roads heading out of Jackson." A second bond issue in 1909 added another forty-one miles of modern roads. Key supported the cost of construction with a story from his own experience. Using the old road, he could haul 1,000 pounds of hay the twelve miles to Jackson with his two mules. The effort required a hard day's work from his animals. On the new road, Key routinely hauled a load of 2,260 pounds (28 bales of hay) using the same mule power. In his estimation, he had "saved on one load one hand and one team one day's work." This savings would cover any increase in taxes, allowing him to "break even if I didn't haul another load this year."[71]

The barrage of Good Roads literature and the enthusiastic speeches from progressive farmers, Farmers' Union lecturers, and agricultural college professors rained down on the heads of farm families. Southern boosters encouraged road building as the mechanism for regional uplift. In the words of one historian of the movement, roads were "to the rural South what skyscrapers were to the urban South." Indeed, many viewed road construction as an urban-rural-business-state partnership. In a speech to the Appalachian Good Roads Convention in Asheville, North Carolina, G. Grosvenor Dawe, managing director of the Southern Commercial Congress, highlighted the cooperative effort between Hamilton County authorities and the Chattanooga and Tennessee River Power Company. The company agreed to construct an "automobile road" from the city to its lock and dam on the Tennessee River. In a cooperative effort with the county, the road would be extended to Monteagle, where it would link with "good roads that will carry the route to Nashville." Many modern firms, including electric companies and railroads, supported road building as a mechanism for promoting enhanced sales through farm diversification and consumer spending.[72]

Women were especially interested in good roads. The opportunities for better education in consolidated schools attracted women concerned about their children's futures. Even opponents of woman suffrage, such as Tennessee's Josephine Anderson Pearson, heard "'the whirr' of the automobile as it climbs the mountains and dashes along" as a vehicle of liberation that would free women "from the mud and isolation" that "breeds insanity." Farm women readily understood Pearson's characterization of rural isolation and used similar language in their correspondence with agricultural officials who sought their views on pending farm legislation. As the Smith-Lever bill made its way through congress, the USDA mailed questionnaires to farm women across the nation seeking to determine how the agency could expand its programs to meet the needs of country wives and daughters. The connected topics of better education, better roads, and access to money of their own topped their lists.[73]

Some farm women did not wait for farmers, legislators, and congressmen to act. Marshaling their own resources and calling upon the networks that bound rural and urban women together in the demands of household work, a group of farm women in Giles County demonstrated the benefits of cooperative marketing over a considerable distance. At the fourth annual meeting of the Middle Tennessee Home-Makers Association, the delegates heard a report on the butter and egg sales of the Pleasant Valley Ladies Aid Society, a women's club in rural Giles County. First organized at a local country church, the Society "developed into a club of diversified interests and activities" that brought together government programs, producers, and consumers in a mutually beneficial exchange of goods and services.[74]

USDA pamphlets instructed farmers on how to standardize their products to meet the demands of urban consumers, and the U.S. Postal Service, through its Rural Free Delivery program, provided the means to move home-produced goods from farm to urban kitchens. Beginning with urban friends, the Pleasant Valley women contracted to supply farm-fresh eggs and butter "of standard requirements." As word spread of the reliability of their products, the women took orders for butter and eggs from customers in Memphis, Decatur, Birmingham, and Montgomery. As one satisfied customer explained, "The enterprise, which started in a missionary spirit was soon established on a business basis." The rural society, made up of the wives of landowners and tenants, recorded sales in 1913 of over $2,000.[75]

Like their wives, progressive farmers increasingly bridged the traditional divisions between government, business, and agriculture as well as those between rural and urban interests to craft a more modern and inclusive approach to southern economic and social problems. The transformation was evident in the biennial reports of the successive agricultural commissioners. A combination of information and boosterism, the reports reveal a shift toward efficiency, a reliance on scientific research, and a business approach to farming that earlier reports lacked. As early as 1903, Commissioner Thomas H. Paine noted the "spirit of progress" manifest in new railroad construction, the expansion of rural free delivery, the good roads movement, and educational development. Statistical data suggest that Paine's enthusiasm was overblown, as the South continued to lag behind the rest of the nation in all the areas where he saw "progress." Nevertheless, the reports and the farmers who supported their conclusions demonstrate the presence of a small but important rural progressive movement that captured the spirit first manifested in the Grange, the Wheel, and the Alliance, and that resolved to see country life enter the modern age.[76]

By the end of the first decade of the twentieth century, Commissioner John Thompson (1909–1910) believed the overlapping agricultural agencies of the state and federal governments had confused Tennessee farmers. Some farmers complained that the Tennessee Department of Agriculture, the College of Agriculture at the University of Tennessee, and the experiment station at the University of Tennessee represented a costly duplication of services. But, as Thompson explained, their duties supported, rather than duplicated one another. "The business of the College of Agriculture is instruction, and the work of the Experiment Station of the College is investigation of the manifold problems of agriculture," he wrote in the biennial report to the governor. The Department of Agriculture executed the agricultural laws enacted by the state legislators. In an increasingly bureaucratic and scientific world, this directive included the compilation of crop reports, inspection of livestock for disease, and certification of fertilizers, feed stock, and seeds for purity. To accomplish the many tasks, the staff of the department expanded over the decade with a commensurate rise in budgets. By 1914, the Department of Agriculture oversaw a paid staff that included three assistant commissioners, a veterinarian, four seed and fertilizer inspectors, a state inspector of apiaries, and a state entomologist. Biennial appropriations increased from $14,448 in 1899–1900 to more than $45,000 in 1913–1914 as the department expanded and professionalized.[77]

At the turn of the century, the primary functions of the department were fertilizer inspection, the compilation of crop reports, and the annual convention of the Farmers' Institutes—functions little changed from the original duties of the Bureau of Agriculture. By 1914, Tennessee agricultural officials performed the services traditionally delegated to the department and oversaw new health and sanitation regulations designed to protect the state's producers and consumers. National markets in meat, dairy, livestock, and fruits and vegetables left buyers vulnerable to fraudulent labeling and unsafe or unsanitary products. At both the state and federal levels, government enacted legislation to address the need for pure food and drugs. Likewise, livestock producers and farmers faced increasing threats to their cattle, hogs, and poultry, as well as crops such as cotton, as the movement of agricultural products exposed their farms and ranches to outbreaks of plant and animal diseases and pests. In one example of modern medicine's effect on agriculture, the state constructed an anti-hog-cholera serum plant and devoted an entire issue of *Tennessee Agriculture* to an explanation of cholera epidemiology and the efficacy of protective vaccination.[78]

Increasingly, agricultural professionals promoted progressive agricultural practices and diversification through demonstration fields. Tennessee's first county demonstration agents were local progressive farmers who recruited demonstration plots from a variety of county farms, a practice that put demonstration work within the reach of virtually all farmers. The agents selected the farmers who would participate in the program, outlined the methods to be used, and oversaw the implementation of the demonstration. In addition, they organized Boys' Corn Clubs and other local farm groups. They were required to devote three days per week to demonstration work and received a modest salary of $37 per month. This local demonstration work, both in terms of staffing and land use, no doubt eroded the defenses of many farmers who harbored a strong suspicion against agricultural "book learning." The establishment of experiment stations in Jackson (West Tennessee) and Columbia (Middle Tennessee) addressed the ongoing concerns of agricultural leaders, but the continuous presence of local Farmers' Institutes, demonstration work, and boys' and girls' agricultural clubs brought progressive ideas home. Familiar faces were now trying to change old agricultural practices, and resistance became more difficult.[79]

Southern agriculture had been historically unaccepting of progressive changes and diversification. Much of that resistance found root in the lack of capital to initiate change. Blooded livestock were costly, and modern urban tastes demanded new standards in texture and taste for meat, milk, butter, and cheese that were obtained only through improved animal breeding. Cotton and tobacco still paid enough to discourage transformation of fields to fruit and vegetable production. And for farmers with marginal lands, the shift toward commercial agriculture carried unacceptable risks. Farmers decided to embrace or reject progressive agriculture on the basis of a complex calculus of soil fertility, risk, family and communal culture, and regional history. But progressivism was not sufficient to transform the region; only the Great Depression and World War II could accomplish such change.

Historians generally assert that the Progressives and New Dealers implemented Alliance and Populist demands. The direct election of U.S. senators, a federal income tax, the Federal Reserve system, and the New Deal agricultural programs all claim a heritage reaching back to the Alliance reformers. But most histories of the Alliance and Populism end with the election of 1896, and Progressive or New Deal histories note the similarities to earlier demands while suggesting an unbridgeable divide between Populism and Progressivism. Following the threads that tied Populism and Progressive agriculture together provides an interesting and complex tale of conflict and compromise.

The political defeat of Populism did not end agricultural problems. Low commodity prices, a decline in owner-operator farms, the lack of access to capital to make improvements or diversify production, and the economic and social consequences associated with tobacco and cotton monoculture combined to perpetuate poverty and localism. Dissuaded from engaging in mass political movements, farmers organized more narrowly, allying themselves by commodity or by locality. Generally these associations pursued limited goals through a combination of legal and bureaucratic channels and economic boycotts. Focused inwardly, these associations demanded adherence to rigid community standards to enforce solidarity and protect a way of life they believed to be at risk. As a result, confrontations between farmers and "trusts" were often violent, as in the Black Patch and Reelfoot Lake episodes. The principal exception was the organization of cotton farmers in the Farmers' Union. Pursuing a broader agenda that included economic advancement and rural uplift, Union men became embedded in the emerging agricultural bureaucracy and thereby influenced farm legislation at the state and federal levels.

One of the first post-Populist efforts to protect agrarian interests coalesced in the dark-fired tobacco region of Tennessee and Kentucky. Organized and directed by the largest planters in the Black Patch, the Planters' Protective Association attacked the monopolistic buying practices of the American Tobacco Company and the Italian Regie. The PPA built on the buying and marketing cooperativism made familiar by earlier agrarian organizations and promised to sustain traditional rural networks of communal and familial relations even as it expected to transform market relations. Largely unsuccessful in altering buying practices or transforming the market, the efforts of the PPA demonstrated a tantalizing blend of older forms of community vigilantism and newer judicial and bureaucratic pressures to gain stated objectives.

Other men, with weaker ties to commercial agriculture, fought for the perpetuation of the common rights in a second example of agrarian upheaval. The conflicts between free-ranging hog and scrub cattle raisers and their neighbors with vested interests in protecting crops and blooded livestock shaped the politics of the Gilded Age. The violent defense of common rights to land use reemerged in the area surrounding Reelfoot Lake in 1908. Similar to activities in the Black Patch, the Reelfoot Lake events are remembered primarily for the episodes of night riding. But like the tobaccomen, lakers pursued their goals through both traditional and modern mechanisms.

A third example more clearly engaged a significant number of former Wheel and Alliance leaders in a Progressive effort to transform agricultural markets and uplift the open country. These men and women joined the Farmers' Union, supported Farmers' Institutes, and campaigned for consolidated schools, good roads, and social reforms such as prohibition. As rural Progressives, they influenced legislative and regulatory changes, not through attempts to alter the political process but as part of the emerging bureaucracies of the day. Embedded in the educational and agricultural bureaucracy, they served as inspectors for regulatory agencies; demonstration farmers for the fledgling extension service; officers for county, regional, and state Farmers' Institutes; and expert witnesses in testimony before legislative committees at the state and national levels. Their expertise provided local farmers with practical plans for diversifying agricultural production. Their ideas and demands were written into Progressive-era legislation because they had a hand in writing the enabling acts and the regulatory rules.

The influence of these Progressive farmers extended into the future. Their sons and daughters became the teachers who taught at the rural consolidated schools, the public health doctors and nurses who tackled hookworm and pellagra, the home demonstration agents and extension agents who improved rural households and farms, and the Tennessee Valley Authority engineers who brought electricity to the countryside. The most notable of the Alliance and Populist descendents was James McGill Buchanan, the grandson of John P. Buchanan. Born in Murfreesboro and reared on the farm his grandfather established, Buchanan won the 1986 Nobel Prize for economics for "his development of the contractual and constitutional bases for the theory of economic and political decision making." A conservative economic theorist, this distinguished fellow of the American Economic Association and fellow of the American Academy of Arts and Sciences enjoys an international reputation for his articulation of public choice theory.[80]

Despite their cautious conservatism and their commitment to communal ties, Tennessee farmers of the late nineteenth century were not romantic agrarians.

They recognized the changing economy of the period and demanded the tools to participate in that change as independent producers, not hired hands. They questioned the image of progress that trampled men underfoot in the scramble for power and wealth. In their view, a few millionaires did not represent true progress. Unless all the people moved forward, progress became a hollow vessel. The success of the Bourbons and the New South proponents in retaining political and economic control stifled the voices of the Wheel and Alliance farmers in Tennessee, but it did not end their struggle to effect change. Indeed, when the doors to elective office closed, farmers found new means to exert their influence over changes in the countryside. As bureaucrats, not as legislators, they added their voices to the Progressive reshaping of American life. Recapturing the agrarian critique of American industrial society demonstrates the degree to which their demand for economic justice, their struggle with social change, and their questioning of the definition of progress has resonated throughout the twentieth century.

NOTES

Abbreviations

TSLA Tennessee State Library and Archives, Nashville
SHSC Southern Historical Society Collection, University of North Carolina

Introduction

1. Brice, "*I Had a Real Good Time*," 172.

2. Quote from Moore, "Agrarianism and Populism in Tennessee." Robison, *Bob Taylor*; Hart, *Redeemers, Bourbons, and Populists*. Scholars who view the agrarian reformers more favorably include Connie L. Lester (see, for example, "'Let Us Be Up and Doing'") and J. A. Sharp.

3. Hofstadter, *Age of Reform*; Goodwyn, *Democratic Promise*.

4. Barrett, *Mission, History, and Times*, 97.

Chapter One. Surveying the Barnyards

1. Diary of Archelaus M. Hughes (1867–1884) (hereafter cited as Hughes Diary), TSLA.

2. *DeBow's Review* (1866): 615–616; McKenzie, *One South or Many?* 88; Goldin and F. D. Lewis, "Economic Cost of the American Civil War"; Ransom and Sutch, "Impact of the Civil War and Emancipation on Southern Agriculture."

3. Killebrew, *Introduction to the Resources of Tennessee*; S. B. Smith, "Joseph Buckner Killebrew."

4. Corlew, *Tennessee*; West, *Tennessee Encyclopedia of History and Culture*.

5. McKenzie, *One South or Many?*; Winters, *Tennessee Farming*; Ash, *Middle Tennessee Society Transformed*; McDonald and McWhiney, "Antebellum Southern Herdsman."

6. *DeBow's*, 618; *Niles Weekly Register*, 8 May 1841.

7. J. H. Moore, *Agriculture in Ante-Bellum Mississippi*, 107; Winters, *Tennessee Farming*, 180; Gray, *History of Agriculture*; Burnett, "Hog Raising." Toll keepers along the French Broad River placed the annual number of hogs driven to market at 150,000 to 175,000.

8. Bolton, *Poor Whites of the Antebellum South*; Salstrom, *Appalachia's Road to Dependency*; Elizabeth Fortson Arroyo, "Poor Whites, Slaves, and Free Blacks in Tennessee, 1796–1861," in West, *Tennessee History*, 101–112.

9. Ash, *Middle Tennessee Society Transformed* and *When the Yankees Came*; Fisher, *War at Every Door*; Groce, *Mountain Rebels*; McDonough and Connelly, *Five Tragic Hours*; Connelly, *Civil War Tennessee.*

10. Killebrew, *Introduction to the Resources of Tennessee*, 353. Recent studies suggest that, although planters suffered economic losses with the destruction of slavery, they retained the same relative positions in terms of wealth within the context of local and state economic parameters. In Tennessee, the "bottom rail" did not move to the top. See McKenzie, *One South or Many?* chapter 3.

11. U.S. Bureau of the Census, *Report of the Statistics of Agriculture in 1890*, 100, 101, 103, 104, 106. Economic historians placed southern recovery from the Civil War at 1909, whereas northern recovery had been accomplished by 1879. See Goldin and F. D. Lewis, "Economic Cost of the American Civil War."

12. *Weekly Toiler* (hereafter *Toiler*), 10 Oct. 1888 and 23 April 1890.

13. Killebrew, *Introduction to the Resources of Tennessee*; Tenn. *Report of Bureau of Agriculture of Tennessee, 1889–90*; Tenn. *Bureau of Agriculture, Statistics, and Mines, 1891 and 1892.*

14. U.S. Bureau of the Census, *Report of the Statistics of Agriculture in 1890*, 57–58.

15. Senate, *Letters to Committee on Agriculture and Forestry*, 397. Cotton prices can be found in Sigafoos, *Cotton Row to Beale Street*, 347–348.

16. Killebrew, "Tobacco Growing in Tennessee," in Tenn. *Biennial Report of the Commissioner of Agriculture, 1899–1900*, 68–70; Waldrep, *Night Riders*, 22; T. A. Campbell, *Politics of Despair.*

17. Killebrew, "Tobacco Growing in Tennessee," 69–70.

18. Ibid., 73–78.

19. Ibid., 79.

20. Tenn. *Report of Bureau of Agriculture of Tennessee, 1889–1890*, 31–32; *Toiler*, 27 June, 11 July, and 7 Nov. 1888.

21. Tenn. *Biennial Report of the Bureau of Agriculture, Statistics, and Mines, 1891 and 1892*, 133–137.

22. Diary, Nannie Haskins Williams Papers, 17 May and 30 Dec. 1887, SHSC.

23. Many diaries discuss the production of butter, eggs, and chickens, and their importance to the family economy. See Robert Freeman Diary, 1885–1886, 18 April 1885, TSLA; Mary Kirkpatrick Tinsley Diary, 1891–1894, TSLA; Cartmell Diary, 8 March 1891, TSLA. J. B. Killebrew published many articles in the *Rural Sun* in support of poultry raising, see especially 28 Aug. 1873. D. Fink, *Open Country, Iowa*, 57–62, offers an excellent discussion of the importance of poultry and eggs for farm women. For another view, see L. A. Jones, *Mama Learned Us to Work.*

24. For examples, see Killebrew, *Introduction to the Resources of Tennessee*, 112–154; D. S. William, "Dairy Husbandry," Paper read before the Middle Tennessee Farmers' Institute, 12 Dec. 1900, in *Biennial Report, 1899–1900*, 80–85.

25. Tenn. *Report of the Bureau of Agriculture, 1889–90*, 45–46; *Toiler*, 8 Jan. 1890; Cartmell Diary, 8 March 1891, TSLA.

26. Hahn, *Roots of Southern Populism*; Turner, "Understanding the Populists."

27. Hahn, *Roots of Southern Populism*; Kirwan, *Revolt of the Rednecks*; Hackney, *Populism to Progressivism in Alabama*; Turner, "Understanding the Populists"; and Mayhew, "Reappraisal."

28. Killebrew, *Introduction to the Resources of Tennessee*, 361; Dyersburg *State Gazette*, 2 May 1874; J. T. Essary, "Does Farming Pay in Tennessee? If Not, Why Not?" *Proceedings of the Twenty-third Annual Session of the East Tennessee Farmers' Convention, held at Knoxville, May 17 and 18, 1898*, in Tenn. *Biennial Report of the Bureau of Agriculture, 1897–1898*, 120–126.

29. McKenzie, *One South or Many?* 151–152. As McKenzie viewed the change, he suggested that "the combined effect of these changes was to eliminate entirely the disparity in farm size that had distinguished the state's major regions before emancipation."

30. U.S. Bureau of the Census, *Historical Statistics of the United States: Colonial Times to 1970*, 461.

31. Harold Woodman, "Class, Race, Politics, and the Modernization of the Postbellum South," *Journal of Southern History* 63 (1997): 3–22; Claire Strom, "Texas Fever and the Dispossession of the Southern Yeoman Farmer," *Journal of Southern History* 66 (2000): 49–74.

32. U.S. Bureau of the Census, *Report of the Statistics of Agriculture in 1890*, 117–118; J. T. Moore, "Agrarianism and Populism in Tennessee."

33. U.S. Bureau of the Census, *Report on Population of the United States in 1890*; D. B. Dodd and W. S. Dodd, *Historical Statistics of the South*.

34. McKenzie, *One South or Many?* 154; U.S. Bureau of the Census, *Agriculture, 1909 and 1910*, vol. 5, 126.

35. U.S. Bureau of the Census, *Report on the Productions of Agriculture in 1880*; U.S. Bureau of the Census, *Report on Population of the United States in 1890*; U.S. Bureau of the Census, *Agriculture in 1900*; U.S. Bureau of the Census, *Agriculture, 1909 and 1910*; G. Wright, *Old South, New South*, 118.

36. Senate, *Report of the Committee on Agriculture and Forestry on Condition of Cotton Growers*, vol. 1; U.S. Bureau of the Census, *Report of the Statistics of Agriculture in 1890*. Tenancy included both cash renters and sharecroppers.

37. *Toiler*, 5 March 1890; Nannie Haskins Williams Diary, 6 Aug. 1888, SHSC; Kyriakoudes, "Country Comes to Town."

38. Bush, *Dorie*.

39. Keith, *Country People in the New South*, 22–23; *Carthage Courier*, 9 and 23 April and 21 May 1925.

40. Civil War Veterans' Questionnaires, TSLA.

41. Cash, *Mind of the South*, 161.

42. Woodman, "Class, Race, Politics," 8, emphasis in the original; McKenzie, *One South or Many?* 119.

43. McKenzie, *One South or Many?* 162; Cash, *Mind of the South*, 161.

44. Cartmell Diary, 7 June and 18 July 1890, TSLA; emphasis in the original.

45. Testimony of J. F. Frank, as reported in Senate, *Report of the Committee on Agriculture and Forestry on Condition of Cotton Growers*, vol. 1, 1473–1474.

46. J. O. Porter, *Problems Relating to the Development*, 33.

47. Henry Grady's "The Farmer and the Cities Speech," presented at Elberton, Georgia, 23 July 1889, reprinted from the *Atlanta Constitution*, 26 July 1889, in Bryan, *Henry Grady or Tom Watson?* 114–116; *Toiler*, 25 July 1888.

48. Hughes Diary, TSLA; *Toiler*, 28 Nov. 1888.

49. Bryan, *Henry Grady or Tom Watson?* 114–116; *Toiler*, 19 Dec. 1888. No doubt southern farmers were familiar with the problems of Irish tenant farmers. Democratic Party platforms routinely included a plank condemning the British for the so-called Irish Question.

50. *Toiler*, 4 Dec. 1889, 22 Jan. 1890, and 29 Jan. 1890.

51. *National Economist*, 2 May 1891. Critics of the Wheel and Alliance attempted to undermine the effect of the mortgage question by asserting that rural people found the census questions about mortgage indebtedness intrusive and resented being asked about their financial status.

52. U.S. Bureau of the Census, *Report on Farms and Homes in 1890*.

53. U.S. Bureau of the Census, *Report on Cotton Production in the United States*, 11–105; Killebrew, *Introduction to the Resources of Tennessee*, 1015–1193; F. A. Bailey, *Class and Tennessee's Confederate Generation*; McKenzie, *One South or Many?*; U.S. Bureau of the Census, *Report on Farms and Homes in 1890*, 622.

54. Hughes Diary, 14 March 1875, 10 April 1875, 31 Oct. 1875, 13 Dec. 1875, 2 July 1876, 8 Oct. 1876, and 15 May 1877, TSLA. Farmers likely recalled the hard times of the 1870s when sales of land for taxes were common. See Dyersburg's *Neal's State Gazette*, 7 March and 4 July 1874. Stock, "Real Estate Mortgages, Foreclosures, and Midwestern Agrarian Unrest," 89–106; Snowden, "Mortgage Rates"; Eichengreen, "Mortgage Interest Rates." Interest payments on mortgages in Tennessee in 1890 averaged $41 per farm. Not surprisingly, many of the agrarian reform leaders fell into the age range most likely to have mortgaged indebtedness. The lack of a manuscript census for 1890, however, makes it merely speculative to assert that Tennessee agrarian leaders were debtors.

55. *Toiler*, 26 Dec. 1888; *National Economist*, 2 May 1891; Eichengreen, "Mortgage Interest Rates," 1008–1009.

56. Dun, *Dun's Mercantile Agency Book*. For a discussion of post-Reconstruction banking, see James, "Development"; Sylla, "Federal Policy."

57. McKenzie, *One South or Many?* 188; *Toiler*, 5 Sept. and 12 Dec. 1888.

58. Hughes Diary, 27 Aug. 1875, TSLA.

59. For a discussion of farm life in antebellum Tennessee, see F. A. Bailey, *Class and Tennessee's Confederate Generation*; Winters, *Tennessee Farming*; McKenzie, *One South or Many?*; B. H. Clark, *Tennessee Yeomen*; Owsley, *Plain Folk of the Old South*.

60. Mayhew, "Reappraisal," 474–475; Hughes Diary, 27 Jan. 1875, TSLA.

61. Nannie Haskins Williams Diary, 6 Aug. 1888, SHSC, emphasis in the original.

62. Ibid. 15 Sept. 1888, 8 Nov. 1888, 14 Jan. 1889, and 27 Oct. 1889; emphasis in the original.

63. Hahn, *Roots of Southern Populism*; *Toiler*, 19 Sept. 1888 and 28 Nov. 1888; Response to Questionnaire by Attwood Pierson in Senate, *Letters to Committee on Agriculture and Forestry*, 400.

64. Cartmell Diary, 31 Dec. 1890 and 2 Feb. 1891, TSLA; emphasis in the original.

65. For a discussion of the national business changes and the growth of railroads, see Kolko, *Triumph of Conservatism*; Chandler, *Visible Hand*; Lloyd, *Wealth Against Commonwealth*; G. Porter, *Rise of Big Business*; Zunz, *Making America Corporate*.

66. Corlew, *Tennessee*, 241–268.

67. Hart, *Redeemers, Bourbons, and Populists*, 28–56; Duggins, "Background for Regulation," 13–32.

68. Herr, *Louisville and Nashville Railroad*; Edward A. Johnson, "Railroads," in West, *Tennessee Encyclopedia of History and Culture*, 771–773.

69. Killebrew Papers, "Recollections," TSLA; Stover, *Railroads in the South*; Burt, "Four Decades."

70. *Toiler*, 4 and 11 July 1888; *National Economist*, Nov. 1892.

71. *Toiler*, 30 Jan. 1889.

72. B. B. Davis, *Gates, Tennessee.*

73. McDowell's schedule appeared frequently in the *Toiler*; Keith, *Country People in the New South*, 76–89; Hughes Diary, 27 Jan. 1875, TSLA.

74. Duggins, "Background for Regulation," 49–97; *Tribune*, 15 March 1877, 17 Dec. 1882, and 4 March 1883; J. B. Killebrew, the former Agricultural Commissioner and a railroad man, attributed the failure of the Railroad Commission to the demagoguery of John H. Savage, a common tactic for discrediting political adversaries. See Killebrew Papers, "Recollections," TSLA; *Acts of the State of Tennessee, 1883*, 199. Also see Burt, "Four Decades," 99–130.

75. Duggins, "Background for Regulation," 93; *Daily American*, 1 March 1884.

76. Hart, *Redeemers, Bourbons, and Populists*, 72–79; Duggins, "Background for Regulation," 95; R. H. White, *Messages of the Governors*, 7:110–115; *Journal of the Tennessee House of Representatives, 1885*, 666; *Journal of the Tennessee Senate, 1885*, 334–335; *First and Second Annual Reports of the Railroad Commission for the State of Tennessee*; *Acts of the State of Tennessee, 1885*, 155.

77. *Tribune*, 21 April 1887.

78. McKenzie, *One South or Many?* 188–189.

Chapter Two. Organizing for the Grand Work

1. Winters, *Tennessee Farming*, 173–74. Also see Appleby, "Commercial Farming"; Burnett, "Hog Raising"; Gray, *History of Agriculture*; T. B. Jones,

"Public Lands of Tennessee"; McDonald and McWhiney, "Antebellum Southern Herdsman"; Owsley, *Plain Folk of the Old South*; Owsley and Owsley, "Economic Structure of Rural Tennessee"; Morton Rothstein, "Antebellum South as a Dual Economy"; Winters, "Agricultural Ladder," "Farm Size and Production Choices," and "Plain Folk of the Old South Reexamined"; Woodman, *King Cotton and His Retainers*; G. Wright, *Old South, New South* and *Political Economy of the Cotton South.*

2. Winters, *Tennessee Farming*, 174–178.

3. S. B. Smith, "Joseph Buckner Killebrew," 61–68 (quote from page 65); Killebrew Papers, "Recollections," TSLA.

4. S. B. Smith, "Joseph Buckner Killebrew," 41.

5. S. B. Smith, "Joseph Buckner Killebrew," 61–68; Joseph B. Killebrew, *Address Before the Montgomery County Farmers' Club*, White's Chapel, Montgomery County Tennessee, 12 March 1870 (n.p., n.d.), 1–11, in Killebrew Papers, TSLA; Killebrew Papers, "Recollections," TSLA.

6. Killebrew Papers, "Recollections," 232–288, TSLA; S. B. Smith, "Joseph Buckner Killebrew," 65. In December 1871, as a result of agitation across the state, spurred in no small part by Killebrew, the Tennessee General Assembly established a Bureau of Agriculture. Made up of two appointed citizens from each of the state's three divisions, the bureau published crop reports and provided information about the state's timber and mineral resources. In 1875, the state legislature reorganized the cumbersome bureau and appointed Killebrew as the first commissioner of agriculture, a position he held until 1880.

7. S. B. Smith, "Joseph Buckner Killebrew," 65–67.

8. Ibid., 68–69; Nashville *Union and American*, 1, 4, and 5 Oct. 1871.

9. S. B. Smith, "Joseph Buckner Killebrew," 71–75; *Union and American*, 28, 29, 30, and 31 May 1873; Nashville *Daily American*, 16 Sept. 1875 and 16 May 1879; Dunning, *Farmers' Alliance History*, 298–302.

10. Buck, *Granger Movement*; Woods, *Knights of the Plow*; Saloutos, *Farmer Movements in the South*, 31–43; Saloutos, "The Grange in the South"; McMath, *American Populism*, chapter 2.

11. F. A. Bailey, "Patrons of Husbandry" in West, *Tennessee Encyclopedia of History and Culture*, 723–24; Saloutos, "The Grange in the South," 479–485.

12. S. B. Smith, "Joseph Buckner Killebrew"; *Declaration of Principles as Set Forth by the National Grange at Its Seventh Annual Meeting*, a pamphlet found in "Minutes of the Wilson Grange, Clarksville, Tennessee, 1873–1876," TSLA.

13. "Letter to O. P. Temple from J. A. Turley," box 2, f. 53, O. P. Temple Papers, Hoskins Library.

14. Saloutos, "The Grange in the South," 478–479.

15. Hughes Diary, 27 Jan., 27 Aug., and 5 Sept. 1875, TSLA.

16. Marti, *Women of the Grange*, 1; Columbus, Miss., *Patrons of Husbandry*, 1880, Mississippi Valley Collection.

17. "Minutes of the Wilson Grange," TSLA; "Minutes of the Mill Valley Creek Grange," Tennessee State Museum.

18. Hart, *Redeemers, Bourbons, and Populists*, 111; Saloutos, "The Grange in the South," 487.

19. McBride and Robison, *Biographical Directory*, 392–393; W. S. Morgan, *History of the Wheel*, 254, 269, 16; *Toiler*, 25 July 1888.

20. The seven founders of the organization were W. A. Suit, W. T. McBee, J. W. Walls, W. W. Tedford, H. B. Lakey, J. T. Thrasher, and J. W. McBee. Six of the founders were landowners, all were Democrats, and their motive in organizing was "self protection." W. S. Morgan, *History of the Wheel*, 16, 67, 82. Morgan was a member of the Agricultural Wheel and later the Farmers' Alliance. His history has served as a source of information for most studies dealing with the Wheel. Dunning, *Farmers' Alliance History*, 197. The biblical reference to Ezekiel's vision of the wheels is to Ezekiel 10. For a discussion of Jacksonian ideology, see Cooper, *The South and the Politics of Slavery*; Thornton, *Politics and Power*; Hahn, *Roots of Southern Populism*; H. L. Watson, *Liberty and Power*. For a discussion of Tennessee politics in the Jacksonian era, see Bergeron, *Antebellum Politics in Tennessee*, and Atkins, *Parties, Politics*.

21. W. S. Morgan, *History of the Wheel*, 65–76; Dunning, *Farmers' Alliance History*, 197–215. The five Tennessee delegates at the 1886 Litchfield convention were J. R. Miles of Ralston, W. T. Davis of Gleason, A. C. Gardner of Dresden, C. C. Adams of Raulston, and J. A. Welch of Dukedom.

22. W. S. Morgan, *History of the Wheel*, 109–114; *Toiler*, 1 Aug. 1888.

23. *Toiler*, 25 July 1888.

24. W. S. Morgan, *History of the Wheel*, 254, 269, 16; *Toiler*, 25 July 1888.

25. W. S. Morgan, *History of the Wheel*, 144–145. For a discussion of the organization of American society as a buttress against the disruptions of industrial life, see Hofstadter, *Age of Reform*; Wiebe, *Search for Order*; Hays, *Response to Industrialism*; Trachtenberg, *Incorporation of America*; De Santis, *Shaping of Modern America*; Cashman, *America in the Gilded Age*; Zunz, *Making America Corporate*.

26. W. S. Morgan, *History of the Wheel*, 24–28; Palmer, *"Man Over Money."*

27. W. S. Morgan, *History of the Wheel*, 123–125.

28. Ibid.

29. Lawrence Goodwyn, "The Alliance Develops a Movement Culture," in Holmes, *American Populism*, 26; McMath, *American Populism*, 53–55.

30. Goodwyn, "The Alliance Develops a Movement Culture," 27–34.

31. McMath, *American Populism*, 77–81.

32. Address by President Charles W. Macune to the National Farmers' Alliance and Cooperative Union, meeting at Shreveport, La., 12 Oct. 1887, as reported in Dunning, *Farmers' Alliance History*, 69; Schwartz, *Radical Protest and Social Structure*, 91–104.

33. W. S. Morgan, *History of the Wheel*, 295; Dunning, *Farmers' Alliance History*, 247; *Toiler*, 4 Dec. 1889 and 3 Dec. 1891; *National Economist*, 14 March 1889.

34. *Toiler*, 18 July 1888.

35. Ibid., 15 Jan. 1890.

36. Ibid., 27 June 1890.

37. Ibid., 22 Jan. 1890.

38. Ibid., 27 June and 19 Sept. 1888. Dunning, *Farmers' Alliance History*, 104. McDowell was born near Trenton in Gibson County in 1844. He attended St. Andrews College before leaving to serve in the Confederate ranks. He first joined Company H, 12th Tennessee Infantry, but was discharged as under age. He later served under Van Dorn and in Forrest's cavalry. He married Emma Sandeford of Gibson County, and they had seven children. Sometime after 1867, he moved to Arkansas, where he was a cotton farmer. On his return to Obion County, he again farmed until elected to the general assembly. He served in both the house (1883–1885, 1905–1907) and senate (1885–1889). He was a member of the Grange, the Agricultural Wheel, and the Farmers' Alliance. In the Wheel and Alliance, he held national office as state vice president. When John P. Buchanan was elected governor of Tennessee in 1890, McDowell became president of the Tennessee State Alliance. Martin VanBuren Ingram was born in 1832 in Montgomery County. At age seventeen, he took charge of the family's land and slaves. He supported Douglas in the presidential race of 1860. During the Civil War, he served as a member of Hawkins's battalion, attached to Maney's brigade. He was discharged for disability after Shiloh. He married Annie L. Farmer of Springfield, and they had seven children. In 1865, he launched his newspaper career with the publication of the *Register* at Springfield. In the 1890s he published the Clarksville *Tobacco-Leaf*. Ingram's interests in tobacco and journalism were still evident during the formation of the Planters' Protective Association and the Black Patch War of the early twentieth century.

39. *Toiler*, 3 Oct. 1888; *Gibson County Herald*, 7 Sept. 1888.

40. "Official Proceedings of the State Wheel," published in the *Toiler*, 1 Aug. 1888.

41. *Toiler*, 19 Sept. 1888.

42. Ibid., 19 Sept., 3 Oct., and 10 Oct. 1888. Ruff continued to publish his paper under the title of *State Wheel* until December 1888, when a fire at his Jackson office destroyed his presses. McDowell reported the loss, noting that the blaze had been set by an arsonist. See *Toiler*, 9 Jan. 1889.

43. *Toiler*, 1 Aug., 5 Sept. 1888, and 27 Nov. 1889.

44. Ibid., 12 Sept. 1888.

45. Schwartz, *Radical Protest and Social Structure*, 91–102; Hart, *Redeemers, Bourbons, and Populists*, 129–154.

46. *Pulaski Citizen*, 13 Sept. 1888; Cartmell Diary, 12 Sept. 1892, TSLA.

47. *Toiler*, 12 Oct. 1888.

48. Hart, *Redeemers, Bourbons, and Populists*, 126; *Nashville Tennessean*, 15 May 1930; West, *Tennessee's Historic Landscapes*, 419; *Toiler*, 12 Sept. 1888; McBride and Robison, *Biographical Directory*, 65; *Toiler*, 8 Aug. 1888.

49. Union City *Independent*, 7 May 1896; Obion County Tax Records, 1890, tenth district, Obion County Courthouse, Union City.

50. *Civil War Veterans' Questionnaires*, TSLA.

51. *Toiler*, 3 Oct. 1888.

52. Ibid., 31 July 1889 and 10 Feb. 1890.

53. Ibid., 11 July 1888; 7, 14, 21, and 28 Nov. 1888; 20 Feb. 1889; 27 March 1889. Salaries for state officers remained surprisingly low. Members of the Executive Committee received $2 per day while on Wheel or Alliance business. The state president received a $300 annual salary and $600 in travel expenses. See *Toiler*, 7 Aug. 1889.

54. Dunning, *Farmers' Alliance History*, 93–105. The financial woes of the Tennessee Alliance continued to the very end. See *Toiler*, 4 Dec. 1891. When the Alliance met in Cincinnati in 1891, the Tennessee delegation could not participate in the balloting because the state organization was not current in its membership dues.

55. *Toiler*, 4 Dec. 1889.

56. L. Fink, *Workingmen's Democracy*; McLaurin, *Knights of Labor in the South*; Shapiro, *New South Rebellion*; and Voss, *Making of American Exceptionalism*.

57. *Toiler*, 11 July, 7 Nov., and 21 Nov. 1888; 20 March 1889.

58. See Daniel, "Tennessee Convict War"; Hutson, "Overthrow of the Convict Lease System," 82–103; and Shapiro, *New South Rebellion*.

59. Gaston, *New South Creed*, 88, 173; Cash, *Mind of the South*, 183.

60. Wilson, *Baptized in Blood*; Gerster and Cords, *Myth and Southern History*; Simpson, *S. A. Cunningham and the Confederate Heritage*; Cox, *Dixie's Daughters*.

61. Rosenburg, *Living Monuments*, 162; Skocpol, *Protecting Soldiers and Mothers*; Cartmell Diary, 8 Jan. 1891, TSLA.

62. Rosenburg, *Living Monuments*, 37; *Acts of the State of Tennessee, 1887*, 105–106; *Acts of the State of Tennessee, 1889*, 344.

63. *Civil War Veterans' Questionnaires*, TSLA; McBride and Robison, *Biographical Directory*, 128, 109, 580–81; John Hugh McDowell Daybook in the McDowell Papers, Mississippi Valley Collection. Some histories refer to McDowell as General McDowell. This was not an earned military rank but an honorary rank bestowed by the Confederate veterans' organization in recognition of his leadership.

64. *Civil War Veterans' Questionnaires*, TSLA; Forrester, *Glory and Tears*, 143–160; E. H. Marshall, *History of Obion County*, 18–21.

65. Forrester, *Glory and Tears*, 143–160; E. H. Marshall, *History of Obion County*, 18–21; *Obion County History*, 1:100, 243; McBride and Robison, *Biographical Directory*, 128–29; Union City *Independent*, "The Populist Convention," 7 May 1896, Obion County Historical Society, Union City. For an example of the bitterness that accompanied Reconstruction politics, see Lufkin, "Forgotten Controversy," 37–50.

66. *Obion County History*, 3, 20; Christopher Losson, "Monuments, Civil War" in West, *Tennessee Encyclopedia of History and Culture*, 640–641.

67. F. A. Bailey, *Class and Tennessee's Confederate Generation*, 160, 112.

68. *Toiler*, 9 Jan. 1889; E. S. Evans, "Progressive Party in Tennessee," 85–87.

69. F. A. Bailey, *Class and Tennessee's Confederate Generation*, 58–76, 157–158.

70. Cartmell Diary, 30 Aug. 1892, TSLA; Hart, *Redeemers, Bourbons, and Populists*, 179–198; Simpson, *S. A. Cunningham and the Confederate Heritage*.

71. Jenkins, "An 1890s Tour of Chattanooga"; *Toiler*, 11 July 1890, 5 June 1889, and 3 July 1889; *Appeal-Avalanche*, 15–19 Nov. 1892; *National Economist*, 3 and 10 Dec. 1892; E. S. Evans, "Progressive Party in Tennessee," 85–87.

72. *Toiler*, 11 Dec. 1889.

73. Ibid., 31 July 1889; As Charles Reagan Wilson noted, most southerners eventually found a rationale for reconciliation after 1900 and the American experience in the Spanish-American War. Wilson, *Baptized in Blood*, 161–182.

74. Jeffrey, "Women in the Southern Farmers' Alliance," 75. Both Jeffrey and McMath, *Populist Vanguard*, chapter 5, depict Southern Alliance women as conservative in their views and hesitant to enter the public sphere. Also see Wagner, "Farms, Families, and Reform"; Watkins, "Political Activism."

75. Watkins, "Political Activism," 197–213; Wagner, "Farms, Families, and Reform," 361; Jeffrey, "Women in the Southern Farmers' Alliance," 75.

76. Mrs. C. A. Mitchell to the *Toiler*, 17 Oct. 1888; "L.E.R." to the *Toiler*, 16 April 1890; "Phoeba" to the *Toiler*, 19 March 1890.

77. Anonymous to the *Toiler*, 23 July 1890; "Wheeler's Wife" to the *Toiler*, 23 Jan. 1889. As the editor of the *Toiler* explained, women were "both members and honorary members since they did not pay assessments and dues." They could "get all the secret work and are initiated like male members. They can hold office, but do not vote unless the subordinate lodge, in making its by-laws, grants them that privilege. Ladies [cannot] be fined for non-attendance unless they accept an office, if the by-laws made by the sub union impose a fine for non-attendance." *Toiler*, 2 April 1890.

78. "Welcoming Address" presented to the Carroll County Wheel, by Miss Cora McLean, October 1889, printed in the *Toiler*, 4 Dec. 1889; essay by Miss Maggie Gwin, ibid.; Nancy Tarpley to the *Toiler*, 14 May 1890 and 8 May 1889.

79. *Toiler*, 27 March and 24 July 1889.

80. Minnie Herrell to the *Toiler*, 9 April 1890; "Mama's Boy" to the *Toiler*, 10 July 1889.

81. Fannie Davis to the *Toiler*, 24 July 1889.

82. *Toiler*, 31 July 1889, 24 Oct. 1888, 26 June 1889, and 17 July 1889; Dunning, *Farmers' Alliance History*, 310.

83. Gaither, *Blacks and the Populist Revolt*.

84. Ibid., chapters 2 and 3. Tennessee claimed 100,000 members of the Colored Alliance in 1890.

85. *Toiler*, 8 Aug. 1888, 19 Sept. 1888, and 18 Dec. 1889.

86. Ibid., 11 July 1888.

87. *Toiler*, 27 Feb. 1889. Although whites wanted the support of black farmers, they also wanted control over them. Black Wheelers seemed to be more dependent upon whites, for example to help establish Colored Wheels, while black Alliancemen

attempted to establish their autonomy in organizations in alliance with whites rather than subject to white control.

88. Although the Colored Wheels refused white help in organizing local units, the Colored Alliance actively solicited white agents to help in establishing suballiances. The differences between the two organizations' willingness to accept white help may have influenced McDowell's position in his conflict with Bernard.

Chapter Three. Building Cooperativism

1. *Report of the Commission on Country Life*; Wiebe, *Search for Order*. Numerous agricultural historians have pointed to the importance of rural communal networks, including Barron, *Mixed Harvest*; C. Clark, *Roots of Rural Capitalism*; Henretta, "Families and Farms"; Kulikoff, *From British Peasants to Colonial American Farmers*; Neth, *Preserving the Family Farm*.

2. Friedman, *Enclosed Garden*, 3; Rutman and Rutman, *Place in Time*; R. Isaac, *Transformation of Virginia*; McCurry, *Masters of Small Worlds*.

3. *Toiler*, 30 April 1890.

4. Obion County Historical Society, *Obion County History*, 1:452, 243, 158, 114; *Confederate Veterans' Questionnaires*, TSLA; McBride and Robison, *Biographical Directory*, 570–71.

5. *Toiler*, 4 Dec. and 25 July 1888; Union City *Independent*, 26 March and 7 May 1896; Obion County Tax Records, 1888–1896.

6. Obion County Historical Society, *Obion County History*, 1:452, 243, 158, 114.

7. U.S. Census, *Population, 1890*.

8. *Toiler*, 4 Dec., 27 June, and 9 Jan. 1889; DeFiore, "'Come and Bring the Ladies'"; Kinchloe, "Similarities in Crowd Control Techniques."

9. A brush arbor meeting is a summer revival held in a temporary structure covered with leafy branches. A dinner-on-the-grounds is a midday church picnic, often potluck.

10. For a discussion on the rural networks and the transition to capitalism, see Henretta, "Families and Farms"; Lemon, "Comment on James A. Henretta's 'Families and Farms'"; Rothberg, *From Market Places to Market Economy*; S. A. Mann, *Agrarian Capitalism in Theory and Practice*.

11. *Toiler*, 22 Aug. 1888. Corlew, *History of Dickson County*. For a larger discussion on the role of churches in rural communities, see Faragher, *Sugar Creek*.

12. *Toiler*, 25 July and 27 June 1888.

13. Ibid., 27 June 1888 and 17 July 1889.

14. Ibid., 5 Feb. and 19 March 1890.

15. Ibid., 23 April 1890.

16. Marty G. Bell, "Landmarkism" in West, *Tennessee Encyclopedia of History and Culture*, 522–523; Wamble, "Landmarkism"; Norton, *Religion in Tennessee*; Crews, *Church of God*; Spain, *At Ease in Zion*; Hill, *South and the North in American Religion*.

17. Soden, "Social Gospel in Tennessee"; W. H. Smith, "William Jennings Bryan and the Social Gospel"; Goode, "Godly Insurrection"; Bode, "Religion and Class Hegemony"; Luker, "Social Gospel"; Swanson, "Country Life Movement."

18. *Toiler*, 31 July 1889 and 19 Sept. 1888.

19. Ibid., 19 Sept. 1888.

20. Soden, "Social Gospel in Tennessee"; R. H. White, *Messages of the Governors*, 7:370; Kazin, *Populist Persuasion*.

21. *Toiler*, 18 July 1888.

22. Ibid., 18 July 1888, 1 Aug. 1888, 15 Jan. 1890, 19 Feb. 1890, 26 Feb. 1890, and 2 April 1890.

23. Ibid., 12 March 1890.

24. Dunning, *Farmers' Alliance History*, 330; *National Economist*, 29 June 1889; Wheel and Alliance bylaws. For a discussion of the importance of political education in the efforts of the Alliance, see Mitchell, *Political Education*.

25. *Toiler*, 22 Jan. 1890, 23 April 1890, 11 July 1888, and 13 Feb. 1889.

26. Ibid., 8 Aug. 1888.

27. Ibid.

28. Ibid., 23 July 1889. Historian Lawrence Goodwyn viewed the development of strong cooperative systems as the critical forerunner to third-party formation. See Goodwyn, *Democratic Promise*.

29. *Toiler*, 18 July 1888. The Safe Deposit Trust and Banking Company remained in operation until the Depression of 1893, when it closed permanently. See W. Waller, *Nashville in the 1890s*, 100. When the West Tennessee Agency began operations, a second account was opened in Memphis. George Albert Gowan was elected to the position of State Wheel business agent in July 1888 and continued in his position until he was appointed commissioner of agriculture in 1891. *Toiler*, 5 Dec. 1888 and 24 April 1889.

30. *Toiler*, 27 June and 25 July 1888; *Acts of the State of Tennessee, 1889*.

31. *Toiler*, 27 June, 25 July, and 24 Oct. 1888. See W. Waller, *Nashville in the 1890s*, 239–240.

32. *Toiler*, 3 April 1889. *Journal of the Tennessee House of Representatives, 1889*, 464; *Journal of the Tennessee Senate*, 1889, 227; *Acts of the State of Tennessee, 1889*, 346–347.

33. *Toiler*, 28 Nov. 1888. Macune acknowledged some of the financial pitfalls of developing a state agency in his frequent speeches before the national Alliance. Dunning, *Farmers' Alliance History*, 65–196. Also see Goodwyn, *Democratic Promise*, and McMath, *Populist Vanguard*, chapters 2 and 4.

34. *Toiler*, 28 Nov. 1888.

35. Ibid.

36. Ibid., 5 Sept. 1888.

37. Ibid., 7 Nov. 1888.

38. Ibid., 7 Nov. 1888, 24 April 1889.

39. Ibid., 14 and 28 Nov. 1888.

40. Ibid. White was replaced by T. J. Galloway. Although concern over the payment of assessments continued, nothing more was said about the Memphis agent's salary.

41. *Toiler*, 24 April 1889.

42. *Gibson County Herald*, 11 April 1889.

43. *Toiler*, 8 May and 9 Jan. 1889; McMath, *Populist Vanguard*, 51; Sharp, "Entrance of the Farmers' Alliance," 78n.

44. *Toiler*, 8 May 1889.

45. Cocke County Farmers' Alliance Proceedings, McClung Collection, January 1889, April 1890, July 1890. Also see A. C. Smith, "Southern Violence Reconsidered."

46. *Toiler*, 3 July 1889.

47. Ibid., 10 July 1889.

48. Ibid. Information about the Wheel community can be found in the Bedford County history in West, *Tennessee Encyclopedia of History and Culture*.

49. *Toiler*, 12 Sept. 1888, 12 Dec. 1888, and 30 Jan. 1889. A letter from the Chattanooga company was published in the *Toiler*, 20 Feb. 1889.

50. *Toiler*, 29 May 1889.

51. Ibid., 14 Nov. 1888.

52. Ibid., 5 June 1889. On the changing American economy, see Chandler, *The Visible Hand*; Lamoreaux, *Great Merger Movement in American Business*. For a discussion of farmers and the problems of trusts, see Sanders, *Roots of Reform*.

53. For a review of the antimonopoly rhetoric of the antebellum era, see Thornton, *Politics and Power*; H. L. Watson, *Liberty and Power*; Temin, *Jacksonian Economy*; G. Wright, *Political Economy*; Sellers, *Market Revolution*; Cutler, "Jackson, Polk, and Johnson."

54. Holmes, "Southern Farmers' Alliance." More limited discussions of the jute boycott can be found in Goodwyn, *Democratic Promise*, and McMath, *Populist Vanguard* and *American Populism*.

55. *Toiler*, 8 Aug. 1888.

56. Ibid., 12 Sept., 26 Sept., 3 Oct., and 17 Oct. 1888. Culp and Ross, *Gibson County Past and Present*, 323; Sigafoos, *Cotton Row to Beale Street*, appendix B: National and Local Cotton Market, 1865–1977, p. 347.

57. *Toiler*, 7 Nov. 1888.

58. McMath, *Populist Vanguard*, 57.

59. *Toiler*, 5 Sept. 1888.

60. Ibid.

61. Ibid.; Holmes, "Southern Farmers' Alliance," 65; McMath, *Populist Vanguard*, 57.

62. *Toiler*, 12 Sept. 1888, 24 Oct., and 7 Nov. 1888. Though no action was taken on the suggestion to use convict labor to produce jute bagging in Tennessee, it surfaced several times during the course of the boycott and provides an interesting insight into agrarian thinking in light of subsequent problems with convict leasing.

63. *Toiler*, 12 Sept. 1888; Holmes, "Southern Farmers' Alliance," 65–80.

64. Memphis *Avalanche*, 27 Oct. 1888; *Gibson County Herald*, 2 Nov. 1888. For a history of the Memphis Cotton Exchange, see Lax, "Memphis Cotton Exchange," and Sigafoos, *Cotton Row to Beale Street*, chapter 6, "The Power Brokers of Front Street."

65. *Avalanche*, 27 Oct. 1888; *Toiler*, 23 Jan. 1889; "Minutes of the Gibson County Wheel," January 1889, Gibson County Historical Society.

66. *Gibson County Herald*, 28 Sept. and 9 Nov. 1888. Among the agents of the Cotton Association was E. Richardson, the proprietor of the Wheel Store in Trenton. Richardson's actions point up the recurring problems experienced by wheels and alliances when they contracted with local merchants to obtain goods at a discount rather than establishing a county agency system.

67. *Toiler*, 24 April 1889.

68. Ibid., 29 May 1889; McMath, *Populist Vanguard*, 54.

69. *Toiler*, 29 May 1889; Holmes, "Southern Farmers' Alliance," 71–74.

70. *Toiler*, 2 Oct. 1889.

71. Ibid.

72. Ibid.

73. This second front in the southern war on monopoly has received no attention by historians of the era. Tobacco historians focus on the so-called Black Patch War of the early twentieth century, an uprising that encompassed the same area. For studies of tobacco culture, see Daniel, *Breaking the Land*; Kulikoff, *Tobacco and Slaves*; Breen, *Tobacco Culture*. For a history of the Black Patch and the Black Patch War, see T. A. Campbell, *Politics of Despair*; Waldrep, *Night Riders*; S. Marshall, *Violence in the Black Patch*; Gregory, "Desperate Farmers." For a contemporary view of tobacco farming, see Killebrew Papers, *Tobacco*, TSLA.

74. *Toiler*, 1 Aug. and 12 Dec. 1888.

75. Ibid.

76. Ibid., 1 Aug. 1888. The leadership in the Tobacco Congress and the Tobacco Association included a number of men who were founders of the Farmer's Association, an organization of upper-class planters. Among the members of both organizations was M. V. Ingram, former editor of the *Toiler*, who edited the *Tobacco Cultivator* during this period.

77. *Toiler*, 12 Dec. 1888.

78. Ibid.

79. Ibid., 10 and 17 Oct. 1888. Only a month later, Bozart (or Bogard) was having second thoughts. At the meeting of the Tobacco Congress, he "spoke as a speculator," saying that he had "already rented out his land . . . [and] was carrying now, about forty families, and did not see how he could keep it up." Although he "hoped the scheme would go through," he did not believe "the movement had begun soon enough." *Clarksville Evening Chronicle*, 29 Nov. 1888.

80. *Toiler*, 24 Oct. 1888.

81. Ibid., 24 Oct. and 28 Nov. 1888.

82. Ibid., 1 Aug., 5 Sept., and 12 Sept. 1888.

83. Ibid., 6 Feb. and 5 June 1889.

84. Ibid., 28 Nov. 1888.

85. Ibid., 22 May and 5 June 1889. Warfield turned up in another war of farmers and buyers. When the Black Patch erupted in a war to defeat the American Tobacco Company and Regie, Warfield became an important participant. See T. A. Campbell, *Politics of Despair*, 35.

86. *Toiler*, 1 and 22 May 1889.

87. *Toiler*, 22 May and 19 June 1889.

88. T. A. Campbell, *Politics of Despair*, 30–52.

Chapter Four. Farm Issues in the Political Economy

1. Winters, *Tennessee Farming*, 55–60; Killebrew, *Introduction to the Resources of Tennessee*, 360–415; Cronon, *Nature's Metropolis*; Mayhew, "Reappraisal."

2. For lists of members of the general assembly, see Miller, *Official and Political Manual*. Roger Hart counts the number of Alliance members in the 1889 legislature as between 41 and 52. See Hart, *Redeemers, Bourbons, and Populists*, 133–134. McDowell counted 46 members in the 1889 general assembly. See *Toiler*, 3 April 1889.

3. For a discussion on the role of republicanism in nineteenth-century political thought, see Atkins, *Parties, Politics*; H. Bergeron, *Antebellum Politics in Tennessee*; Ford, *Origins of Southern Radicalism*; Benning, *Jeffersonian Persuasion*; H. L. Watson, *Liberty and Power*; Thornton, *Politics and Power*; Shalhope, "Toward a Republican Synthesis."

4. Hicks, *Populist Revolt*; Hofstadter, *Age of Reform*; Goodwyn, *Democratic Promise*. Voss-Hubbard, "'Third Party Tradition,'" 124; Ryan, *Women in Public*; Clemens, *People's Lobby*; Kazin, *Populist Persuasion*.

5. Hart, *Redeemers, Bourbons, and Populists*.

6. Hofstadter, *American Political Tradition*, 36n; Peterson, *Jefferson Image in the American Mind*, 82.

7. Senator Warren Smith, "A Tribute to Labor," *Toiler*, 25 July 1888.

8. Quoted in F. J. Bryan, *Henry Grady or Tom Watson?* 111.

9. *Toiler*, 11 July 1888, 31 July 1889, and 19 Sept. 1888. Catherine McNicol Stock recently argued that such views characterized agrarian radicals from Bacon's Rebellion in 1676 to recent outbreaks of violence in the western states, which have been attributed to so-called militiamen. See McNicol Stock, *Rural Radicals*.

10. S. M. Evans and Boyte, *Free Spaces*; *Toiler*, 31 Oct. 1888.

11. *Toiler*, 7 May 1890, 11 July 1888, 4 Dec. 1889, and 11 July 1888.

12. Dunning, *Farmers' Alliance History*, 59, 72, 80–82; typed copy, *Proceedings of the National Farmers' Alliance and Co-operative Union of America*, Shreveport, La., Oct. 12, 13, and 14, 1887, p. 16, original in the possession of John Brannon, Union City, Tennessee; copy, *National Farmers' Alliance and Co-operative Union of America,*

Constitution and By-laws (Dallas, Tex.: Mercury Job Office, 1887); original in the possession of John Brannon.

13. *Toiler*, 27 June 1888.

14. Ibid., 25 July 1888.

15. Ibid., 16 Jan. 1889.

16. Ibid.

17. Ostler, *Prairie Populism*.

18. McKinney, *Southern Mountain Republicans*; Hart, *Redeemers, Bourbons, and Populists*, 28–55; Queener, "Decade," "East Tennessee Republicans as a Minority Party," and "East Tennessee Republicans in State and Nation"; P. E. Isaac, *Prohibition and Politics*.

19. Hart, *Redeemers, Bourbons, and Populists*, 1–84; Robison, *Bob Taylor*; Ball, "Public Career of Colonel A. S. Colyar"; Howell, "Editorials of Arthur S. Colyar." State Party platforms of the period can be found in R. H. White, *Messages of the Governors*, vols. 6 and 7; W. S. Morgan, *History of the Wheel*, 254.

20. McKinney, *Southern Mountain Republicans*, 143–144; Archer, "Life of John Chiles Houk"; Gentry, "Public Career of Leonidas Campbell Houk"; Seehorn, "Life and Public Career of Henry Clay Evans."

21. De Santis, "President Arthur"; Cartwright, *Triumph of Jim Crow*, 66; McKinney, *Southern Mountain Republicans*, 144.

22. Cartwright, *Triumph of Jim Crow*, 78, 85; Nashville *American*, 18 March 1884.

23. Miller, *Official and Political Manual*, 238–244; *Chattanooga Times*, 6 and 8 Aug. 1886; Nashville *American*, 28 June and 24 Aug. 1888. See also Couto, *Lifting the Veil*, 42, 61, 68–69.

24. Hart, *Redeemers, Bourbons, and Populists*, 56–83.

25. Woodward, *Origins of the New South*, 140–174; Billings, *Planters and the Making of a "New South"*; Wiener, *Social Origins of the New South*; Cobb, "Beyond Planters and Industrialists."

26. Napoleon Hill Folder, Memphis Public Library, Memphis, Tenn.; Hart, *Redeemers, Bourbons, and Populists*, 12–28; Belissary, "Rise of Industry"; T. W. Davis, "Arthur S. Colyar and the New South"; Ball, "Public Career of Colonel A. S. Colyar," 23–47, 106–128, 213–238; Howell, "Editorials of Arthur S. Colyar," 262–276; Sigafoos, *Cotton Row to Beale Street*.

27. *Rural Sun*, 12 June 1873.

28. P. E. Isaac, *Prohibition and Politics*, 15–17; *Constitution of Tennessee, 1870*, in Miller, *Official and Political Manual*, 126–148; Fussell Collection, TSLA.

29. *Chattanooga Times*, 8 Aug. 1886; Hart, *Redeemers, Bourbons, and Populists*, 64–65; Speer, *Sketches of Prominent Tennesseans*, 226–230; McBride and Robison, *Biographical Directory*.

30. *Toiler*, 18 July 1888.

31. Ibid., 11 July 1888.

32. Queener, "Decade of East Tennessee Republicanism," 61–89; Queener, "East Tennessee Republicans as a Minority Party," 49–73; Cartwright, *Triumph of Jim*

Crow, 210–220; Woodward, *Reunion and Reaction*; Brice, *"I Had a Real Good Time,"* 148–151; Hutton, "Election of 1896"; Terral, *Newell Sanders*, 81–82; Seehorn, "Life and Public Career of Henry Clay Evans," 52–72.

33. Letter, Leonidas C. Houk to G. P. M. Turner, 9 Jan. 1889, Houk Papers, McClung Collection.

34. *Toiler*, 18 July 1888.

35. Granberry, "When the Rabbit Foot Was Worked."

36. R. L. Taylor, "Demagoguery, Personality, and the Gospel of Democracy"; Robison, *Bob Taylor*.

37. Hart, *Redeemers, Bourbons, and Populists*, 100; *Toiler* quoted in Knoxville *Tribune*, 6 May 1888.

38. Hart, *Redeemers, Bourbons, and Populists*, 102; Memphis *Avalanche*, 10 May 1888; *Banner*, 9 May 1888; *American*, 10 and 11 May 1888.

39. Hart, *Redeemers, Bourbons, and Populists*, 267. Tennessee became the first southern state to enact disfranchising legislation. See also Woodward, *Origins of the New South*; Key, *Southern Politics in State and Nation*; Kousser, *Shaping of Southern Politics*; and Perman, *Struggle for Mastery*, 48–59.

40. For a thorough discussion of the Dortch Law, see Kousser, *Shaping of Southern Politics*, 110–113; Perman, *Struggle for Mastery*, 54–55; *Avalanche*, 28 Dec. 1888, 3 April 1889, and 27 March 1889; *Toiler*, 26 March 1890.

41. *Acts of the State of Tennessee, 1889*, 414, 416–417; *Acts of the State of Tennessee, 1890*, extra sess., 59–60.

42. *Acts of the State of Tennessee, 1889*, 437–438. For a discussion of the Lodge Elections Bill, see Williams, *Years of Decision*, chapter 1, "The Party Setting, 1888," and chapter 2, "The Republicans Govern, 1889–1891"; and Keller, *Affairs of State*, 456, 527, 563.

43. Sharp, "Farmers' Alliance and Tennessee Politics," 91; Miller, *Official and Political Manual*, 260, 261, 280, 281; Article 4, sect. 1 of the Tennessee State Constitution, *Journal of the Proceedings of the Convention of Delegates Elected by the People of Tennessee to Amend, Revise, or Form and Make a New Constitution for the State* (Nashville, 1870), 426, emphasis in the original.

44. "Protests," in *Proceedings of the Convention of Delegates*, 397–398.

45. Andrew Johnson to George W. Nixon, 4 March 1870, Papers of Andrew Johnson, Library of Congress, microfilm, Presidential Center, Hoskins Library.

46. F. B. Williams, "Poll Tax"; *Toiler*, 19 March 1890.

47. Sharp, "Farmers' Alliance and Tennessee Politics," 85; Hart, *Redeemers, Bourbons, and Populists*, 271–273.

48. *Chattanooga Times*, 18 Jan. and 10 June 1890; *Appeal*, 5 Nov. 1890; and Nashville *Herald*, 16 March and 13 July 1890. Kousser, *Shaping of Southern Politics*, 104.

49. Sanford, *Blount College and the University of Tennessee*, 36–39; Montgomery, Folmsbee, and Greene, *To Foster Knowledge*, 15–22.

50. Sanford, *Blount College and the University of Tennessee*, 112–117; R. H. White, *Messages of the Governors*, 6:87.

51. R. H. White, *Messages of the Governors*, 6:87–90.

52. Ibid., 90–92, emphasis in the original.

53. Ibid., 93–98, 104–107.

54. *Toiler*, 5 February 1890.

55. *National Economist*, 11 July 1891; Charles W. Dabney Papers, SHSC; Johnston, "Quills into Plowshares"; Montgomery, Folmsbee, and Greene, *To Foster Knowledge*, 132–145; Merrian, *Higher Education in Tennessee*. See also Neth, *Preserving the Family Farm*; Beaulieu, *Rural South in Crisis*; Mitchell and Lowe, "To Sow Contentment."

56. *Toiler*, 5 Feb. 1890; R. H. White, *Messages of the Governors*, 7:311.

57. *Toiler*, 5 Feb. 1890.

58. Ibid., 18 July 1888.

59. *National Economist*, 13 Dec. 1890.

60. West, *Tennessee Encyclopedia of History and Culture*, 1013–1014; Watts, "Knowledge for the Tennessee Farmer," 67.

61. *Acts of the State of Tennessee, 1871*, 105–106; R. H. White, *Messages of the Governors* 6:464–465, 6:471–472; S. B. Smith, "Joseph Buckner Killebrew."

62. *Rural Sun*, 3 Oct. 1872; Killebrew Papers, "Recollections," 247, TSLA.

63. Killebrew Papers, "Recollections," 258–266, TSLA.

64. Ibid.; *Journal of the Tennessee Senate, 1877*, 589; *Memphis Daily Appeal*, 3 Jan. 1877; *American*, 8 March 1877; McBride and Robison, *Biographical Directory*, 324–325, 551, 625.

65. *Toiler*, 6 March 1889; R. H. White, *Messages of the Governors*, 7:446–447; *Acts of the State of Tennessee, 1891*, 331.

66. Caldwell, *Studies in the Constitutional History of Tennessee*, 160.

67. Kantor and Kousser, "Common Sense or Commonwealth?" 201–241; Steven Hahn, "A Response"; Kantor and Kousser, "A Rejoinder." A survey of the legislature in the 1880s shows that lawyers constituted the largest professional group in both the house and senate in the 1880s. In 1881, there were 44 lawyers in the general assembly and only 9 farmers. By 1885, the farmers had improved their numbers to 25, but were still outnumbered by the 45 lawyers in the house and senate. More than one-third of the men who listed their occupation as farming were classified as large planters, an indication that their interests may not have coincided with those of the smaller, middle-level farmers. The impact of the election of 1890 can be measured in the number of legislators claiming law as their profession. Thirty-three members of the two houses were lawyers. The effect was felt most strongly in the house, where the number of lawyers dropped from the usual 30+ to 20. Sixty-six men, of whom 54 belonged to the Alliance, claimed agriculture as their occupation. In addition to the persistent strength of the legal profession in the legislature, a growing number of physicians were elected to public office. See Miller, *Official and Political Manual*, 238–248; McBride and Robison, *Biographical Directory*.

68. "Proceedings of the Twenty-third Annual Session of the East Tennessee Farmers' Convention, 1898" in Tenn. *Biennial Report of the Bureau of Agriculture*,

1897–1898, 83–86; Charles B. Fort, "Fence Law," address delivered at the Clarksville Farmers' Institute, November 1900, and quoted in Tenn. *Biennial Report of the Commissioner of Agriculture, 1899–1900*, 138–144.

69. Cartmell Diary, 4 Jan. 1892, TSLA; Hughes Diary, 19 Feb. 1875, TSLA; *Toiler*, 27 Feb. 1889.

70. Tennessee produced 427,623 sheep, valued at $832,330 in 1890, according to the report of the state agricultural commissioner. See Tenn. *Biennial Report of the Bureau of Agriculture, 1889–1890*, 40; *Toiler*, 25 July and 8 Aug. 1888.

71. *Rural Sun*, 8 Feb. 1877; Hughes Diary, 7 Jan. 1877, TSLA.

72. *Toiler*, 27 Feb. 1889.

73. Ibid.

74. Durrill, "Producing Poverty," 764–781; Keith, *Country People in the New South*; Barron, *Mixed Harvest*.

75. *Toiler*, 27 Feb. 1889.

76. Ibid., 27 Feb., 6 March, 13 March, and 27 March 1889; Durrill, "Producing Poverty," 764–781.

77. R. H. White, *Messages of the Governors*, 7:224, 284–286, 307; *Toiler*, 11 July 1888.

78. *Journal of the Tennessee House of Representatives, 1883*, 60; R. H. White, *Messages of the Governors*, 1:57.

79. Hyman, *Anti-Redeemers*, 99–100.

80. *Acts of the State of Tennessee, 1887*, 24–56; R. H. White, *Messages of the Governor*, 7:285.

81. "Minutes of the Gibson County Wheel and Alliance, 1888–1892," 3 Jan. 1889, photocopy in the possession of the Gibson County Historical Society; *Toiler*, 27 June 1888, 4 July 1888, 21 Nov. 1888, 4 July 1888, 23 Jan. 1888, 13 Feb. 1889, 27 Feb. 1889, 13 March 1889, and 20 March 1889.

82. *Toiler*, 20 March 1889.

Chapter Five. From Nonpartisan Politics to Populism

1. R. Gates, *Farmers, the Conservative Force in the Body Politic* (n.p., 1911) TSLA.

2. Sharp, "Farmers' Alliance and Tennessee Politics"; Sharp, "Entrance of the Farmers' Alliance," 9:77–92; Sharp, "Downfall of the Radicals in Tennessee," 105–124; Hart, *Redeemers, Bourbons, and Populists*, 234–235.

3. *Herald*, 1 Jan. 1890; *Toiler*, 15 Jan., 26 March, and 11 June 1890; *Banner*, 17 Jan. 1890; Speer, *Sketches of Prominent Tennesseans*, 570–572; Hart, *Redeemers, Bourbons, and Populists*, 136; Sharp, "Farmers' Alliance and Tennessee Politics," 43–44.

4. Hart, *Redeemers, Bourbons, and Populists*, 136–137; Sharp, "Farmers' Alliance and Tennessee Politics," 45; Speer, *Sketches of Prominent Tennesseans*, 266–268; *Avalanche*, 5 June 1890; *Journal*, 7 May and 19 July 1890.

5. *American*, 16 July 1890; *Toiler*, 30 April and 11 June 1890; Sharp, "Farmers' Alliance and Tennessee Politics," 51; Hart, *Redeemers, Bourbons, and Populists*, 144.

6. Sharp, "Farmers' Alliance and Tennessee Politics," 32, 47–48; Hart, *Redeemers, Bourbons, and Populists*, 135, 138; *Journal of the Tennessee House of Representatives, 1889*, 464; *Journal of the Tennessee Senate, 1889*, 227; *Acts of the State of Tennessee, 1889*, 346–347; *Toiler*, 3 April 1889, 26 March 1890, and 7 May 1890; *American*, 18 July 1890; *Banner*, 18 July 1890.

7. *Avalanche*, 25 June 1890.

8. Ibid., 15 July 1890.

9. *Toiler*, 19 March and 16 April 1890.

10. *Toiler*, 2 April 1890; *American*, 16 July 1890.

11. *Toiler* 23 and 25 June 1890; *Tribune*, 22 June 1890; Sharp, "Farmers' Alliance and Tennessee Politics," 49–50.

12. *American*, 15 July 1890.

13. Ibid., 19 July 1890; *Toiler*, 23 July 1890.

14. *American*, 15–19 July 1890; *Chattanooga Times*, 16–19 July 1890; Albert L. McNeal to Edward W. Carmack, 23 July 1890, in Carmack Papers, box 6, f. 4, TSLA.

15. *Toiler*, 12 Feb. 1890.

16. Miller, *Official and Political Manual*, 106–108, 174–179; McBride and Robison, *Biographical Directory*, 1:790–791; Sharp, "Farmers' Alliance and Tennessee Politics," 80; *Banner*, 10 Sept. 1890.

17. *Banner*, 10–13 Sept. 1890; Sharp, "Farmers' Alliance and Tennessee Politics," 82.

18. *Banner*, 24–25 Oct. 1890; *American*, 24–25 and 31 Oct. 1890; Sharp, "Entrance of the Farmers' Alliance," 82–83.

19. *Toiler*, 23 July and 12 Aug. 1890; *Banner*, 5 Aug. 1890.

20. *American*, 26 Sept.–1 Oct. 1890; *Appeal*, 25 Sept.–1 Oct. 1890.

21. McBride and Robison, *Biographical Directory*, 767–768; *American*, 16 July 1890; *Toiler*, 18 July 1888; U.S. Congress, 50th Cong., 2nd sess., *Congressional Record* 20 (1889), 254.

22. McBride and Robison, *Biographical Directory*, 956–957; *Biographical Directory*, 2013; *Toiler*, 26 Feb. and 30 April 1890; *American*, 3 July 1890; West, *Tennessee's Historic Landscapes*, 326.

23. McBride and Robison, *Biographical Directory*, 271–272, 968; *Toiler*, 29 Jan., 7 May, and 21 May 1890.

24. McBride and Robison, *Biographical Directory*, 583–584; Sharp, "Farmers' Alliance and Tennessee Politics," 88–89; Keith, *County People in the New South*, 12–25.

25. McBride and Robison, *Biographical Directory*, 987; Sharp, "Farmers' Alliance and Tennessee Politics," 89–96; Sigafoos, *Cotton Row to Beale Street*, 73, 143; Seehorn, "Life and Public Career of Henry Clay Evans."

26. McBride and Robison, *Biographical Directory*, 1647; *American*, 3 July 1890; *Toiler*, 21 May, 25 June, 2 July, and 13 Aug. 1890.

27. *Banner*, 3 Sept. 1890.

28. Ibid., 4 Sept. 1890.

29. Williams, *Years of Decision*, 1–70. Following the election losses of 1892, Populists continued to argue their party's relevancy: "There can be but two political parties in this country. The Republican party is dead. The Democratic party will not do anything on the silver question, and by the time of the next election things will be ripe for the People's party." See interview with L. K. Taylor of the *Toiler* in *Avalanche*, 15 Nov. 1892.

30. *National Economist*, 11 April 1891.

31. *Toiler*, 26 Oct. 1892.

32. Correspondence of the Governors, John P. Buchanan, TSLA; *Appeal-Avalanche*, 21 and 27 Jan. 1891; *American*, 27 Jan. 1891; *Banner*, 27 Jan. 1891; Hart, *Redeemers, Bourbons, and Populists*, 155–157.

33. *Banner*, 1 Jan. 1891; *American*, 11 Feb. 1891; *Tribune*, 4 March 1891.

34. *Journal of the Tennessee House of Representatives, 1891*, 295–296; Hart, *Redeemers, Bourbons, and Populists*, 161.

35. *Journal of the Tennessee Senate, 1891*, 34, 220, 366–367. Eight Democrats and one Republican supported the bill; twelve Democrats and seven Republicans opposed its passage.

36. *American*, 20 Jan., 30 Jan., and 4 Feb. 1891.

37. *Banner*, 19, 20, 21, and 30 March 1891; *Appeal-Avalanche*, 20, 21, and 24 March 1891; *Tribune*, 21 March 1891; *Journal*, 21 March 1891.

38. *Banner*, 20 March and 4 April 1891; *Appeal-Avalanche*, 21 March 1891. Tansail, a farmer from Weakley County, not associated with the Alliance, defended the tax rates by pointing to increasing expenditures. The measures he specifically mentioned as inflating the state budget had Alliance approval: the construction of the Confederate Home and the pension bill, and the proposed expenditures on repairs to the capitol.

39. *Appeal-Avalanche*, 24 March 1891; Sharp, "Entrance of the Farmers' Alliance," 123.

40. Shapiro, *New South Rebellion*; Daniel, "Tennessee Convict War"; Hutson, "Coal Miners' Insurrections"; Hutson, "Overthrow of the Convict Lease System"; Tennessee *Special Report of the Commissioner of Labor and the Inspector of Mines to His Excellency Governor John P. Buchanan*, hereafter cited as Tenn. *Special Report of the Commissioner of Labor*. For a study of the convict labor in the antebellum period, see C. P. White, "Early Experiments with Prison Labor."

41. Hutson, "Coal Miner's Insurrections," 105; Daniel, "Tennessee Convict War," 274; Tenn. *Special Report of the Commissioner of Labor*, 73–74. Hutson estimated that "Tennessee's convict labor from 1870 to 1890 netted a total profit of $771,391.48." He further noted that "this sum lacked but approximately $176,000 of repaying the state for the expenses of its penal institutions from their beginning in 1829." Hutson, "Coal Miners' Insurrections," 105. For studies on Arthur S. Colyar, see Ball, "Public Career of Colonel A. S. Colyar"; Howell, "Editorials of Arthur S. Colyar"; T. W. Davis, "Arthur S. Colyar and the New South."

42. Felton, "Convict Lease System of Georgia"; Hutson, "Coal Miners' Insurrection," 104; Daniel, "Tennessee Convict War," 279; Shapiro, *New South Rebellion*, 49–52.

43. Tenn. *Special Report of the Commissioner of Labor*, 8–16; *Tribune*, 5 Aug. 1891; *Journal*, 16 and 18 July 1891.

44. Tenn. *Special Report of the Commissioner of Labor*, 19–21; *Journal*, 17 and 18 July 1891.

45. Tenn. *Special Report of the Commissioner of Labor*, 22–25; *Journal*, 18 July 1891; Daniel, "Tennessee Convict War," 275–277; Hutson, "Coal Miners' Insurrection," 109–110.

46. Tenn. *Special Report of the Commissioner of Labor*, 24–25; *Journal*, 18–20 July 1891: Shapiro, *New South Rebellion*, 83–84; *Appeal-Avalanche*, 16–23 July 1891; *Tribune*, 18–25 July 1891.

47. Tenn. *Special Report of the Commissioner of Labor*, 29–32; *Journal*, 21 July 1891; Daniel, "Tennessee Convict War," 278–279; Hutson, "Coal Miners' Insurrection," 113–115; Shapiro, *New South Rebellion*, 86–89.

48. *Journal*, 25 July 1891.

49. *National Economist*, 15 Aug. 1891; *American*, 14 Aug. 1891; *Journal*, 14 Aug. 1891.

50. *Journal of the Tennessee House of Representatives, 1891*, extra sess., 61–62; Hart, *Redeemers, Bourbons, and Populists*, 173–175; Hutson, "Overthrow of the Convict Lease System," 82–86.

51. Cartmell Diary, 3 Nov. 1892, TSLA; Hutson, "Overthrow of the Convict Lease System," 86–91; Daniel, "Tennessee Convict War," 285–287. Although historians generally praise the decision during Turney's administration to end convict leasing, the legislation promised more than it delivered. Prisoners continued to mine coal for state use at the new Brushy Mountain Prison. Early in the next century, industrialists once again used convict labor to build their economic empires. Jere Baxter used convict labor to construct the Tennessee Central Railroad across the Cumberland Plateau. See Jeanette Keith, "Jere Baxter," in West, *Tennessee Encyclopedia of History and Culture*, 51–52.

52. *Journal of the Tennessee House of Representatives, 1891*, 152; R. H. White, *Messages of the Governors*, 7:370, 376, 380, 381–382, 387–388.

53. *Acts of the State of Tennessee, 1891*, 295; *Journal of the Tennessee House of Representatives, 1891*, 175, 190.

54. *Acts of the State of Tennessee, 1891*, 150, 48, 360, and 322; Tenn. *Biennial Report of the Bureau of Agriculture, 1891–1892*, 97–126; R. H. White, *Messages of the Governors*, 7:439–440.

55. Letter, M. V. Ingram to Edward Ward Carmack, Carmack Papers, box 5, f. 12, 11 May 1891, TSLA.

56. Ibid.

57. Ibid.

58. Ibid.

59. *American*, 9 April 1891.

60. *American*, 25 June 1891; *Banner*, 25 June 1891; Fitzgerald, *Union League Movement.*

61. *Banner*, 25 June 13 and 14 July 1891; *Appeal-Avalanche*, 25 June, 13 July, and 14 July 1891; Hart, *Redeemers, Bourbons, and Populists*, 169–172.

62. *Toiler*, 21 October 1891; *Gibson County Herald*, 18 Nov. 1891; *Pulaski Citizen*, 14 July 1892.

63. *Journal*, 15 Sept. 1892; Speer, *Sketches of Prominent Tennesseans*, 472–74; Robison, *Bob Taylor*, 178–179; R. H. White, *Messages of the Governors*, 7:451–453.

64. *Times*, 16 Aug. 1888: *Birmingham Herald*, 16 Aug. 1888.

65. *Journal*, 10 March 1892; *Banner*, 1 March 1892. Democrats countered that in 1880, Albert S. Marks had refused to run for a second term when he could not support the Democratic platform of 1880, and Buchanan should put the party ahead of his own views and do the same.

66. *Banner*, 5 May 1891, 29 Feb. 1892, and 23 May 1892; Hart, *Redeemers, Bourbons, and Populists*, 63–66, 177.

67. *Banner*, 10 March 1892; *Weekly Tribune*, 23 March 1892.

68. *Banner*, 4 April 1892.

69. Ibid., 5 April and 4 May 1892; *Gallatin Examiner*, 3 May 1892.

70. *Herald*, 11 May 1891; Fayetteville *Sun*, 16 Sept. 1891; Sparta *Expositor*, 23 Oct. and 6 Nov. 1891.

71. *Herald*, 7 May 1891; *American*, 15 Aug. 1891; *Tribune*, 13 Sept. 1891; Sharp, "Entrance of the Farmers' Alliance," 161–162. All seventeen Democratic dailies had expressed their opposition to Buchanan by March 1892 (Hart, *Redeemers, Bourbons, and Populists*, 181).

72. *Tribune*, 25 and 18 Sept. 1891; *Banner*, 11 March 1892; Cartmell Diary, 26 March 1892, TSLA. Among the well-known Democrats who stumped for Turney were Robert L. Taylor, Isham Harris, William B. Bate, Josiah Patterson, Benton McMillin, J. D. C. Atkins, James D. Porter, Edward W. Carmack, James D. Richardson, Joseph E. Washington, B. A. Enloe, W. P. Caldwell, Frank Wilson, and Frank P. Bond.

73. Sharp, "Farmers Alliance and Tennessee Politics," 179–180.

74. Voting tallies and county primary and convention news appeared daily in both the *Banner* and the *American*. Voting strengths changed daily, and the Nashville editors anticipated a Turney victory, but recognized how evenly divided many counties were. See *Banner*, 7 and 9 May 1892.

75. *Banner*, 10 and 11 May 1892; Hart, *Redeemers, Bourbons, and Populists*, 185–186. Despite Taylor's support for the People's Party, the Tennessee party traced its origins to the earlier Union Labor Party, men who were not members of the Alliance, but entered the political conflict through association with the Populists nationally. See Cincinnati *Enquirer*, 20 May 1891; *Appeal-Avalanche*, 21 May 1891; *Herald*, 16 May 1892.

76. *Banner*, 10 and 11 May 1892.

77. Ibid., 11 May 1892; *National Economist*, 19 March 1892.

78. *Banner*, 17, 18, and 21 May 1892; *American*, 18 May 1892; *Herald*, 12 May 1892; *Chattanooga Press*, 13 May 1892.

79. *American*, 26 and 27 May 1892; *Banner*, 26 and 27 May 1892. Hart notes that historians frequently misrepresent Buchanan's political strength in the early primaries and conventions, since his votes did not show up in the August state convention. As he points out, as late as May, Buchanan carried mass meetings in Warren, Cheatham, and Campbell counties, and lost in Sumner County by only 17 votes. See Hart, *Redeemers, Bourbons, and Populists*, 186–187.

80. *National Economist*, 2 July 1892; Hart, *Redeemers, Bourbons, and Populists*, 28–55.

81. *American*, 31 July, 7 Aug., and 10 Aug. 1892; *Banner*, 9 and 10 Aug. 1892; *Tribune*, 7 and 10 Aug. 1892.

82. *Pulaski Citizen*, 4 Aug. 1892.

83. Ibid., 18 and 19 Aug. 1892.

84. *Journal*, 14 Aug. 1892; Daniel, "Tennessee Convict War"; Hutson, "Overthrow of the Convict Lease System."

85. *Journal*, 15–18 Aug. 1892; Daniel, "Tennessee Convict War," 287–289; Hutson, "Overthrow of the Convict Lease System," 94–95.

86. *Banner*, 19 Aug. 1892; *American* 14, 16–18, 20, and 21 Aug. 1892.

87. *Banner*, 19 Aug. 1892; *American* 14, 16–18, 20, and 21 Aug. 1892.

88. *Banner*, 19 Aug. 1892; *American* 14, 16–18, 20, and 21 Aug. 1892.

89. Daniel, "Tennessee Convict War," 291–292.

90. Ordinary Democrats, like Robert H. Cartmell in Madison County, expressed the same views as the Nashville mass meeting. "Gov. Buchanan is not the man to act in an emergency like this," Cartmell wrote in 1892. He described the Alliance legislature of 1891 as "a poor one" and hoped "we will have no more of it." He noted Buchanan's endorsement by the Populists and claimed that the Madison County delegates to the convention were "no better than the radicals [Republicans]." He believed the party leaders were "pd for the part they acting" and doubted "if there is one of them that could have been elected to any office from constable up before the people." Cartmell assessed the Populist leaders as office-seekers, who "care nothing for the people or the good of government." Cartmell's observations so closely paralleled the Nashville meeting that they probably indicate the extent to which the Bourbon criticism of the Alliance and Populists drove the political debate of the late summer and fall. Cartmell Diary, 20 Aug. 1892, TSLA.

91. *American*, 23 and 24 Oct. 1892; *Appeal-Avalanche*, 23 Oct. 1892; *Commercial*, 23 Oct. 1892; *Chattanooga Times*, 23 Oct. 1892; *Journal*, 23–26 Oct. 1892; Hart, *Redeemers, Bourbons, and Populists*, 196; Sharp, "Farmers Alliance and Tennessee Politics," 196–202; Queener, "East Tennessee Republicans As a Minority Party," 70–72; McKinney, *Southern Mountain Republicans*, 146.

92. *Tribune*, 26 Oct. 1892.

93. *Journal of the Tennessee Senate, 1893*, 15; *Journal of the Tennessee House of Representatives, 1893*, 21; R. H. White, *Messages of the Governors*, 7:469–471.

94. Hart, *Redeemers, Bourbons, and Populists*, 208; *Commercial*, 17 June, 18 June, 6 Aug., and 16 Aug. 1893; *Banner*, 16 Aug. 1893, 11, 12, and 14 April 1894.

95. *Banner*, 11 and 12 April 1894; *Journal*, 12 April 1894; Hart, *Redeemers, Bourbons, and Populists*, 207–208, 263–265; R. H. White, *Messages of the Governors*, 7:517–518.

96. *American* 16 Aug. 1894; *Banner*, 15 Aug. 1894; Hart, *Redeemers, Bourbons, and Populists*, 209; R. H. White, *Messages of the Governors*, 7:518–520; J. E. Lewis, "Tennessee Gubernatorial Campaign and Election of 1894," 111–115.

97. *Commercial Appeal*, 15 Aug. 1894; *Banner*, 15 Aug. 1894; *American*, 15 Aug. 1894; J. E. Lewis, "Tennessee Gubernatorial Campaign and Election of 1894," 111–115; R. H. White, *Messages of the Governors*, 7:518–521, Hart, *Redeemers, Bourbons, and Populists*, 209.

98. *Banner* and *American*, 21 and 22 Aug. 1894; J. E. Lewis, "Tennessee Gubernatorial Campaign and Election of 1894," 111–119; R. H. White, *Messages of the Governors*, 7:522; Hart, *Redeemers, Bourbons, and Populists*, 209–211; Queener, "East Tennessee Republicans in State and Nation," 99–128; McKinney, *Southern Mountain Republicans*, 138–141; Terral, *Newell Sanders*, 80–81.

99. J. E. Lewis, "Tennessee Gubernatorial Campaign and Election of 1894," 224–237; Terral, *Newell Sanders*, 81–82.

100. *Speech of Hon. H. Clay Evans, Republican Nominee for Governor of Tennessee, delivered at Memphis, Tenn., Wednesday, Oct. 10th, 1904*, pamphlet (n.p., n.d.), TSLA; J. E. Lewis, "Tennessee Gubernatorial Campaign and Election of 1894," 224–243.

101. J. E. Lewis, "Tennessee Gubernatorial Campaign and Election of 1894," 234–236; *Chattanooga Times*, 10 Oct. 1894.

102. *Contest for Governor. Peter Turney, Contestant vs. H. Clay Evans, Contestee. Minority Report of General Committee of Investigation* (n.p., n.d), TSLA, 20; R. H. White, *Messages of the Governors*, 7:542–567, quoted on p. 564, emphasis in the original.

103. *Journal of the Tennessee Senate, 1893*, 15; *Journal of the Tennessee House of Representatives, 1893*, 21; *Journal of the Tennessee Senate, 1895*, 175–177.

104. "Congressional Directory," printed in *Senate Journal, 1895*.

105. *A Plea for the Right*, pamphlet (n.p, n.d.), TSLA, 1.

106. Eisenberg, "House Divided," 92–93; Hutton, "Election of 1896"; *Commercial Appeal*, 12 June 1895.

107. *Commercial Appeal*, 13 and 14 June 1895; W. J. Bryan, *First Battle*, 162; Eisenberg, "House Divided," 94–96.

108. *Commercial Appeal*, 13 June 1895.

109. W. J. Bryan, *First Battle*, 188–258; Goodwyn, *Populist Moment*, 247–256.

110. W. J. Bryan, *First Battle*, 260–286; Goodwyn, *Populist Moment*, 256–263; McMath, *American Populism*, 201.

111. *Commercial Appeal*, 14 Oct. 1894 and 31 July 1896; Hart, *Redeemers, Bourbons, and Populists*, 218–219; Hutton, "Election of 1896," 49, 68, 117–118.

112. W. J. Bryan, *First Battle*, 525; *Banner*, 21 July, 27 July, 5 Aug., 14–16 Oct. 1896; *Commercial Appeal*, 6 Oct. 1896.

113. R. H. White, *Messages of the Governors*, 7:640, 646–647.

114. Charles B. Simonton Papers, box 1, f. 5, Mississippi Valley Historical Collection; Letter from J. H. Crass, 16 May 1906, box 4, f. 4; and letter from William Hall, 16 May 1906, box 5, f. 2 in Carmack Papers, TSLA.

115. Sharp, "Farmers Alliance and Tennessee Politics," 200–205; E. S. Evans, "Progressive Party," 86–87.

116. Buchanan also addressed the Farmers' Institute with Governor Patterson in December 1908. See Tenn. *Biennial Report of the Department of Agriculture, 1907–1908*, 51–68.

Chapter Six. Up from the Mudsills of Hell

1. Tenn. *Biennial Report of the Bureau of Agriculture, Statistics, and Mines, 1891 and 1892*, 59.

2. *Biennial Report, Tennessee Bureau of Agriculture, 1893–1894*; Watts, "Knowledge," 86–91.

3. Tenn. *Biennial Report of the Commissioner of Agriculture, 1899–1900*, 199–201, 208–209, 215–218.

4. Justi, *Official History of the Tennessee Centennial Exposition*; *Banner*, 1 May 1897; Tennessee Centennial Exposition Papers, TSLA; Doyle, *Nashville*, 144–156; Doyle, *New Men*, 206–207; W. Waller, *Nashville in the 1890s*, 103–116.

5. Justi, *Official History*, 13–41.

6. Union City *Independent*, 6 Feb. 1896.

7. Ibid.

8. Justi, *Official History*, 360.

9. Ibid., 74.

10. U.S. Bureau of the Census, Agriculture, 1900 (12th) census.

11. U.S. Bureau of the Census, Agriculture, 1909 and 1910 (13th) census, vol. 5.

12. Ibid., 5:567–581; ibid., vol. 8.; Sigafoos, *Cotton Row to Beale Street*, 347–348.

13. U.S. Bureau of the Census, Agriculture, 1909 and 1910 (13th) census; U.S. Bureau of the Census, 1900 (12th) census, part 2, vol. 6; T. A. Campbell, *Politics of Despair*, 38.

14. See Gregory, "Desperate Farmers," 68–69. For a listing of average tobacco prices over a forty-year period from 1866–1905, see T. A. Campbell, *Politics of Despair*, 18–19.

15. Although all histories of the so-called Black Patch wars include accounts of night riding, the most important recent works are T. A. Campbell, *Politics of Despair*; S. Marshall, *Violence in the Black Patch*; and Waldrep, *Night Riders*. For a discussion of southern Progressivism, see Grantham, *Southern Progressivism*; Link, *Paradox of Southern Progressivism*; and Woodward, *Origins of the New South*.

16. T. A. Campbell, *Politics of Despair*, table 4, p. 25.

17. Ibid., 25–29.

18. For ledger books and sharecropper contracts, see Washington Family Papers, TSLA. See also Waldrep, *Night Riders*, 39; McBride and Robison, *Biographical Directory*, 300–301, 905–906, 955–957; and T. A. Campbell, *Politics of Despair*, 45.

19. Clarksville *Leaf-Chronicle*, 24 and 26 Sept. 1904; *Banner*, 26 Sept. 1904; Waldrep, *Night Riders*, 42–50; T. A. Campbell, *Politics of Despair*, 30–32; Gregory, "Desperate Farmers," 51–59, 73–79; Dyson, *Farmers' Organizations*, 281. Tennessee counties sending official representatives to this first meeting included Cheatham, Dickson, Houston, Montgomery, Obion, Robertson, and Weakley.

20. Waldrep, *Night Riders*, 40; Gregory, "Desperate Farmers," 78 (quote), 79, 93.

21. Waldrep, *Night Riders*, chapter 4, "Local Problems, Federal Help," 52–62.

22. Ibid., 56–62; Gregory, "Desperate Farmers," 219–228; *Banner*, 4 June, 28 June, 4 July, and 6 Aug. 1904; 21 June, 5, 7, 9, 11, 13, and 17 July 1906.

23. As early as 1874, Joseph Killebrew complained about tobacco taxes in his *Resources of Tennessee*, 100; Gregory, "Desperate Farmers," 215–216; Irish, *Joseph W. Byrns of Tennessee*, 44.

24. T. A. Campbell, *Politics of Despair*, 12–13, 51–52.

25. Ibid., 78–79; Waldrep, *Night Riders*, 70.

26. For a general discussion on southern violence, see Cash, *Mind of the South*; Reed, *Enduring South*; Woodward, *Origins of the New South*; Wyatt-Brown, *Southern Honor*; and Hackney, "Southern Violence." For discussions on violence in Black Patch during the Civil War and Reconstruction periods, see Ash, *Middle Tennessee Society Transformed* and *When the Yankees Came*; S. Marshall, *Violence in the Black Patch*; Waldrep, *Night Riders*; Gildrie, "Lynch Law"; and Kathleen R. Zebley, "Unconditional Unionist: Samuel Mayes Arnell and Reconstruction in Tennessee," in West, *Tennessee History*, 180–200. *Senate Journal of the Extra Session of the 35th General Assembly of the State of Tennessee*, convened 27 July 1868; Lufkin, "Forgotten Controversy."

27. Waldrep, *Night Riders*, 95–101.

28. Killebrew, *Introduction to the Resources of Tennessee*, 1118–1125, 1148–1160; Obion County Historical Society, *Obion County History*, 1:21–22; Vanderwood, *Night Riders of Reelfoot Lake*, 8–9.

29. *Toiler*, 31 Oct. 1888; Brice, *"I Had a Real Good Time,"* 182–185. Accounts differ on Harris's origins (Kentucky or North Carolina) and his time of arrival in Lake County (1865 or 1873).

30. McBride and Robison, *Biographical Directory*, 385; Brice, *"I Had a Real Good Time,"* 184–185.

31. Tennessee General Assembly, "Report of the Special Committee Appointed to Investigate Conditions at Reelfoot Lake," *Journal of the Tennessee House of Representatives, 1909*, 493; *Banner*, 20 Oct. 1908.

32. *Webster v. Harris* (1903).

33. Ibid.; Vanderwood, *Night Riders of Reelfoot Lake*, 10–11.

34. *Webster v. Harris* (1903); Vanderwood, *Night Riders of Reelfoot Lake*, 10–11.

35. Vanderwood, *Night Riders of Reelfoot Lake*, 12–13.

36. Ibid., 13; *Journal of the Tennessee House of Representatives, 1909*, 494–495; *Banner*, 17 March 1909.

37. *Banner*, 20 Oct. and 22 Dec. 1908; *Chattanooga Times*, 22 Dec. 1908; *Louisville Courier-Journal*, 5 Oct. 1908; Vanderwood, *Night Riders of Reelfoot Lake*, 26–28, 40–41.

38. Marquis, "Menace of the Mask," 15–21; *Tennessean*, 22 Dec. 1908; Vanderwood, *Night Riders of Reelfoot Lake*, 33–40.

39. *Banner*, 20 Oct. 1908; *Commercial-Appeal*, 20 Dec. 1908; Vanderwood, *Night Riders of Reelfoot Lake*, 46–47, 53–55.

40. *Banner*, 20 and 21 Oct. 1908; *Commercial-Appeal*, 20 Oct. 1908.

41. *New York Tribune*, 9 Jan. 1909; *Atlanta Constitution*, 1 Nov. 1908; Vanderwood, *Night Riders of Reelfoot Lake*, 48–52.

42. *Banner*, 19–21 Nov. 1908, 17 Dec. 1908–9 Jan. 1909, and 4 July 1909.

43. *Commercial-Appeal*, 5 July 1909; *Atlanta Constitution*, 6 July 1909; *Trenton Herald Democrat*, 8 July 1909; *Fayette Falcon*, 9 July 1909; *McNairy County Independent*, 9 July 1909; Vanderwood, *Night Riders of Reelfoot Lake*, 140.

44. *Banner*, 29 Jan., 13 Feb., 16 Feb., 9 April, and 16 April 1909; "Report of the Special Committee Appointed to Investigate Conditions at Reelfoot Lake"; *State, ex. Rel. Charles T. Cates, Jr., v. West Tennessee Land Company et al.* (1913); Humphreys, "Formation of Reelfoot Lake."

45. "Less Cotton Movement" in Tenn. *Biennial Report of the Bureau of Agriculture, Statistics, and Mines, 1891 and 1892*, 93–126; Saloutos, *Farmer Movements in the South*, 133–134; C. C. Taylor, *Farmers' Movement*; Burkett and Poe, *Cotton*.

46. Sigafoos, *Cotton Row to Beale Street*, 65–97; Boney Wrenn, *Cinderella of the New South*, 10–120.

47. Sanders, *Roots of Reform*. Sanders argues effectively that much of the Progressive legislation enacted during the Wilson administration owed its final shape to the influence of demands from the agricultural states in the Plains and the South. While not entirely new, her argument benefits from a clear delineation of economic regions and vote count analysis. For an earlier study of the long-term effects of agrarian demands, see Saloutos, *Farmer Movements in the South*, 213–236.

48. *Commercial Appeal*, 22 Jan. 1896; McBride and Robison, *Biographical Directory*, 17, 340–341; Saloutos, *Farmer Movements in the South*, chapter 10.

49. New Orleans *Daily Picayune*, 25 Jan. 1905; Saloutos, *Farmer Movements in the South*, 153–166; Robson, "Farmers' Union in Mississippi," 4–20.

50. New Orleans *Daily Picayune*, 25 Jan. 1905; Saloutos, *Farmer Movements in the South*, 153–166; Robson, "Farmers' Union in Mississippi," 4–20.

51. Barrett, *Mission, History and Times*, 170–183, 226–229; Saloutos, *Farmer Movements in the South*, 212.

52. Barrett, *Mission, History and Times*, 226; Tenn. *Biennial Report of the Department of Agriculture, 1913–1914*, 25.

53. McBride and Robison, *Biographical Directory*, 2; Barrett, *Mission, History and Times*, 383, 386.

54. Barrett, *Mission, History and Times*, 228, 387; McBride and Robison, *Biographical Directory*, 569–570; Tenn. *Biennial Report of the Department of Agriculture, 1909–1910*, 10–15.

55. Barrett, *Mission, History and Times*, 383–385; McBride and Robison, *Biographical Directory*, 72–73; T. J. Brooks folder, University Archives, Mitchell Memorial Library, Mississippi State University.

56. Saloutos, *Farmer Movements in the South*, 210–211; "T. J. Brooks, Pioneer in Marketing," typed manuscript by Charles E. Treas and T. J. Brooks, Jr., in T. J. Brooks folder, Mitchell Memorial Library; Brooks, *Markets and Rural Economics*; Brooks, *Business Side of Farming, Bulletin of Markets and Rural Economics*. George Robert Hightower was a Mississippi state senator, president of the Mississippi Farmers' Union, and president of Mississippi A&M College. See Robson, "Farmers' Union in Mississippi."

57. Treas and Brooks, "T. J. Brooks"; West, *Tennessee Encyclopedia of History and Culture*, 96–98.

58. Barrett, *Mission, History and Times*; quote from the Jackson *Mississippi Union Advocate and Southern Farm and Home*, 25 Dec. 1907, as found in Robson, "Farmers' Union in Mississippi," 23. John H. McDowell is pictured with the same bushy beard that invited Edward Carmack's scathing jibes.

59. Barrett, *Mission, History and Times*, 134–138. For an example of the provisions the Farmers' Union made with local cotton warehouses, see "letter from W. C. Moore," box 2, f. 57 in Charles P. Simonton Papers, Mississippi Valley Collection, Ned McWherter Library, University of Memphis; Lax, "Memphis Cotton Exchange," 49.

60. Barrett, *Mission, History and Times*, 226–227; Doyle, *Nashville in the New South*; Doyle, *New Men, New Cities, New South*; Sigafoos, *Cotton Row to Beale Street*.

61. Tenn. *Biennial Report of the Department of Agriculture, 1913–1914*, 10, 25.

62. Ibid., 10–11.

63. Danbom, *The Resisted Revolution*, quote on pp. vii–viii; Hoffschwelle, *Rebuilding the Rural Southern Community*; Keith, *Country People in the New South*. Also see Mitchell and Lowe, "To Sow Contentment"; Scott, *Reluctant Farmer*. For recent examples of the reevaluation of these initiatives, see L. A. Jones, *Mama Learned Us to Work*, and Peters and P. A. Morgan, "Country Life Commission."

64. *Report of the Commission on Country Life*, 41.

65. Barrett, *Mission, History and Times*, 198.

66. *Report of the Commission on Country Life*, 49–58; *Sentinel*, 12 Nov. 1908.

67. *Sentinel*, 13 Nov. 190.

68. *Tennessee Agriculture* 3:8 (Aug. 1914): 313–316.

69. Tenn. *Biennial Report of the Department of Agriculture, 1913–1914*, 62; *Tennessee Agriculture* 2:3 (April 1913), 117; ibid. 2:8 (Sept. 1913), 310–311; ibid. 2:9 (Oct. 1913),

337, 345; ibid. 2:11 (Dec. 1913), 539–544; ibid. 3:1 (Jan. 1914), 63; Barrett, *Mission, History and Times*, 87–89; Watts, "Knowledge," 138–147.

70. *Tennessee Agriculture* 2:11 (Dec. 1913), 419–420, 476–483; ibid. 3:4 (April 1914), 178; ibid. 3:11 (Nov. 1914), 411; James Key, "Good Roads," in Tenn. *Biennial Report of the Department of Agriculture, 1909–1910*, 314–319; John H. Caldwell, "How to Provide Money for Road Construction," Tenn. *Biennial Report of the Department of Agriculture, 1909–1910*, 320–326; Barron, *Mixed Harvest*, 35; Preston, *Dirt Roads to Dixie.*

71. Tenn. *Biennial Report of the Department of Agriculture, 1909–1910*, 314–319.

72. Preston, *Dirt Roads to Dixie*, 5; Dawe, "Good Roads Movement in the South."

73. Preston, *Dirt Roads to Dixie*, 17; unpublished typed speech delivered at 9th Annual Convention of the Southern Appalachian Good Roads Congress, 19 Oct. 1917, at Nashville, Tenn., Josephine Anderson Pearson Papers, TSLA; U.S. Department of Agriculture, "Domestic Needs of Farm Women," Rep. 104, "Economic Needs of Farm Women," Rep. 106, and "Educational Needs of Farm Women," Rep. 105.

74. Abernathy, "Organized Home-Makers of a Country Neighborhood."

75. Ibid.

76. *Handbook of Tennessee Containing Descriptive and Statistical Matter Relative to the Resources and Products of the State*, 2.

77. Tenn. *Biennial Report of the Commissioner of Agriculture, 1899–1900*; ibid., *1909–1910*.

78. *Tennessee Agriculture* 2:10 (Nov. 1913), 355–400.

79. Watts, "Knowledge," 141–142.

80. *Obion County History*, vol. 1; "James McGill Buchanan" in West, *Tennessee Encyclopedia of History and Culture*, 100–101.

BIBLIOGRAPHY

Primary Sources

MANUSCRIPTS (BY REPOSITORY OR COLLECTION)

John Brannon, Personal Collection, Union City, Tennessee. "Constitution and By-Laws of the Farmers' State Alliance of Texas, Adopted at Cleburne, 1886." *Proceedings of the National Farmers' Alliance and Co-operative Union of America*, Shreveport, Louisiana, 1877. "Ritual of the Farmers' Alliance, 1886."

Chattanooga/Hamilton County Bicentennial Library, Chattanooga. H. Clay Evans Scrapbooks. Newell Sanders Papers.

Gibson County Historical Society, Trenton, Tennessee. Minutes of the Gibson County Wheel and Alliance, 1888–1892.

Hoskins Library, Special Collections. University of Tennessee, Knoxville. O. P. Temple Papers. Presidents' Papers.

McClung Collection, Lawson McGhee Library, Knoxville. Cocke County Farmers' Alliance Proceedings, 1889–1891. Hall-Stakely Papers. John C. Houk Papers. Leonidas and John C. Houk Papers.

Memphis Public Library, Memphis. Napoleon Hill folder.

Mississippi Valley Collection, Ned McWherter Library, University of Memphis. John Hugh McDowell Daybook and Scrapbook. Patrons of Husbandry, Membership Certificate. Charles B. Simonton Papers. Jones-Black Family Papers, 1842–1919.

Murray State University, Special Collections. Ollie M. James Papers. Pinkard Papers. Clarence L. Timberlake Papers.

National Archives, Atlanta, Georgia. *United States v. J. T. Caldwell*. Case 1850, 6th Circuit of Federal District Court (Memphis).

Obion County Courthouse, Union City, Tennessee. *Record of Elections*, 1890, 1892, 1894, 1896. *Tax Records*, 1890, 1892.

Tennessee State Library and Archives (TSLA), Nashville. Agricultural and Mechanical Society, Macon County, 1890–1910. Luther Meade Blackman Papers. Dark Tobacco Growers Protective Association of Clarksville, Tennessee: Minutes. Robertson Yeatman Johnson Collection. Farmers' Alliance Price List. Fort Family Papers. James Douglas Anderson Papers. Edward W. Carmack Papers. Robert Cartmell Diary. Walter Chandler Memoir. Samuel A. Champion Correspondence. Civil War Veterans' Questionnaires. Correspondence of the Governors: John P. Buchanan; Isham G. Harris; Malcolm Patterson; Robert L. Taylor; Peter Turney. Jacob McGavock Dickinson Papers. Robert Freeman Diary. Fussell

Collection. Diary of William J. Hine. Archelaus M. Hughes Diary. Papers of Joseph B. Killebrew: "Recollections of My Life: An Autobiography." Polly Langford Diary, 1888–1889. Liberty Wheel Minute Book, Decatur, Tennessee, 1886–1891. William McClinchy Diary. Nichols-Britt Collection. Josephine Anderson Pearson Papers. Flournoy Rivers Scrapbook. Mrs. Frank Robbins Diary, 1897–1898. Ruskin Cooperative Association Papers, 1896–1963. John H. Savage Papers. Emma Schindler Diary, 1890–1916. Tennessee Centennial Exposition. Minutes of the Thompson's Chapel Grange, 1873–1874. Mary Kirkpatrick Tinsley Diary, 1891–1894. George Augustine Washington Papers. Joseph Edwin Washington Papers. Paul Welch Memoir. Wilson Grange Minute Book, Clarksville, Tennessee, 1873–1876. Yeatman-Polk Papers.

Tennessee State Museum, Nashville. Mill Valley Creek Grange Minutes, 1873–1874. Tennessee Historical Society Collection.

Southern Historical Society Collection (SHSC). University of North Carolina, Chapel Hill. Charles W. Dabney Papers. Leonidas Lafayette Polk Papers. Nannie Haskins Williams Papers.

University Archives, Mitchell Memorial Library, Mississippi State University. T. J. Brooks folder.

NEWSPAPERS

Bolivar Bulletin, 1890.

Carroll County, *Tennessee Republican*, 1889, 1901.

Carthage, *Record*, 1894–1896.

Chattanooga Press, 1892.

Chattanooga Times, 1885–1892.

Clarksville Evening Chronicle, 1888.

Clarksville, *Tobacco-Leaf Chronicle*, 1888–1889.

DeBow's Review, 1866, 1867, 1869.

Dyersburg, *Neal's State Gazette*, 1880–1907.

Fayetteville Sun, 1891.

Gallatin Examiner, 1892.

Knoxville Journal, 1885–1892.

Knoxville Sentinel, 1890.

Knoxville Tribune, 1885–1892.

Lincoln County News, 1894, 1904.

McMinnville, *Southern Standard*, 1905–1906.

Memphis Appeal, 1888–1890.

Memphis Appeal-Avalanche, 1892.

Memphis Avalanche, 1888–1890.

Memphis, *Tennessee Baptist*, 1886–1888.

Milan (Gibson County) *Exchange*, 1889.

Nashville Banner, 1889–1910.

Nashville, *Baptist and Reflector*, 1888–1890.

Nashville, *Christian Advocate*, 1888–1890.
Nashville, *Confederate Veteran*, 1893–1896.
Nashville, *Daily American*, 1882–1910.
Nashville Democrat, 1888.
Nashville, *Rural Sun*, 1871–1873.
Nashville, *Tennessee Farmer*, 1889.
Nashville, *Union and American*, 1871.
Nashville, *Weekly Toiler*, 1888–1892.
Niles Weekly Register, 1841.
Obion County *Union*.
Pulaski Citizen, 1887–1892.
Raleigh, North Carolina, *Progressive Farmer*, 1888–1892.
Sparta *Expositor*, 1891.
Trenton, *Gibson County Herald*, 1885–1892.
Trenton Herald Democrat, 1909.
Union City, *Independent*, 1896.
Union City, *Obion Democrat*, 1895.
Washington, D.C., *National Economist*, 1888–1892.

U.S. GOVERNMENT DOCUMENTS

U.S. Bureau of the Census. *Historical Statistics of the United States: Colonial Times to 1970. Part 1*. White Plains, N.Y.: Kraus International Publications, 1989.

———. *Agriculture, 1909 and 1910: Compiled from the Original Returns of the Thirteenth Census*. Washington, D.C.: Government Printing Office, 1914.

———. *Agriculture in 1900: Compiled from the Original Returns of the Twelfth Census. Part 1: Farms, Live Stock, and Animal Products. Part 2: Crops and Irrigation*. Washington, D.C.: U.S. Census Office, 1902.

———. *Report on Farms and Homes in 1890: Proprietorship and Indebtedness in the United States: Compiled from the Original Returns of the Eleventh Census*. Washington, D.C.: Government Printing Office, 1896.

———. *Report of the Statistics of Agriculture in 1890: Compiled from the Original Returns of the Eleventh Census*. Washington, D.C.: Government Printing Office, 1895.

———. *Report on Cotton Production in the United States: Compiled from the Original Returns of the Tenth Census. Part 1*. Eugene W. Hilgard, special agent in charge. Washington, D.C.: Government Printing Office, 1884.

———. *Report on Population of the United States in 1890: Compiled from the Original Returns of the Eleventh Census. Part 1*. Washington, D.C.: Government Printing Office, 1895.

———. *Report on the Productions of Agriculture in 1880: Compiled from the Original Returns of the Tenth Census*. Washington, D.C.: Government Printing Office, 1883.

———. *Special Reports: Occupations at the Twelfth Census*. Washington, D.C.: Government Printing Office, 1904.

U.S. Congress. *Congressional Record: Debates and Proceedings in the Congress of the United States*. Washington, D.C., 1890–1897.

———. 1914. *Rural Credits: Joint Hearings before the Subcommittee of the Committees on Banking and Currency of the Senate and House of Representatives Charged with the Investigation of Rural Credit*. 63rd Cong., 2nd sess.

———. Senate. 1895. *Agricultural Depression: Causes and Remedies: Report by Mr. Pfeffer Submitted to The Senate Committee on Agriculture and Forestry*. 53rd Cong., 3rd sess., S. Rep. no. 986, serial 3288.

———. Senate. 1895. *Letters to Committee on Agriculture and Forestry*. 53rd Cong., 3rd sess., S. Rep. no. 986, serial 3290–3291.

———. Senate. 1895. *Report of the Committee on Agriculture and Forestry on Condition of Cotton Growers in the United States, the Present Prices of Cotton, and the Remedy; on Cotton Consumption and Production*. 53rd Cong., 3rd sess., S. Rep. no. 986, serial 3290–3291.

———. *Senate Journal*. 54th Cong., 1895, ser. 3361. Washington, D.C.: 1896.

———. Senate. 1912. *Preliminary Report on Land and Agricultural Credit in Europe*. 62nd Cong., 3rd sess., S. Doc. 967.

———. Senate. 1913. *Agricultural Co-operation and Rural Credits in Europe*. 63rd Cong., 3rd sess., S. Doc. 114.

———. Senate. 1913. *The Landschaften System of Rural Credits*. 63rd Cong., 1st sess., S. Doc. 114.

———. Senate. 1913. *Southern Commercial Congress. American Commission for the Study of the Application of the Co-operative System to Agricultural Production, Distribution & Finances in European Countries*. 62nd Cong., 3rd sess., S. Doc. 1071.

U.S. Department of Agriculture. "Domestic Needs of Farm Women." Rep. no. 104. Washington, D.C.: Government Printing Office, 1915.

———. "Educational Needs of Farm Women." Rep. no. 105. Washington, D.C.: Government Printing Office, 1915.

———. "Economic Needs of Farm Women." Rep. no. 106. Washington, D.C.: Government Printing Office, 1915.

———. *Yearbook of the United States Department of Agriculture, 1901*. Washington, D.C.: Government Printing Office, 1902.

———. *Yearbook of the United States Department of Agriculture, 1903*. Washington, D.C.: Government Printing Office, 1904.

———. *Yearbook of the United States Department of Agriculture, 1904*. Washington, D.C.: Government Printing Office, 1905.

———. *Yearbook of the United States Department of Agriculture, 1908*. Washington, D.C.: Government Printing Office, 1909.

TENNESSEE GOVERNMENT DOCUMENTS

Acts of the State of Tennessee, 1873–1891. Nashville.

Biennial Report of the Bureau of Agriculture, Statistics, and Mines, 1891 and 1892. Nashville: Marshall & Bruce, Printers to the State, 1893.

Biennial Report, Tennessee Bureau of Agriculture, 1893–1894. Nashville: Franc. M. Paul, Printer to the State, 1895.

Biennial Report of the Bureau of Agriculture, 1897–1898. Nashville: Brandon Printing, 1899.

Biennial Report of the Commissioner of Agriculture, 1899–1900. Nashville: Gospel Advocate Publishing, 1901.

Biennial Report of the Department of Agriculture, 1905–1907. Nashville: B. L. Foster, 1907.

Biennial Report: Department of Agriculture, 1907–1908. Nashville: Folk-Keeling Printing Co., 1909.

Biennial Report of the Department of Agriculture, 1909–1910. Nashville: Brandon Printing, 1911.

Biennial Report of the Department of Agriculture, 1913–1914. Nashville: Foster & Parker, 1915.

Biennial Report, Board of Prison Commissioners of Tennessee, 1897–1908. Nashville.

First and Second Annual Reports of the Railroad Commission for the State of Tennessee. Nashville, 1884.

Handbook of Tennessee. Nashville, 1882. Hand-book of Tennessee: prepared by A. W. Hawkins, commissioner of agriculture, statistics, mines and immigration, assisted by Henry E. Colton, practical geologist and mining engineer. Nashville, 1882.

Handbook of Tennessee Containing Descriptive and Statistical Matter Relative to the Resources and Products of the State. Nashville: McQuiddy Printing Co., 1903.

Journal of the Proceedings of the Convention of Delegates Elected by the People of Tennessee to Amend, Revise, or Form and Make a New Constitution for the State. Nashville: 1870. Assembled in the city of Nashville, January 10, 1870. Tennessee.

Journal of the Tennessee House of Representatives, 1873–1909. Nashville.

Journal of the Tennessee Senate, 1873–1909. Nashville.

Report of Bureau of Agriculture of Tennessee, 1889–1890. [Nashville?]

Revised Handbook of Tennessee. Nashville, 1885.

Senate Journal of the Extra Session of the 35th General Assembly of the State of Tennessee, convened July 27, 1868.

Special Report of the Commissioner of Labor and the Inspector of Mines to His Excellency Governor John P. Buchanan. Extra sess., August 31, 1891. Nashville, 1891.

Tennessee Bureau of Agriculture, Statistics, and Mines: Bulletin of Farms, Residences, etc., for Sale in Tennessee. January 1, 1882–July 31, 1883.

Tennessee General Assembly, Joint Select Committee to Investigate the State Debt, 1879. Nashville, 1879.

OTHER PRIMARY SOURCES

Anderson, James Douglas. *Contest for Governor Pending before the Forty-ninth General Assembly of Tennessee* (pamphlet).

Ash, Stephen V., ed. *Messages of the Governors of Tennessee, 1907–1933.* Vols. 9 and 10. Nashville: Tennessee Historical Commission, 1990.

Bailey, L. H. *The Country-Life Movement in the United States*. New York: MacMillan, 1911.

Barrett, Charles Simon. *The Mission, History and Times of the Farmers' Union: A Narrative of the Greatest Industrial-Agricultural Organization in History and Its Makers*. Nashville: Marshall & Bruce, 1909.

Brice, James M. *"I Had a Real Good Time": The Making of a Country Editor*. Edited by Rebel C. Forrester and Betty B. Wood. Union City, Tenn.: Lanzer Printing, 1984.

Brooks, T. J. *Bulletin of Markets and Rural Economics*. Meridian, Miss.: Wortham Printing, 1914.

———. *Business Side of Farming*. Meridian, Miss.: Wortham Printing, 1914.

———. *Markets and Rural Economics: Science of Commerce and Distribution: An Investigation of Agricultural Production, and the Economics of Distribution; Cooperation in Marketing; Rural Credits; Agencies of Impending Change; Present Conditions and Tendencies; Future Possibilities and Opportunities*. New York: Shakespeare Press, 1914.

Burkett, Charles William, and Clarence Hamilton Poe. *Cotton: Its Cultivation, Marketing, Manufacture, and the Problems of the Cotton World*. New York: Doubleday, Page & Co., 1906.

Bryan, William J. *The First Battle: A Story of the Campaign of 1896*. Chicago: W. B. Conkey, 1896.

Caldwell, Joshua W. *Studies in the Constitutional History of Tennessee*. Cincinnati: n.p., 1895.

Carlton, Frank T. "The South During the Last Decade." *Sewanee Review* 12 (1904): 174–181.

Carroll, William H., *The Gubernatorial Contest* (pamphlet).

Dickinson, Jacob McGavock. *Address to Southern Club of Chicago on Progress of the South* (pamphlet). 10 April 1909.

———. *Present Financial and General Condition of the South, 1899* (pamphlet).

Dun, R. G., & Co. *Dun's Mercantile Agency Book: Tennessee*. New York: R. G. Dun & Co., 1886.

Dunning, N. A. *The Farmers' Alliance History and Agricultural Digest*. Washington, D.C.: Farmers' Alliance, 1889.

Elliott, J. R. *American Farms: Their Condition and Future*. New York: G. P. Putnam's Sons, 1890.

Felton, Rebecca. "The Convict Lease System of Georgia." *Forum* (1887): 484–490.

Fortune, T. Thomas. *Black and White: Land, Labor, and Politics in the South*. New York: Fords, Howard, & Hulbert, 1884. Reprint, New York: Arno Press and the New York Times, 1968.

Foster, Florence J. "The Grange and the Co-operative Enterprises in New England." *Annals of the American Academy of Political and Social Sciences* 4 (1893–1894): 798–805.

Gates, Robert. *Farmers, the Conservative Force in the Body Politics: An Address Delivered Before the Farmers' Institute of Middle Tennessee, December 6, 1911* (pamphlet). 1911.

Godwin, D. G. *Tennessee: Its Resources, Capabilities, and Development Illustrated by Graphics*. [Chicago: Rand, McNally, 1892.]

Hibbard, Benjamin H. "Tenancy in the Southern States." *Quarterly Journal of Economics* 27 (May 1913).

Holman, D. A. *The Taxpayers vs. Monopolies* (pamphlet). Louisville, Ky.: 1875.

Holmes, George K. "A Decade of Mortgages." *Annals of Political and Social Sciences* 4 (1893–1894): 904–918.

———. "The Peons of the South." *Annals of the American Academy of Political and Social Science* 4 (1893–1894): 265–274.

———. "Tenancy in the United States." *Quarterly Journal of Economics* 10 (1895): 34–53.

Justi, Herman, ed. *Official History of the Tennessee Centennial Exposition*. Nashville: Brandon Printing Co., 1898.

Killebrew, Joseph B. *Agricultural Products, Live Stock and Grazing Land, Poultry Raising Along the Line of the Nashville, Chattanooga & St. Louis Railway* (pamphlet). Nashville.

———. *Introduction to the Resources of Tennessee*. Prepared under the direction of the (Tennessee) Bureau of Agriculture. Nashville: Tavel, Eastman & Howell, 1874.

———. *The South: Its Opportunities for the Investment of Capital* (pamphlet). An address given at a convention in Southern Pines, North Carolina, 19 April 1899.

———. *Tobacco: Its Culture in Tennessee with Statistics of Its Commercial Importance* (pamphlet). [Nashville?]: Tavel, Eastman & Howell, 1876.

———. *West Tennessee: Its Resources and Advantages—Cheap Homes for Immigrants* (pamphlet). Nashville: Tavel, Eastman & Howell.

Killebrew, Joseph B., Robert Gates, and M. V. Richards. *Tennessee: Its Advantages, Resources,and Possibilities* (pamphlet). Nashville: Marshall Bruce, no. 86.

Lloyd, Henry Demarest. *Wealth Against Commonwealth*. Washington, D.C.: National Home Library Foundation, 1936. First published 1896.

Mercantile Credit Book, Tennessee. N.p.: R. G. Dun & Co., 1886.

Merrian, Lucius Salisbury. *Higher Education in Tennessee*. Washington, D.C.: U.S. Government Printing Office, 1893.

Miller, Charles A. *The Official and Political Manual of the State of Tennessee*. Nashville: Marshall & Bruce, Stationers, 1890.

Morgan, W. Scott. *History of the Wheel and Alliance and the Impending Revolution*. Hardy, Ark.: n.p., 1889.

Page, Walter Hines. "A Journey through the Southern States: The Changes of Ten Years." *World's Work* 14 (June 1907): 9026–9035.

A Plea for the Right: Opinions of the Democratic Press (pamphlet).

Poe, Clarence. *How Farmers Co-operate and Double Profits: First-Hand Reports on All the Leading Forms of Rural Co-operation in the United States and Europe—Stories That Show How Farmers Can Co-operate by Showing How They Have Done It and Are Doing It*. New York: Orange Jud, 1915.

Proceedings of the Joint Investigating Committee (pamphlet).

Report of the Commission on Country Life, With an Introduction by Theodore Roosevelt. Sturgis & Walton, 1911. Reprint, Chapel Hill: Univ. of North Carolina Press, 1944.
Sanford, Edward T. *Blount College and the University of Tennessee: An Historical Address Delivered Before the Alumni Association and Members of the University of Tennessee, June 12th, 1894* (pamphlet). Knoxville: Univ. of Tennessee.
Speech of the Honorable H. Clay Evans, October 10, 1894 (pamphlet).
Speer, William S. *Sketches of Prominent Tennesseans: Containing Biographies and Records of Many of the Families Who Have Attained Prominence in Tennessee.* Nashville: Albert B. Tavel, 1888.
Tennessee General Committee of Investigation, Turney vs. Evans (pamphlet).
Walker, C. S. "Farmers' Movement." *Annals of the American Academy of Political and Social Science* 4 (1893–1894): 790–798.
Washington, Booker T. "The Negro as a Farmer." *North American Review* 114 (Feb. 1912): 175–181.
Watson, Thomas E. *The People's Party Campaign Book, 1892.* Washington, D.C.: National Watchman Publishing, n.d. Facsimile reprint, New York: Arno Press, 1975.
White, Robert H., ed. *Messages of the Governors of Tennessee, 1796–1907.* Vols. 1–8. Nashville: Tennessee Historical Commission, 1952–1972.
Youngman, Anna. "The Tobacco Pools of Kentucky and Tennessee." *Journal of Political Economy* 18 (1910): 34–49.

Secondary Sources

Abernathy, Elizabeth Denty. "Organized Home-Makers of a Country Neighborhood." *Tennessee Agriculture* 3:11 (Dec. 1914): 475.
Abrams, Richard M., ed. *Issues of the Populist and Progressive Eras, 1892–1912.* New York: Harper & Row, 1969.
Aldrich, Mark. "A Note on Railroad Rates and the Populist Uprising." *Agricultural History* 54 (1980): 424–432.
Alexander, Mai. *Dyer County: The Garden Spot of the World.* Dyersburg, Tenn.: Wallace Printing, 1974.
Alexander, Thomas B. "Persistant Whiggery in the Confederate South, 1860–1877." *Journal of Southern History* 27 (1961): 305–329.
———. "Whiggery and Reconstruction in Tennessee." *Journal of Southern History* 16 (1950): 291–305.
Appleby, Joyce. *Capitalism and a New Social Order: The Republican Vision of the 1790s.* New York: New York Univ. Press, 1984.
———. "Commercial Farming and the 'Agrarian Myth' in the Early Republic." *Journal of American History* 68 (1982): 833–849.
Archer, Claude Jackson. "The Life of John Chiles Houk." M.A. thesis, University of Tennessee, 1941.
Ash, Stephen V. *Middle Tennessee Society Transformed, 1860–1870: War and Peace in the Upper South.* Baton Rouge: Louisiana State Univ. Press, 1988.

———. "Postwar Recovery: Montgomery County, 1865–70." *Tennessee Historical Quarterly* 36 (1977): 208–221.

———. *When the Yankees Came: Conflict and Chaos in the Occupied South, 1861–1865.* Chapel Hill: Univ. of North Carolina Press, 1995.

———, ed. *Secessionists and Other Scoundrels: Selections from Parson Brownlow's Book.* Baton Rouge: Louisiana State Univ. Press, 1999.

Atack, Jeremy. "Tenants and Yeomen in the Nineteenth Century." *Agricultural History* 62 (1988): 6–32.

Atack, Jeremy, and Peter Passell. *An Economic View of American History: From Colonial Times to 1940.* 2nd ed. New York: W. W. Norton, 1994.

Atkins, Jonathan M. *Parties, Politics, and the Sectional Conflict in Tennessee, 1832–1861.* Knoxville: Univ. of Tennessee Press, 1997.

Ayers, Edward L. *The Promise of the New South: Life After Reconstruction.* New York: Oxford Univ. Press, 1992.

Bailey, Clay. "Looking Backward in the Ruskin Cooperative Association." *Tennessee Historical Quarterly* 53 (1994): 101–113.

Bailey, Fred A. *Class and Tennessee's Confederate Generation.* Chapel Hill: Univ. of North Carolina Press, 1987.

———. "The Poor, Plain Folk, and Planters: A Social Analysis of Middle Tennessee Respondents to the Civil War Veterans' Questionnaires." *West Tennessee Historical Society Papers* 36 (1982): 5–24.

———. "Tennessee's Antebellum Common Folk." *Tennessee Historical Quarterly* 55 (1996): 40–55.

Ball, Clyde L. "The Public Career of Colonel A. S. Colyar." *Tennessee Historical Quarterly* 12 (1953): 23–47, 106–128, 213–238.

Barnes, Donna A. *Farmers in Rebellion: The Rise and Fall of the Southern Farmers' Alliance and People's Party in Texas.* Austin: Univ. of Texas Press, 1984.

Barron, Hal S. *Mixed Harvest: The Second Great Transformation in the Rural North, 1870–1930.* Studies in Rural Culture. Chapel Hill: Univ. of North Carolina Press, 1997.

Barthelme, Marion K. *Women in the Texas Populist Movement: Letters to the Southern Mercury.* Centennial Series of the Association of Former Students, Texas A&M University. College Station: Texas A&M Univ. Press, 1997.

Bartlett, Richard A., ed. *The Gilded Age: America, 1865–1900.* Reading, Mass.: Addison-Wesley, 1969.

Beard, William E. "Albert Virgil Goodpasture: Gentleman and Scholar." *Tennessee Historical Quarterly* 2 (1943): 50–58.

Beasely, Gaylon Neil. *True Tales of Tipton: Historical Accounts of Tipton County, Tennessee.* Covington: Tipton County Historical Society, 1981.

Beatty, Bess. *Alamance: The Holt Family and Industrialization in a North Carolina County, 1837–1900.* Baton Rouge: Louisiana State Univ. Press, 1999.

Beaulieu, Lionel J. *The Rural South in Crisis: Challenges for the Future.* Boulder, Colo.: Westview Press, 1988.

Belissary, Constantine G. "The Rise of Industry and the Industrial Spirit in Tennessee, 1865–1885." *Journal of Southern History* 19 (1953): 193–215.

Benning, Lance. *Jeffersonian Persuasion: Evolution of a Party Ideology*. Ithaca, N.Y.: Cornell Univ. Press, 1978.

Bergeron, Paul. *Antebellum Politics in Tennessee*. Lexington: Univ. Press of Kentucky, 1982.

Billings, Dwight B., Jr. *Planters and the Making of a "New South": Class, Politics, and Development in North Carolina, 1865–1900*. Chapel Hill: Univ. of North Carolina Press, 1979.

Binder, John J., and David T. Brown. "Bank Rates of Return and Entry Restrictions, 1869–1914." *Journal of Economic History* 51 (1991): 47–66.

Bland, Gaye K. "Populism in the First Congressional District of Kentucky, 1892." *The Filson Club History Quarterly* 51 (1977): 31–43.

Bode, Frederick A. "Religion and Class Hegemony: A Populist Critique in North Carolina." *Journal of Southern History* 38 (1971): 417–438.

Boles, John B., and Evelyn Thomas Nolan, eds. *Interpreting Southern History: Historiographical Essays in Honor of Sanford W. Higginbotham*. Baton Rouge: Louisiana State Univ. Press, 1987.

Bolton, Charles C. *Poor Whites of the Antebellum South: Tenants and Laborers in Central North Carolina and Northeast Mississippi*. Durham, N.C.: Duke Univ. Press, 1994.

Boney Wrenn, Lynette. *Cinderella of the New South: A History of the Cottonseed Industry, 1855–1955*. Knoxville: Univ. of Tennessee Press, 1995.

Bonner, Edmund Cody. "Hog Raising and Hog Driving in the Region of the French Broad River." *Agricultural History* 20 (1946): 87–103.

Boutwell, Lane L. "The Oratory of Robert Love Taylor." *Tennessee Historical Quarterly* 9 (1950): 10–45.

Breen, T. H. *Tobacco Culture: The Mentality of the Great Tidewater Planters on the Eve of Revolution*. Princeton, N.J.: Princeton Univ. Press, 1985.

Breisach, Ernst A. *American Progressive History: An Experiment in Modernization*. Chicago: Univ. of Chicago Press, 1993.

Brundage, W. Fitzhugh. *Lynching in the New South: Georgia and Virginia, 1880–1930*. Urbana: Univ. of Illinois Press, 1993.

Bryan, Ferald J. *Henry Grady or Tom Watson? The Rhetorical Struggle for the New South, 1880–1890*. Macon, Ga.: Mercer Univ. Press, 1994.

Buck, Solon. *The Granger Movement*. Cambridge, Mass.: Harvard Univ. Press. 1913.

Burckel, Nicholas C. "William Goebel and the Campaign for Railroad Regulation in Kentucky, 1888–1890." *Filson Club Quarterly* 48 (1974): 43–60.

Burnett, Edmund Cody. "Big Creek's Response to the Coming of the Railroad: 'Old Buncombe' Promotes the Better Life in a Rural Community." *Agricultural History* 21 (July 1947): 129–148.

———. "Hog Raising and Hog Driving in the Region of the French Broad River." *Agricultural History* 20 (April 1946): 86–103.

———. "Shingle Making on the Lesser Waters of the Big Creek of the French Broad River." *Agricultural History* 20 (Oct. 1946): 225–235.

Burt, Jesse C., Jr. "Four Decades of the Nashville, Chattanooga and St. Louis Railway, 1873–1916." *Tennessee Historical Quarterly* 9 (1950): 99–130.

Bush, Florence Cope. *Dorie: Woman of the Mountains*. Knoxville: Univ. of Tennessee Press, 1992.

Calhoun, Charles W., ed. *The Gilded Age: Essays on the Origins of Modern America*. Wilmington, Del.: Scholarly Resources, 1996.

Campbell, Ballard C., ed. *The Human Tradition in the Gilded Age and Progressive Era*. Wilmington, Del.: Scholarly Resources, 2000.

Campbell, Tracy A. "The Limits of Agrarian Action: The 1908 Kentucky Tobacco Strike." *Agricultural History* 66 (1992): 76–97.

———. *The Politics of Despair: Power and Resistance in the Tobacco Wars*. Lexington: Univ. Press of Kentucky, 1993.

Carter, Dan T. *The Politics of Rage: George Wallace, The Origins of New Conservatism, and the Transformation of American Politics*. New York: Simon & Schuster, 1995.

———. *When the War Was Over: The Failure of Self-Reconstruction in the South, 1865–1867*. Baton Rouge: Louisiana State Univ. Press, 1985.

Cartwright, Joseph H. *The Triumph of Jim Crow: Tennessee Race Relations in the 1880s*. Knoxville: Univ. of Tennessee Press, 1976.

Cash, W. J. *The Mind of the South*. New York: Alfred A. Knopf, 1941.

Cashin, Joan E. "The Structure of Antebellum Planter Families: 'The Ties that Bound Us Was Strong.'" *Journal of Southern History* 56 (1990): 55–70.

Cashman, Sean Dennis. *America in the Gilded Age: From the Death of Lincoln to the Rise of Theodore Roosevelt*. 3rd ed. New York: New York Univ. Press, 1993.

Chandler, Alfred D., Jr. *The Visible Hand: The Managerial Revolution in American Business*. Cambridge, Mass.: Harvard Univ. Press, 1977.

Cimpich, John. *Slavery's End in Tennessee, 1861–1865*. University: Univ. of Alabama Press, 1985.

Clark, Blanche Henry. *Tennessee Yeomen, 1840–1860*. Nashville: Vanderbilt Univ. Press, 1942.

Clark, Christopher. *Roots of Rural Capitalism: Western Massachusetts, 1780–1860*. Ithaca, N.Y.: Cornell Univ. Press, 1990.

Clark, Thomas D. *Three Paths to the Modern South: Education, Agriculture, Conservation*. Baton Rouge: Louisiana State Univ. Press, 1965.

Clemens, Elisabeth S. *The People's Lobby: Organizational Innovation and the Rise of Interest Group Politics in the United States, 1890–1925*. Chicago: Univ. of Chicago Press, 1997.

Clements, Kendrick A. *William Jennings Bryan: Missionary Isolationist*. Knoxville: Univ. of Tennessee Press, 1982.

Cobb, James C. "Beyond Planters and Industrialists: A New Perspective on the New South." *Journal of Southern History* 54 (1988): 45–68.

———. *Industrialization and Southern Society, 1877–1984*. Lexington: Univ. Press of Kentucky, 1984. Chicago: Dorsey Press, 1988.

———. *The Most Southern Place on Earth: The Mississippi Delta and the Roots of Regional Identity*. New York: Oxford Univ. Press, 1992.

Connelly, Thomas L. *Civil War Tennessee: Battles and Leaders*. Knoxville: Univ. of Tennessee Press in cooperation with the Tennessee Historical Commission, 1979.

Cooper, William J., *The South and the Politics of Slavery, 1828–1856*. Baton Rouge: Louisiana State Univ. Press, 1978.

Corlew, Robert E. *The History of Dickson County, Tennessee*. Nashville: Tennessee Historical Commission and Dickson County Historical Society, 1956.

———. *Tennessee: A Short History*. Knoxville: Univ. of Tennessee Press, 1990.

Coté, Joseph A., "Clarence Hamilton Poe: The Farmers' Voice, 1899–1964." *Agricultural History* 53 (1979): 30–41.

Cotham, Perry C. *Toil, Turmoil, and Triumph: A Portrait of the Tennessee Labor Movement*. Franklin, Tenn.: Hillsboro Press, 1995.

Courtwright, David T. "Opiate Addiction as a Consequence of the Civil War." *Civil War History* 24 (1978): 101–111.

Couto, Richard A. *Lifting the Veil: A Political History of Struggles for Emancipation*. Knoxville: Univ. of Tennessee Press, 1993.

Cox, Karen L. *Dixie's Daughters: The United Daughters of the Confederacy and the Preservation of Confederate Culture*. Gainesville: Univ. Press of Florida, 2003.

Crampton, John A. *The National Farmers Union: Ideology of a Pressure Group*. Lincoln: Univ. of Nebraska Press, 1965.

Cresswell, Stephen. *Multi-party Politics in Mississippi, 1877–1802*. Jackson: Univ. Press of Mississippi, 1995.

Crews, Mickey. *The Church of God: A Social History*. Knoxville: Univ. of Tennessee Press, 1990.

Cronon, William, *Nature's Metropolis: Chicago and the Great West*. New York: W. W. Norton, 1991.

Crow, Jeffrey J., Paul D. Escott, and Charles L. Flynn Jr. *Race, Class, and Politics in Southern History*. Baton Rouge: Louisiana State Univ. Press, 1989.

Culp, Frederick M., and Mrs. Robert E. Ross. *Gibson County Past and Present: The First General History of One of West Tennessee's Pivotal Counties*. Trenton, Tenn.: Gibson County Historical Society, 1961.

Cunningham, Raymond J. *The Populists in Historical Perspective*. Boston: D. C. Heath, 1968.

Cutler, Wayne. "Jackson, Polk, and Johnson: Defenders of the Moral Economy." *Tennessee Historical Quarterly* 54 (1995): 178–189.

Dalecki, Michael G., and C. Milton Coughenour. "Agrarianism in American Society." *Rural Sociology* 57 (1992): 48–64.

Danbom, David B. *Born in the Country: A History of Rural America*. Baltimore: Johns Hopkins Univ. Press, 1995.

———. *The Resisted Revolution: Urban America and the Industrialization of Agriculture, 1900–1930*. Ames: Iowa State Univ. Press, 1979.

Daniel, Pete. *Breaking the Land: The Transformation of Cotton, Tobacco, and Rice Culture Since 1880*. Urbana: Univ. of Illinois Press, 1985.

———. "The Tennessee Convict War." *Tennessee Historical Quarterly* 34 (1975): 273–292.

Darrah, Marsha Young. "Political Career of Col. William B. Stokes of Tennessee." M.A. thesis, Tennessee Technological Univ., 1968.

Davis, Bettie B. *Gates, Tennessee: Now and Then*. Ripley, Tenn.: n.p., 1986.

Davis, Lance E. "The Investment Market, 1870–1914: The Evolution of a National Market." *Journal of Economic History* 25 (1965): 355–399.

Davis, Ronald L. F. *Good and Faithful Labor: From Slavery to Sharecropping in the Natchez District, 1860–1890*. Westport, Conn.: Greenwood Press, 1982.

Davis, Thomas Woodrow. "Arthur S. Colyar and the New South, 1860–1905." Ph.D. diss., Univ. of Missouri, 1962.

Dawe, G. Grosvenor. "The Good Roads Movement in the South." *Taylor and Trotwood Magazine* (Jan. 1910): 350.

DeFiore, Jayne Crumpler. "'Come and Bring the Ladies': Tennessee Women and the Politics of Opportunity During the Presidential Campaigns of 1840 and 1844." *Tennessee Historical Quarterly* 51 (1992): 197–212.

De Santis, Vincent P. "President Arthur and the Independent Movements in the South in 1882." *Journal of Southern History* 19 (1953): 346–363.

———. *The Shaping of Modern America, 1877–1916*. New York: Forum Press, 1973.

Diner, Steven J. *A Very Different Age: Americans in the Progressive Era*. New York: Hill & Wang, 1998.

Dobson, John M. *Politics in the Gilded Age: A New Perspective on Reform*. New York: Praeger, 1972.

Dodd, Donald B., and Wynell S. Dodd. *Historical Statistics of the South, 1790–1970*. University: Univ. of Alabama Press, 1973.

Doster, James F. "Were Populists Against Railroad Corporations? The Case of Alabama." *Journal of Southern History* 20 (1954): 385–399.

Downing, Marvin. "Woodley Farm, Martin, Tennessee." *West Tennessee Historical Society Papers* 38 (1984): 51–68.

Doyle, Don H. *Nashville in the New South, 1880–1930*. Knoxville: Univ. of Tennessee Press, 1985.

———. *New Men, New Cities, New South: Atlanta, Nashville, Charleston, Mobile, 1860–1910*. Chapel Hill: Univ. of North Carolina Press, 1990.

Dubofsky, Melvyn. *Industrialism and the American Worker, 1865–1920*. 2nd ed. Arlington Heights, Ill.: Harlan Davidson, 1985.

Duggins, Edward Cameron. "Background for Regulation of the Railroads in Tennessee." M.A. thesis, University of Tennessee, 1939.

Durden, Robert F. *The Climax of Populism: The Election of 1896*. Lexington: Univ. Press of Kentucky, 1965.

Durrill, Wayne K. "Producing Poverty: Local Government and Economic Development in a New South County, 1874–1884." *Journal of American History* 71 (June 1984–March 1985): 764–781.

Dyson, Lowell K. *Farmers' Organizations. The Greenwood Encyclopedia of American Institutions*. New York: Greenwood Press, 1986.

Eichengreen, Barry. "Mortgage Interest Rates in the Populist Era." *American Economic Review* 74 (1984): 995–1014.

Eisenberg, John M., Jr. "A House Divided: Silver Democrats and Their Party." *West Tennessee Historical Society Papers* 29 (1975): 86–99.

Elkins, F. Clark. "The Agricultural Wheel in Arkansas, 1887." *Arkansas Historical Quarterly* 40 (1982): 249–260.

Ellsworth, Clayton S. "Theodore Roosevelt's Country Life Commission." *Agricultural History* 34 (Oct. 1960): 155–172.

Escott, Paul D. "Yeoman Independence and the Market: Social Status and Economic Development in Antebellum North Carolina." *North Carolina Historical Review* 66 (1989): 275–300.

Evans, Edith Snyder. "The Progressive Party in Tennessee in 1912." M.A. thesis, University of Tennessee, 1933.

Evans, Sara M., and Harry C. Boyte. *Free Spaces: The Sources of Democratic Change in America*. New York: Harper & Row, 1988.

Faragher, John Mack. *Sugar Creek: Life on the Illinois Prairie*. New Haven, Conn.: Yale Univ. Press, 1986.

Ferguson, James S. "The Grange and Farmer Education in Mississippi." *Journal of Southern History* 8 (1942): 497–512.

Ferleger, Louis. "Self-Sufficiency and Rural Life on Southern Farms." *Agricultural History* 58 (1984): 314–329.

Fine, Sidney. *Laissez Faire and the General-Welfare State: A Study of Conflict in American Thought, 1865–1901*. Ann Arbor: Univ. of Michigan Press, 1956; reprint 1967.

Finger, John R. *Tennessee Frontiers: Three Regions in Transition*. History of the Trans-Appalachian Frontier Series. Bloomington: Indiana Univ. Press, 2001.

Fink, Deborah. *Agrarian Women: Wives and Mothers in Rural Nebraska, 1880–1940*. Chapel Hill: Univ. of North Carolina Press, 1992.

———. *Open Country, Iowa: Rural Women, Tradition and Change*. Chapel Hill: Univ. of North Carolina Press, 1986.

Fink, Leon. *Workingmen's Democracy: The Knights of Labor and American Politics*. Urbana: Univ. of Illinois Press, 1983.

Fisher, Noel C. *War at Every Door: Partisan Politics and Guerrilla Violence in East Tennessee, 1860–1869*. Chapel Hill: Univ. of North Carolina Press, 1997.

Fite, Gilbert. "The Agricultural Trap in the South." *Agricultural History* 60 (1986): 38–52.

Fitzgerald, Michael W. *The Union League Movement in the Deep South: Politics*. Baton Rouge: Louisiana State Univ. Press, 1989.

Flynt, Wayne. *Poor But Proud: Alabama's Poor Whites*. Tuscaloosa: The Univ. of Alabama Press, 1989.

Folmsbee, Stanley J., and Lee Seifert Greene. *To Foster Knowledge: A History of the University of Tennessee, 1794–1970*. Knoxville: Univ. of Tennessee Press, 1974.

Ford, Lacy K. *Origins of Southern Radicalism: The South Carolina Upcountry, 1800–1860*. New York: Oxford Univ. Press, 1988.

———. "Rednecks and Merchants: Economic Development and Social Tensions in the South Carolina Upcountry, 1865–1900. *Journal of American History* 71 (1984): 294–318.

Forrester, Rebel C. *Glory and Tears: Obion County, 1860–1870*. Union City, Tenn.: H. A. Lanzer, 1970.

———. "The Making of a Country Editor: James M. Brice and the *News-Banner* of Troy, Tennessee." *West Tennessee Historical Society Papers* 35 (1981): 103–117.

Friedman, Jean E. *The Enclosed Garden: Women and Community in the Evangelical South, 1830–1900*. Chapel Hill: Univ. of North Carolina Press, 1985.

Fuller, Wayne E. "The Populists and the Post Office." *Agricultural History* 65 (1991): 1–16.

———. "The Rural Roots of Progressive Leaders." *Agricultural History* 42 (Jan. 1968): 1–13.

Gaither, Gerald H. *Blacks and the Populist Revolt: Ballots and Bigotry in the "New South."* University: Univ. of Alabama Press, 1977.

———. "The Negro Alliance Movement in Tennessee, 1888–1891." *West Tennessee Historical Society Papers* 27 (1973): 50–62.

Gaston, Paul M. *The New South Creed: A Study in Southern Mythmaking*. New York: Alfred A. Knopf, 1970.

Gates, Paul W. *The Farmer's Age: Agriculture, 1815–1860*. Armonk, N.Y.: M. E. Sharpe, 1960. Previously published as vol. 3 in the Economic History of the United States, New York: Holt, Rinehart & Winston, 1960.

Gates, Warren J. "Modernization as a Function of an Agricultural Fair: The Great Grangers' Picnic Exhibition at Williams Grove, Pennsylvania, 1873–1916." *Agricultural History* 58 (1984): 252–265.

Geertz, Clifford. *The Interpretation of Cultures*. New York: Basic, 1973.

Gentry, Amos Leo. "The Public Career of Leonidas Campbell Houk." M.A. thesis, University of Tennessee, 1939.

Gerber, David A. "Cutting Out Shylock: Elite Anti-Semitism and the Quest for Moral Order in the Mid-Nineteenth-Century American Market Place." *Journal of American History* 69 (1982): 615–637.

Gerster, Patrick, and Nicholas Cords, eds. *Myth and Southern History*. Chicago: Rand McNally College Publishing, 1974.

Gildrie, Richard P. "Lynch Law and the Great Clarksville Fire of 1878: Social Order in a New South Town." *Tennessee Historical Quarterly* 42 (1983): 58–75.

Ginger, Ray. *Age of Excess: The United States from 1877 to 1814*. New York: MacMillan, 1965; reprint 1975.

Glymph, Thavolia, and John J. Kushma, eds. *Essays on the Postbellum Southern Economy*. College Station: Texas A&M Press, 1985.

Goldin, Claudia, and Frank D. Lewis, "The Economic Cost of the American Civil War: Estimates and Implications." *Journal of Economic History* 35 (1975): 299–326.

Goode, Richard C. "The Godly Insurrection in Limestone County: Social Gospel Populism and Southern Culture in the Late Nineteenth Century." *Religion and Culture* (1992): 155–169.

Goodwyn, Lawrence. *Democratic Promise: The Populist Moment in America*. New York: Oxford Univ. Press, 1976.

———. *The Populist Moment: A Short History of the Agrarian Revolt*. New York: Oxford Univ. Press, 1978.

Granberry, Dorothy. "Black Community Leadership in a Rural Tennessee County, 1865–1903." *Journal of Negro History* 83 (1998): 249–257.

———. "When the Rabbit Foot Was Worked and Republican Votes Became Democratic Votes: Black Disfranchisement in Haywood County, Tennessee." *Tennessee Historical Quarterly* 63 (2004): 34–47.

Grantham, Dewey W. "Black Patch War: The Story of the Kentucky and Tennessee Night Riders, 1905–1909." *South Atlantic Quarterly* 59 (1960): 215–225.

———. *Southern Progressivism: The Reconciliation of Progress and Tradition*. Knoxville: Univ. of Tennessee Press, 1983, 1987.

Gray, Lewis C. *History of Agriculture in the Southern United States to 1860*. 2 vols. Washington, D.C.: Carnegie Institution of Washington, 1933.

Gregory, Rick S. "Desparate Farmers: The Dark Tobacco District Planters' Protective Association of Kentucky and Tennessee, 1904–1914." Ph.D. diss., Vanderbilt University, 1989.

———. "Robertson County and the Black Patch War, 1904–1909." *Tennessee Historical Quarterly* 39 (1980): 341–358.

Groce, W. Todd. *Mountain Rebels: East Tennessee Confederates and the Civil War, 1860–1870*. Knoxville: Univ. of Tennessee Press, 1999.

Groneman, Carol, and Mary Beth Norton. *"To Toil the Livelong Day": America's Women at Work, 1780–1980*. Ithaca, N.Y.: Cornell Univ. Press, 1987.

Gutman, Herbert G. *Power and Culture: Essays on the American Working Class*. Edited by Ira Berlin. New York: New Press, 1987.

———. *Work, Culture, and Society in Industrializing America: Essays in American Working-Class and Social History*. New York: Alfred A. Knopf, 1976.

Hackney, Sheldon. *Populism to Progressivism in Alabama*. Princeton, N.J.: Princeton Univ. Press, 1969.

———. "Southern Violence." *American Historical Review* 74 (1969): 906–925.

Hahn, Steven. "A Response: Common Cents or Common Sense?" *Journal of Southern History* 59 (1993): 243–258.

———. *The Roots of Southern Populism: Yeoman Farmers and the Transformation of the Georgia Upcountry, 1850–1890*. New York: Oxford Univ. Press, 1983.

Hahn, Steven, and Jonathan Prude, eds. *The Countryside in the Age of Capitalist Transformation: Essays in the Social History of Rural America*. Chapel Hill: Univ. of North Carolina Press, 1985.

Hair, William Ivy. *Bourbonism and Agrarian Protest: Louisiana Politics, 1877–1900*. Baton Rouge: Louisiana State Univ. Press, 1969.

Hart, Roger L. *Redeemers, Bourbons, and Populists: Tennessee, 1870–1896*. Baton Rouge: Louisiana State Univ. Press, 1975.

Hays, Samuel P. *The Response to Industrialism, 1885–1914*. Chicago: Univ. of Chicago Press, 1957.

Henretta, James. "Families and Farms: *Mentalité* in Pre-Industrial America." *William and Mary Quarterly* 35 (1978): 3–32.

Herr, Kincaid A. *The Louisville and Nashville Railroad, 1850–1963*. Lexington: Univ. Press of Kentucky, 1999.

Hicks, John D. *The Populist Revolt: A History of the Farmers' Alliance and the People's Party*. Minneapolis: Univ. of Minnesota Press, 1931, 1955.

Hill, Samuel S., Jr. *The South and the North in American Religion*. Athens: Univ. of Georgia Press, 1980.

Hoffschwelle, Mary S. *Rebuilding the Rural Southern Community: Reformers, Schools, and Homes in Tennessee, 1900–1930*. Knoxville: Univ. of Tennessee Press, 1998.

Hofstadter, Richard. *The Age of Reform: From Bryan to FDR*. New York: Vintage, 1955.

———. *The American Political Tradition and the Men Who Made It*. New York: Vintage, 1948.

Holmes, William F. "The Demise of the Colored Farmers' Alliance." *Journal of Southern History* 41 (1975): 187–200.

———. "Populism: In Search of Context." *Agricultural History* 64 (1990): 26–58.

———. "The Southern Farmers' Alliance: The Georgia Experience." *Georgia Historical Quarterly* 72 (1988): 627–652.

———. "The Southern Farmers' Alliance and the Jute Cartel." *Journal of Southern History* 60 (1994): 59–80.

———, ed. *American Populism*. Problems in American Civilization. Lexington, Mass.: D. C. Heath, 1994.

Holt, Michael F. *Political Parties and American Political Development from the Age of Jackson to the Age of Lincoln*. Baton Rouge: Louisiana State Univ. Press, 1992.

Howell, Sarah McCanless. "The Editorials of Arthur S. Colyar, Nashville Prophet of the New South." *Tennessee Historical Quarterly* 27 (1969): 262–276.

Humphreys, Cecil C. "The Formation of Reelfoot Lake and Consequent Land and Social Problems." *West Tennessee Historical Society Papers* 14 (1960): 32–72.

Hunt, James L. *Marion Butler and American Populism*. Chapel Hill: Univ. of North Carolina Press, 2003.

Hutson, A. C., Jr. "The Coal Miners' Insurrections of 1891 in Anderson County, Tennessee." *East Tennessee Historical Society Publications* 7 (1935): 103–121.

———. "The Overthrow of the Convict Lease System in Tennessee." *East Tennessee Historical Society's Publications* 8 (1936): 82–103.

Hutton, Marilyn J. "The Election of 1896 in the Tenth Congressional District of Tennessee." Ph.D. diss., Memphis State University, 1976.

Hyman, Michael R. *The Anti-Redeemers: Hill-Country Political Dissenters in the Lower South From Redemption to Populism*. Baton Rouge: Louisiana State Univ. Press, 1990.

Irish, Ann B. *Joseph W. Byrns of Tennessee: A Political Biography*. Knoxville: Univ. of Tennessee Press, 2001.

Irwin, James R. "Farmers and Laborers: A Note on Black Occupations in the Postbellum South." *Agricultural History* 64 (1990): 53–60.

Isaac, Paul E. *Prohibition and Politics: Turbulent Decades in Tennessee, 1885–1920*. Chapel Hill: Univ. of North Carolina Press, 1982.

———. *Tennessee Republicans in the Era of William McKinley, Theodore Roosevelt, and William Howard Taft: Factions, Leaders, and Patronage*. Studies in American History 19. Lewiston, N.Y.: Edwin Mellen Press, 1998.

Isaac, Rhys. *The Transformation of Virginia, 1740–1790*. Chapel Hill: Univ. of North Carolina Press, 1983.

Ivins, Lester S., and A. E. Winship. *Fifty Famous Farmers*. New York: MacMillan, 1925.

Jacobson, Timothy C. "Joseph Buckner Killebrew: Agrarianism in the New South." *Tennessee Historical Quarterly* 33 (1974): 157–174.

James, John A. "The Development of the National Money Market, 1893–1911." *Journal of Economic History* 36 (1976): 878–897.

Jeffrey, Julie Roy. "Women in the Southern Farmers' Alliance: A Reconsideration of the Role and Status in the Late Nineteenth-Century South." *Feminist Studies* 3 (1975): 72–91.

Jenkins, Gary C. "An 1890s Tour of Chattanooga." *Tennessee Historical Quarterly* 52 (1993): 242–255.

Jensen, Richard J. *Grass Roots Politics: Parties, Issues, and Voters, 1854–1983*. Westport, Conn.: Greenwood Press, 1983.

Johnston, Kevin. "Quills into Plowshares: Charles W. Dabney and Technical Education at the University of Tennessee, 1887–1893." M.A. thesis, Middle Tennessee State University, 1993.

Jones, James B., Jr. "Class Consciousness and Worker Solidarity in Urban Tennessee: The Chattanooga Carmen's Strikes of 1899–1917." *Tennessee Historical Quarterly* 52 (1993): 98–112.

———. "Strikes and Labor Organization in Tennessee During the Depression of 1893–1897." *Tennessee Historical Quarterly* 52 (1993): 256–264.

Jones, Lu Ann. *Mama Learned Us to Work: Farm Women in the New South*. Chapel Hill: Univ. of North Carolina Press, 2002.

Jones, Thomas B. "The Public Lands of Tennessee." *Tennessee Historical Quarterly* 27 (1968): 13–26.

Kantor, Shawn Everett, and J. Morgan Kousser. "Common Sense or Commonwealth?: The Fence Law and Institutional Change in the Postbellum South." *Journal of Southern History* 59 (1993): 201–241.

———. "A Rejoinder: Two Visions of History." *Journal of Southern History* 59 (1993): 259–266.

Kazin, Michael. *The Populist Persuasion: An American History*. New York: Basic, 1995.

Keith, Jeanette. *Country People in the New South: Tennessee's Upper Cumberland*. Chapel Hill: Univ. of North Carolina Press, 1995.

Keller, Morton. *Affairs of State: Public Life in Late Nineteenth Century America*. Cambridge, Mass.: Belknap Press of Harvard University, 1977.

Key, V. O., Jr. *Southern Politics in State and Nation*. Knoxville: Univ. of Tennessee Press, 1984. First published 1949 by Alfred A. Knopf.

Kiger, Joseph C. "Social Thought as Voiced in Rural Middle Tennessee Newspapers, 1878–1898." *Tennessee Historical Quarterly* 9 (1950): 131–154.

Kinchloe, Joe L. "Similarities in Crowd Control Techniques of the Camp Meeting and Political Rally: The Pioneer Role of Tennessee." *Tennessee Historical Quarterly* 37 (1978): 155–169.

———. "Transcending Role Restrictions: Women at Camp Meetings and Political Rallies." *Tennessee Historical Quarterly* 40 (1981): 158–169.

Kirby, Jack Temple. *Darkness at the Dawning: Race and Reform in the Progressive South*. Philadelphia: J. P. Lippincott, 1972.

———. *Rural Worlds Lost: The American South, 1920–1960*. Baton Rouge: Louisiana State Univ. Press, 1987.

Kirwan, Albert D. *Revolt of the Rednecks: Mississippi Politics, 1876–1925*. New York: Harper Torchbooks, 1951; reprint 1965.

Kolko, Gabriel. *The Triumph of Conservatism: A Reinterpretation of American History, 1900–1916*. New York: Free Press, 1963.

Kousser, J. Morgan. *The Shaping of Southern Politics: Suffrage Restriction and the Establishment of the One-Party South, 1880–1910*. New Haven, Conn.: Yale Univ. Press, 1974.

Kulikoff, Allan. *The Agrarian Origins of American Capitalism*. Charlottesville: Univ. Press of Virginia, 1992.

———. *From British Peasants to Colonial American Farmers*. Chapel Hill: Univ. of North Carolina Press, 2000.

———. *Tobacco and Slaves: The Development of Southern Cultures in the Chesapeake, 1680–1800*. Chapel Hill: Univ. of North Carolina Press, 1986.

Kyriakoudes, Louis Michael. "Country Comes to Town: Rural Migration to Nashville, Tennessee, 1890–1930." Ph.D. diss., Vanderbilt University, 1997.

Lacy, Eric Russell. "Tennessee Teetotalism: Social Forces and the Politics of Progressivism." *Tennessee Historical Quarterly* 24 (1965): 219–240.

Lamon, Lester D. *Blacks in Tennessee, 1791–1970*. Three Star Series. Knoxville: Univ. of Tennessee Press, 1981.

Lamoreaux, Naomi R. *The Great Merger Movement in American Business, 1895–1904*. New York: Cambridge Univ. Press, 1986.

Lax, Forrest Orren. "The Memphis Cotton Exchange From Beginning to Decline." M.A. thesis, Memphis State University, 1970.

Lemon, James. "Comment on James A. Henretta's 'Families and Farms: *Mentalité* in Pre-Industrial America.'" *William and Mary Quarterly* 37 (1980): 688–696.

Lester, Connie L. "'Let Us Be Up and Doing': Women in the Tennessee Movements for Agrarian Reform, 1870–1892." *Tennessee Historical Quarterly* 54 (1995): 80–97.

Lewis, Frank D. "The Economic Cost of the American Civil War: Estimates and Implications." *Journal of Economic History* 35 (1975): 299–326.

Lewis, J. Eugene. "The Tennessee Gubernatorial Campaign and Election of 1894." *Tennessee Historical Quarterly* 13 (1954): 99–126.

———. "The Tennessee Gubernatorial Campaign and Election of 1894: Evans Stumps the State." *Tennessee Historical Quarterly* 13 (1954): 224–243.

———. "The Tennessee Gubernatorial Campaign and Election of 1894: The Election." *Tennessee Historical Quarterly* 13 (1954): 301–328.

Link, William A. *The Paradox of Southern Progressivism*. Chapel Hill: Univ. of North Carolina Press, 1992.

Lovett, Bobby L., and Linda T. Winn. *Profiles of African Americans in Tennessee*. Nashville: Annual Local Conference on Afro-American Culture and History, 1996.

Lufkin, Charles L. "A Forgotten Controversy: The Assassination of Senator Almon Case of Tennessee." *West Tennessee Historical Society Papers* 39 (1985): 37–50.

Luker, Ralph E. "The Social Gospel and the Failure of Racial Reform, 1877–1898." *Church History* 46 (1977): 80–99.

Lyson, Thomas A., and William W. Falk. *Forgotten Places: Uneven Development in Rural America*. Lawrence: Univ. Press of Kansas, 1993.

Majka, Linda C., and Theo J. Majka. *Farm Workers, Agribusiness, and the State*. Philadelphia: Temple Univ. Press, 1982.

Malone, Ann Patton. "Piney Woods Farmers of South Georgia, 1850–1900: Jeffersonian Yeomen in an Age of Expanding Commercialism." *Agricultural History* 60 (1985): 51–84.

Mandel, Jay R. *The Roots of Black Poverty: The Southern Plantation Economy After the Civil War*. Durham, N.C.: Duke Univ. Press, 1978.

Mann, Ralph. "Mountains, Land, and Kin Networks: Burkes Garden, Virginia, in the 1840s and 1850s." *Journal of Southern History* 58 (1992): 411–434.

Mann, Susan Archer. *Agrarian Capitalism in Theory and Practice*. Chapel Hill: Univ. of North Carolina Press, 1990.

Marquis, Don. "Menace of the Mask, Part 1: Tennessee's Tragic Story of Night Riders Who Lynched at Reelfoot Lake." *Uncle Remus's: The Home Magazine* 24 (Dec. 1908): 15–21.

Marshall, E. H. *History of Obion County: Towns and Communities, Churches, Schools, Farming, Factories, Social and Political, Beginning with Goodspeed's Published in 1887,*

Proceeding with Alexander's Reelfoot Lake in 1925, Personal Effort and Interview. Union City, Tenn.: H. A. Lanzer, 1970.
Marshall, Suzanne. *Violence in the Black Patch of Kentucky and Tennessee.* Columbia: Univ. of Missouri Press, 1994.
Marti, Donald B. "Sisters of the Grange: Rural Feminism in the Late Nineteenth Century." *Agricultural History* 58 (1984): 247–261.
———. *Women of the Grange: Mutuality and Sisterhood in Rural America, 1866–1920.* New York: Oxford Univ. Press, 1991.
Mayhew, Anne, "A Reappraisal of the Causes of Farm Protest in the United States, 1870–1900." *Journal of Economic History* 61 (1972): 464–475.
McBride, Robert M., and Dan M. Robison. *Biographical Directory of the Tennessee General Assembly.* 2 vols. Nashville: Tennessee State Library and Archives, 1979.
McCraw, Thomas K. *Prophets of Regulation: Charles Francis Adams, Louis D. Brandeis, James M. Landis, Alfred E. Kahn.* Cambridge, Mass.: Belknap Press of Harvard University, 1984.
McCurry, Stephanie. *Masters of Small Worlds: Yeoman Households, Gender Relations, and the Political Culture of the Antebellum South Carolina Low Country.* New York: Oxford Univ. Press, 1995.
McDonald, Forrest, and Grady McWhiney. "The Antebellum Southern Herdsman: A Reinterpretation." *Journal of Southern History* 41 (1975): 147–176.
McDonough, James L., and Thomas L. Connelly. *Five Tragic Hours: The Battle of Franklin.* Knoxville: Univ. of Tennessee Press, 1983.
McGerr, Michael E. *The Decline of Popular Politics: The American North, 1865–1928.* New York: Oxford Univ. Press, 1986.
———. "Political Style and Women's Power, 1830–1930." *Journal of American History* 77 (1990): 864–885.
McGuire, Robert A. "Economic Causes of Late-Nineteenth-Century Agrarian Unrest: New Evidence." *Journal of Economic History* 41 (1981): 835–849.
McKenzie, Robert Tracy. "Civil War and Socioeconomic Change in the Upper South: The Survival of Local Agricultural Elites in Tennessee, 1850–1870." *Tennessee Historical Quarterly* 52 (1993): 170–184.
———. "Freedmen and the Soil in the Upper South: The Reorganization of Tennessee Agriculture, 1865–1880." *Journal of Southern History* 59 (1993): 63–84.
———. "From Old South to New South in the Volunteer State: The Economy and Society of Rural Tennessee, 1850–1880." Ph.D. diss., Vanderbilt University, 1988.
———. *One South or Many? Plantation Belt and Upcountry in Civil War–Era Tennessee.* New York: Cambridge Univ. Press, 1994.
———. "Postbellum Tenancy in Fayette County, Tennessee: Its Implications for Economic Development and Persistent Black Poverty." *Agricultural History* 61 (1987): 16–33.
McKinney, Gordon B. "Southern Mountain Republicans and the Negro, 1865–1900." *Journal of Southern History* 41 (1975): 493–516.

———. *Southern Mountain Republicans, 1865–1900: Politics and the Appalachian Community*. Knoxville: Univ. of Tennessee Press, 1998. First published 1978 by Univ. of North Carolina Press.

McLaurin, Melton Alonzo. *The Knights of Labor in the South*. Westport, Conn.: Greenwood Press, 1978.

McMath, Robert C., Jr. *American Populism: A Social History, 1877–1898*. New York: Hill & Wang, 1993.

———. "C. Vann Woodward and the Burden of Southern Populism." *Journal of Southern History* 67 (2001): 741–768.

———. *Populist Vanguard: A History of the Southern Farmers' Alliance*. Chapel Hill: Univ. of North Carolina Press, 1975.

McNicol Stock, Catherine. *Rural Radicals: Righteous Rage in the American Grain*. Ithaca, N.Y.: Cornell Univ. Press, 1996.

McSeveney, Samuel T. *The Politics of Depression: Political Behavior in the Northeast, 1893–1896*. New York: Oxford Univ. Press, 1972.

Merrill, Horace Samuel. *Grover Cleveland and the Democratic Party*. Boston: Little, Brown, 1957.

Mitchell, Theodore R. *Political Education in the Southern Farmers' Alliance, 1887–1900*. Madison: Univ. of Wisconsin Press, 1987.

Mitchell, Theodore R., and Robert Lowe. "To Sow Contentment: Philanthropy, Scientific Agriculture, and the Making of the New South, 1906–1920." *Journal of Social History* 24 (1990): 317–339.

Modelski, Andrew M. *Railroad Maps of North America: The First Hundred Years*. Washington, D.C.: Government Printing Office, 1984. Reprint, New York: Bonanza, 1987.

Montgomery, James Riley. *Threshold of a New Day: The University of Tennessee, 1919–1946*. Knoxville: Univ. of Tennessee Press, 1971.

Montgomery, James Riley, Stanley J. Folmsbee, and Lee Seifert Greene. *To Foster Knowledge: A History of the University of Tennessee, 1794–1970*. Knoxville: Univ. of Tennessee Press, 1974.

Moore, Gary E. "The Involvement of Experiment Stations in Secondary Agricultural Education, 1887–1917." *Agricultural History* 62 (1988): 164–176.

Moore, James Tice. "Agrarianism and Populism in Tennessee, 1886–1896: An Interpretative Overview." *Tennessee Historical Quarterly* 42 (1983): 76–94.

———. "Redeemers Reconsidered: Change and Continuity in the Democratic South, 1870–1900." *Journal of Southern History* 44 (1978): 357–378.

Moore, John Hebron. *Agriculture in Ante-Bellum Mississippi*. New York: Octagon Books, 1971.

Morgan, H. Wayne. *Industrial America: The Environment and Social Problems, 1865–1920*. Chicago: Rand McNally College Publishing, 1974.

Moss, Jeffrey W., and Cynthia B. Lass. "A History of Farmers' Institutes." *Agricultural History* 62 (1988): 150–163.

Nall, James D. *The Tobacco Night Riders of Kentucky and Tennessee, 1905–1909.* Louisville, Ky.: Standard Press, 1939. Reprinted as bicentennial edition, Kuttawa, Ky.: McClanahan, 1995.

Neely, Wayne Caldwell. *The Agricultural Fair.* New York: Columbia Univ. Press, 1935.

Nelson, Richard. "The Cultural Contradictions of Populism: Tom Watson's Tragic Vision of Power, Politics, and History." *Georgia Historical Quarterly* 72 (1988): 1–29.

Neth, Mary. *Preserving the Family Farm: Women, Community, and the Foundations of Agribusiness in the Midwest, 1900–1940.* Baltimore, Md.: Johns Hopkins Univ. Press, 1995.

Niswonger, Richard L. "Arkansas and the Election of 1896." *Arkansas Historical Quarterly* 34 (1975): 41–78.

———. *Arkansas Democratic Politics, 1896–1920.* Fayetteville: Univ. of Arkansas Press, 1990.

Nixon, Herman Clarence. "The Cleavage Within the Farmers' Alliance Movement." *Mississippi Valley Historical Review* 15 (1928–1929): 22–33.

Norton, Herman A. *Religion in Tennessee, 1777–1945.* Three Star Series. Knoxville: Univ. of Tennessee Press, 1981.

O'Brien, Michael. *Rethinking the South: Essays in Intellectual History.* Athens: Univ. of Georgia Press, 1993.

Obion County Historical Society. *Obion County History: Union City, Tennessee.* Vols. 1 and 2. Dallas, Tex.: Taylor Publishing, 1981, 1984.

Osterud, Nancy Grey. "Gender and the Transition to Capitalism in Rural America." *Agricultural History* 67 (1993): 14–29.

Osterweis, Rollin G. *The Myth of the Lost Cause, 1865–1900.* Hamden, Conn.: Archon Books, 1973.

Ostler, Jeffrey Don. *Prairie Populism: The Fate of Agrarian Radicalism in Kansas, Nebraska, and Iowa, 1880–1892.* Rural America. Lawrence: Univ. Press of Kansas, 1993.

Ownby, Ted. "The Defeated Generation at Work: White Farmers in the Deep South, 1865–1890." *Southern Studies: An Interdisciplinary Journal of the South* 23 (1984): 325–347.

———. *Subduing Satan: Religion, Recreation and Manhood in the Rural South, 1865–1920.* Chapel Hill: Univ. of North Carolina Press, 1990.

Owsley, Frank L. *Plain Folk of the Old South.* Baton Rouge: Louisiana State Univ. Press, 1949.

Owsley, Frank L., and Harriet C. Owsley. "The Economic Structure of Rural Tennessee, 1850–1860." *Journal of Southern History* 8 (1942) 161–182.

Palmer, Bruce. *"Man Over Money": The Southern Populist Critique of American Capitalism.* Chapel Hill: Univ. of North Carolina Press, 1980.

Parker, William N. "Quantification in American Agricultural History, 1850–1910: A Re-examination." *Agricultural History* 62 (1988): 113–132.

Parsons, Stanley B., Karen Toombs Parsons, Walter Killilae, and Beverly Borgers. "The Role of Cooperatives in the Development of the Movement Culture of Populism." *Journal of American History* 69 (1983): 866–885.

Perman, Michael. *Struggle for Mastery: Disenfranchisement in the South, 1888–1908*. Chapel Hill: Univ. of North Carolina, 2001.

Peters, Scott J., and Paul A. Morgan. "The Country Life Commission: Reconsidering a Milestone in American Agricultural History." *Agricultural History* 78 (summer 2004): 289–316.

Peterson, Merrill D. *The Jefferson Image in the American Mind*. New York: Oxford Univ. Press, 1960.

Pollock, Norman. *The Populist Response to Industrial America: Midwestern Populist Thought*. Cambridge, Mass.: Harvard Univ. Press, 1962.

Porter, Glen. *The Rise of Big Business, 1860–1920*. 2nd ed. Arlington Heights, Ill.: Harlan Davidson, 1992.

Porter, Jerry O. *Problems Relating to the Development of Northwest Tennessee*. Knoxville: TVA, 1973.

Pressley, Thomas J., and William H. Scofield. *Farm Real Estate Values in the United States by Counties, 1850–1959*. Seattle: Univ. of Washington Press, 1965.

Preston, Howard Lawrence. *Dirt Roads to Dixie: Accessibility and Modernization in the South, 1885–1935*. Knoxville: Univ. of Tennessee Press, 1991.

Pudup, Mary Beth. "The Limits of Subsistence: Agriculture and Industry in Central Appalachia." *Agricultural History* 64 (1990): 59–85.

Queener, Verton M. "A Decade of East Tennessee Republicanism, 1867–1876." *East Tennessee Historical Society Publications* 14 (1942): 61–89.

———. "The East Tennessee Republicans as a Minority Party, 1870–1896." *East Tennessee Historical Society Publications* 14 (1943): 49–73.

———. "The East Tennessee Republicans in State and Nation, 1870–1900." *Tennessee Historical Quarterly* 2 (1943): 99–128.

Ransom, Roger, and Richard Sutch. "Capitalists without Capital: The Burden of Slavery and the Impact of Emancipation." *Agricultural History* 62 (1988): 133–160.

———. "Debt Peonage in the Cotton South After the Civil War." *Journal of Economic History* 32 (1972): 641–669.

———. "The Impact of the Civil War and Emancipation on Southern Agriculture." *Explorations in Southern History* 12 (1975): 1–28.

———. *One Kind of Freedom: The Economic Consequences of Emancipation*. Cambridge: Cambridge Univ. Press, 1977.

Rauchle, Bob Cyrus. "The Political Life of the Germans in Memphis, 1848–1880." *Tennessee Historical Quarterly* 27 (1968): 165–175.

Rea, John C. "The Tennessee Midland Railway." *West Tennessee Historical Society Papers* 35 (1981): 118–121.

Reed, John Shelton. *The Enduring South: Subcultural Persistence in Mass Society*. Lexington, Mass.: D. C. Heath, 1972.

Reidy, Joseph P. *From Slavery to Agrarian Capitalism In the Cotton Plantation South, Central Georgia, 1800–1880*. Chapel Hill: Univ. of North Carolina Press, 1992.

Rischin, Modes, ed. *The American Gospel of Success*. Chicago: Quadrangle, 1965.

Robison, Daniel Merritt. *Bob Taylor and the Agrarian Revolt in Tennessee*. Chapel Hill: Univ. of North Carolina Press, 1935.

———. "The Political Background of Tennessee's War of the Roses." *East Tennessee Historical Society's Publications* 5 (1933): 125–141.

Robson, George Locke, Jr. "The Farmers' Union in Mississippi." M.A. thesis, Mississippi State University, 1963.

Rogers, William Warren. *One-Gallused Rebellion: Agrarianism in Alabama, 1865–1896*. Baton Rouge: Louisiana State Univ. Press, 1970.

———. "Reuben F. Kolb: Agricultural Leader of the New South." *Agricultural History* 32 (April 1958): 109–119.

Rohrbrough, Malcolm J. *The Trans-Appalachian Frontier: People, Societies, and Institutions, 1775–1850*. Belmont, Calif.: Wadsworth, 1990.

Rosenburg, R. B. *Living Monuments: Confederate Soldiers' Homes in the New South*. Chapel Hill: Univ. of North Carolina Press, 1993.

Rothberg, Winifred. *From Market Places to Market Economy: The Transformation of Rural Massachusetts, 1750–1850*. Chicago: Univ. of Chicago Press, 1992.

Rothstein, Morton. "America in the International Rivalry for the British Wheat Market, 1860–1914." *Mississippi Valley Historical Review* 48 (1960): 401–418.

———. "The Antebellum South as a Dual Economy: A Tentative Hypothesis." *Agricultural History* 41 (1967): 373–382.

———. "Farmers' Movements and Organizations: Numbers, Gains, and Losses." *Agricultural History* 62 (1988): 161–181.

———. "The New South and the International Economy." *Agricultural History* 57 (1983): 385–402.

Rutman, Darrett B., and Anita H. Rutman. *A Place in Time: Middlesex County, Virginia, 1650–1750*. New York: W. W. Norton, 1984.

Ryan, Mary P. *Women in Public: Between Banners and Ballots, 1825–1880*. Baltimore: Johns Hopkins Univ. Press, 1990.

Saloutos, Theodore. *Farmer Movements in the South, 1865–1933*. Berkeley: Univ. of California Press, 1960.

———. "The Grange in the South, 1870–1877." *Journal of Southern History* 19 (1953): 473–487.

Salstrom, Paul. *Appalachia's Road to Dependency: Rethinking a Region's Economic History, 1739–1940*. Lexington: Univ. Press of Kentucky, 1994.

Sanders, Elizabeth. *Roots of Reform: Farmers, Workers, and the American State, 1877–1917*. Chicago: Univ. of Chicago Press, 1999.

Saunders, Robert M. "The Southern Populists and the Negro in 1893–1895." *Journal of Negro History* 54 (1969): 240–261.

Schlup, Leonard. "Conservative Counterattack: Adlai E. Stevenson and the Compromise of 1892 with Democrats in Tennessee and the South." *Tennessee Historical Quarterly* 53 (1994): 113–129.

Schwartz, Michael. *Radical Protest and Social Structure: The Southern Farmers' Alliance and Cotton Tenancy, 1880–1890*. New York: Academic Press, 1976.

Schweninger, Loren. "A Vanishing Breed: Black Farm Owners in the South, 1651–1982." *Agricultural History* 63 (1989): 41–57.

Scott, Roy V. *The Reluctant Farmer: The Rise of Agricultural Extension to 1914*. Urbana: Univ. of Illinois Press, 1970.

Seehorn, John B. "The Life and Public Career of Henry Clay Evans." M.A. thesis, University of Tennessee, 1970.

Sellers, Charles. *The Market Revolution: Jacksonian America, 1815–1846*. New York: Oxford Univ. Press, 1991.

Shahan, Joe Michael. "The Limits of Agrarian Reform: The Tennessee Alliance Movement: 1888–92." M.A. thesis, Vanderbilt University, 1973.

———. "Reform and Politics in Tennessee: 1906–1914." Ph.D. diss., Vanderbilt University, 1981.

Shalhope, Robert E. "Thomas Jefferson's Republicanism and the Antebellum Southern Thought." *Journal of Southern History* 42 (1976): 529–556.

———. "Toward a Republican Synthesis: The Emergence of an Understanding of Repulicanism in American Historiography." *William and Mary Quarterly* 29 (1972): 49–80.

Shapiro, Karin A. *A New South Rebellion: The Battle Against Convict Labor in the Tennessee Coalfields, 1871–1896*. Chapel Hill: Univ. of North Carolina Press, 1998.

Sharan, J. M. "The Rhetoric of Reform: The 1906 Gubernatorial Race in Tennessee." *Tennessee Historical Quarterly* 35 (1976): 65–82.

Sharp, Joseph A. "The Downfall of the Radicals in Tennessee." *East Tennessee Historical Society's Publications* 9 (1937): 105–124.

———. "The Entrance of the Farmers' Alliance into Tennessee Politics." *East Tennessee Historical Society's Publications* 9 (1937): 77–92.

———. "The Farmers' Alliance and Tennessee Politics, 1890–1892." M.A. thesis, University of Tennessee, 1931.

Sharpless, Rebecca. "Southern Women and the Land." *Agricultural History* 67 (1993): 30–42.

Shaw, Barton C. *The Wool Hat Boys: Georgia's Populist Party*. Baton Rouge: Louisiana State Univ. Press, 1984.

Shlomowitz, Ralph. "The Origins of Southern Sharecropping." *Agricultural History* 53 (1979): 557–575.

Sigafoos, Robert A. *Cotton Row to Beale Street: A Business History of Memphis*. Memphis, Tenn.: Memphis State Univ. Press, 1979.

Simpson, John A. *S. A. Cunningham and the Confederate Heritage*. Athens: Univ. of Georgia Press, 1994.

Singal, Daniel Joseph. *The War Within: From Victorian to Modernist Thought in the South, 1919–1945*. Chapel Hill: Univ. of North Carolina Press, 1982.

Skocpol, Theda. *Protecting Soldiers and Mothers: The Political Origins of Social Policy in the United States*. Cambridge, Mass.: Harvard Univ. Press, 1992.

Smith, Albert C. "Southern Violence Reconsidered: Arson as Protest in Black-Belt Georgia, 1865–1910." *Journal of Southern History* 51 (1983): 526–564.

Smith, Samuel Boyd. "Joseph Buckner Killebrew and the New South Movement in Tennessee." Ph.D. diss., Vanderbilt University, 1962.

Smith, Sarah Frances. "Post–Civil War Agriculture in West Tennessee." M.A. thesis, Vanderbilt University, 1945.

Smith, William H. "William Jennings Bryan and the Social Gospel." *Journal of American History* 53 (1966): 41–60.

Snowden, Kenneth A. "Mortgage Rates and American Capital Market Development in the Late Nineteenth Century." *Journal of Economic History* 47 (1987): 671–691.

Soden, Dale. "The Social Gospel in Tennessee: Mark Allison Matthews." *Tennessee Historical Quarterly* 41 (1982): 159–170.

Spain, Rufus B. *At Ease in Zion: Social History of Southern Baptists, 1865–1900*. Nashville: Vanderbilt Univ. Press, 1967.

Stock, James H. "Real Estate Mortgages, Foreclosures, and Midwestern Agrarian Unrest, 1865–1920." *Journal of Economic History* 44 (1984): 89–106.

Stover, John F. *Railroads in the South, 1865–1900: A Study in Finance and Control*. Chapel Hill: Univ. of North Carolina Press, 1955.

Strom, Claire. "Texas Fever and the Dispossession of the Southern Yeoman Farmer." *Journal of Southern History* 66 (2000): 49–74.

Summers, Mark Wahlgren. *The Gilded Age, or The Hazard of New Functions*. Upper Saddle River, N.J.: Prentice-Hall, 1997.

Swanson, Merwin. "The Country Life Movement and the Churches." *Church History* 46 (1977): 358–373.

Sylla, Richard. "Federal Policy, Banking Market Structure, and Capital Mobilization in the United States, 1863–1913." *Journal of Economic History* 29 (1969): 657–686.

Szasz, Ferenc Morton. *The Divided Mind of Protestant America, 1880–1930*. University: Univ. of Alabama Press, 1982.

Taylor, Carl C. *The Farmers' Movement, 1620–1920*. New York: American Book, 1953.

Taylor, Jerome G. "Upper Class Violence in Nineteenth Century Tennessee." *West Tennessee Historical Society Papers* 34 (1980): 28–51.

Taylor, Robert L., Jr. "Demagoguery, Personality, and the Gospel of Democracy: Family Influence on Centennial Governor Taylor." *Tennessee Historical Quarterly* 55 (1996): 160–175.

Temin, Peter. *The Jacksonian Economy*. New York: W. W. Norton, 1969.

Terral, Rufus. *Newell Sanders, A Biography*. Privately printed, 1953.

Thornton, J. Mills, III. *Politics and Power in a Slave Society: Alabama, 1800–1860*. Baton Rouge: Louisiana State Univ. Press, 1978.

Trachtenberg, Alan. *The Incorporation of America: Culture and Society in the Gilded Age*. American Century Series, Eric Foner, ed. New York: Hill & Wang, 1982.
Turner, James. "Understanding the Populists." *Journal of American History* 67 (1980): 354–373.
Vanderwood, Paul J. *Night Riders of Reelfoot Lake*. Memphis, Tenn.: Memphis State Univ. Press, 1969.
Voss, Kim. *The Making of American Exceptionalism: The Knights of Labor and Class Formation in the Nineteenth Century*. Ithaca, N.Y.: Cornell Univ. Press, 1993.
Voss-Hubbard, Mark. "The 'Third Party Tradition' Reconsidered: Third Parties and American Public Life, 1830–1900." *Journal of American History* 86 (June 1999): 121–150.
Wagner, MaryJo. "Farms, Families, and Reform: Women in the Farmers' Alliance and Populist Party." Ph.D. diss., University of Oregon, 1986.
Wald, Kenneth D., "The Invisible Empire: The Ku Klux Klan as an Electoral Movement." *Journal of Interdisciplinary History* 11 (1980): 217–234.
Waldrep, Christopher. *Night Riders: Defending Community in the Black Patch, 1890–1915*. Durham: Duke Univ. Press, 1993.
———. "The Reorganization of the Tobacco Industry and Its Impact on Tobacco Growers in Kentucky and Tennessee, 1900–1911." *Mid-America* 73 (1991): 71–81.
Wallace, Louis D. *Makers of Millions*. A Publication of the Tennessee Department of Agriculture, 1951.
Wallenstein, Peter. *From Slave South to New South*. Fred W. Morrison Series in Southern Studies. Chapel Hill: Univ. of North Carolina Press, 1987.
Waller, Altina L. *Feud: Hatfields, McCoys, and Social Change in Appalachia, 1860–1900*. Chapel Hill: Univ. of North Carolina Press, 1988.
Waller, William. *Nashville in the 1890s*. Nashville: Univ. of Tennessee Press, 1970.
Wamble, H. "Landmarkism: Doctrinaire Ecclesiolgy Among Baptists." *Church History* 33 (1964): 429–447.
Watkins, Marilyn P. "Political Activism and Community-Building Among Alliance and Grange Women in Western Washington, 1892–1925." *Agricultural History* 67 (1993): 197–213.
Watson, Harry L. *Liberty and Power: The Politics of Jacksonian America*. New York: Noonday Press, 1990.
Watts, Stanley G., "Knowledge for the Tennessee Farmer: The Development of Agricultural Extension in Tennessee Prior to 1914." M.A. thesis, Mississippi State College, 1967.
Wayne, Michael. *The Reshaping of the Plantation Society: The Natchez District, 1860–1880*. Baton Rouge: Louisiana State Univ. Press, 1983.
Webb, Samuel L. "From Independents to Populists to Progressive Republicans: The Case of Chilton County, Alabama, 1880–1920." *Journal of Southern History* 59 (1993): 707–736.

———. *Two-Party Politics in the One-Party South: Alabama's Hill Country, 1874–1920.* Tuscaloosa: Univ. of Alabama Press, 1997.

West, Carroll Van. *Tennessee Agriculture: A Century Farms Perspective.* Nashville: Tennessee Department of Agriculture, 1986.

———. *Tennessee's Historic Landscapes: A Traveler's Guide.* Knoxville: Univ. of Tennessee Press, 1995.

———. *Tennessee History: The Land, the People, and the Culture.* Knoxville: Univ. of Tennessee Press, 1998.

———, ed. *Tennessee Encyclopedia of History and Culture.* Nashville: Tennessee Historical Society, 1998.

Westphal, Corinne. "The Farmers' Alliance in Tennessee." M.A. thesis, Vanderbilt University, 1929.

Wetherington, Mark V. *The New South Comes to Wiregrass, Georgia, 1860–1910.* Knoxville: Univ. of Tennessee Press, 1994.

White, Charles P. "Early Experiments with Prison Labor in Tennessee." *East Tennessee Historical Society's Publications* 12 (1940): 45–69.

Wiebe, Robert H. *The Search for Order, 1877–1920.* Westport, Conn.: Greenwood Press, 1980. First published 1967 by Hill & Wang.

Wiener, Jonathan M. *Social Origins of the New South: Alabama, 1860–1885.* Baton Rouge: Louisiana State Univ. Press, 1978.

Williams, Frank B., Jr. "The Poll Tax as a Suffrage Requirement in the South, 1870–1901." *Journal of Southern History* 18 (1952): 469–496.

Williams, R. Hal. *Years of Decision: American Politics in the 1890s.* New York: John Wiley & Sons, 1978.

Williams, Samuel C. "The Work of Goodpasture: An Appraisal." *Tennessee Historical Quarterly* 2 (1943): 58–60.

Williamson, Joel. *The Crucible of Race: Black-White Relations in the American South Since Emancipation.* New York: Oxford Univ. Press, 1984.

Willis, John C. *Forgotten Time: The Yazoo-Mississippi Delta after the Civil War.* Charlottesville: Univ. Press of Virginia, 2000.

Wilson, Charles Reagan. *Baptized in Blood: The Religion of the Lost Cause, 1865–1920.* Athens: Univ. of Georgia Press, 1980.

Winters, Donald L. "The Agricultural Ladder in Southern Agriculture: Tennessee, 1850–1870." *Agricultural History* 61 (1987): 36–82.

———. "Farm Size and Production Choices: Tennessee, 1850–1860." *Tennessee Historical Quarterly* 52 (1993): 212–224.

———. "Plain Folk of the Old South Reexamined: Economic Democracy in Tennessee." *Journal of Southern History* 53 (1987): 565–586.

———. "Postbellum Reorganization of Southern Agriculture: The Economics of Sharecropping in Tennessee." *Agricultural History* 62 (1988): 1–19.

———. *Tennessee Farming, Tennessee Farmers: Antebellum Agriculture in the Upper South.* Knoxville: Univ. of Tennessee Press, 1994.

Woodman, Harold D. "Class, Race, Politics, and the Modernization of the Postbellum South." *Journal of Southern History* 63 (1997): 3–22.

———. *King Cotton and His Retainers: Financing and Marketing of the Cotton Crop of the South, 1800–1925*. Columbia: Univ. of South Carolina Press, 1990. First published 1968 by Univ. of Kentucky Press.

———. "Sequel to Slavery: The New History Views the Postbellum South." *Journal of Southern History* 43 (1977): 523–554.

Woods, Thomas A. *Knights of the Plow: Oliver H. Kelley and the Origins of the Grange in Republican Ideology*. Ames: Univ. of Iowa Press, 1991.

Woodward, C. Vann. *The Burden of Southern History*. Baton Rouge: Louisiana State Univ. Press, 1989, 1960.

———. *Origins of the New South, 1877–1913*. 1951. Reprint, Baton Rouge: Louisiana State Univ. Press, 1971.

———. *Reunion and Reaction: The Compromise of 1877 and the End of Reconstruction*. Boston: Little, Brown, 1951.

———. *The Strange Career of Jim Crow*. New York: Oxford Univ. Press, 1974.

———. *Tom Watson: Agrarian Rebel*. New York: MacMillan, 1938. Reprint, New York: Oxford Univ. Press, 1987.

Wright, Gavin. *Old South, New South: Revolutions in the Southern Economy Since the Civil War*. New York: Basic, 1986.

———. *The Political Economy of the Cotton South: Households, Markets, and Wealth in the Nineteenth Century*. New York: Norton, 1978.

Wright, James Edward. *The Politics of Populism: Dissent in Colorado*. New Haven, Conn.: Yale Univ. Press, 1974.

Wyatt-Brown, Bertram. *Southern Honor: Ethics and Behavior in the Old South*. New York: Oxford Univ. Press, 1982.

Young, James Harvey. *The Medical Messiahs: A Social History of Health Quackery in Twentieth Century America*. Princeton, N.J.: Princeton Univ. Press, 1967.

Zunz, Olivier. *Making America Corporate, 1870–1920*. Chicago: Univ. of Chicago Press, 1990.

INDEX